A CRITICAL EVALUATION OF VITAMIN D - BASIC OVERVIEW

Edited by **Sivakumar Gowder**

A Critical Evaluation of Vitamin D - Basic Overview
http://dx.doi.org/10.5772/61867
Edited by Sivakumar Gowder

Contributors

Vera Morsanuto, Claudio Molinari, Francesca Uberti, John Sayer, Dietrich A. Volmer, Lars Kattner, Nikolaos E Koundourakis, Jia Guo, Zhangsuo Liu, Heba Elhusseini, Daria Lizneva, Larisa Gavrilova-Jordan, Noura Eziba, Mohamed Abdelaziz, Soumia Brakta, Sunil Halder, Ayman Al-Hendy, Joanna Jerzyńska, Daniela Podlecka, Iwona Stelmach, Akira Wagatsuma, Jenson Mak, Marta Larrosa, Enrique Casado, Fahimeh Ramezani Tehrani, Alessia Frati, Mercedes Garcia-Gil, Federica Pierucci, Elisabetta Meacci, Khalid Lone

Published by InTech
Janeza Trdine 9, 51000 Rijeka, Croatia

Publishing Process Manager Edi Lipovic
Technical Editor SPi Global
Cover InTech Design team

First published April, 2017
Printed in Croatia
Legal deposit, Croatia: National and University Library in Zagreb

Additional hard copies can be obtained from orders@intechopen.com

A Critical Evaluation of Vitamin D - Basic Overview, Edited by Sivakumar Gowder
p. cm.
Print ISBN 978-953-51-3083-3
Online ISBN 978-953-51-3084-0

For EU product safety concerns:
IN TECH d.o.o., Prolaz Marije Krucifikse Kozulić 3, 51000 Rijeka, Croatia,
info@intechopen.com or visit our website at intechopen.com.

Meet the editor

Dr. Sivakumar Gowder received his academic training and carried out his research in institutions of high academic ranking in India and the US (University of Madras, Chennai, India; All India Institute of Medical Sciences, New Delhi, India; UT Southwestern Medical Center, Dallas, TX, USA; LSH Health Sciences Center, Shreveport, LA, USA and University of Pittsburgh School of Medicine, Pittsburgh, PA, USA). Before working for Qassim University, he worked as a faculty member at the Medical Universities in West Indies. Sivakumar has won prizes and awards in different levels of his academic career. He has developed his own research methods and techniques relevant to his research disciplines and has published several journal articles and book chapters. Dr. Gowder has also edited many books. Currently, he serves as an author and editor of books, editor in chief for an international journal, editorial member and reviewer for journals, fellow and advisory board member of international organizations and external examiner of doctoral thesis work for international universities. He has also served as an invited speaker and chairperson for international conferences.

Contents

Preface

Vitamin D, a fat-soluble vitamin also known as the 'sunshine vitamin', is derived mostly from sun exposure and food. For normal activation, it has to undergo two hydroxylation reactions. Vitamin D affects more than 2000 genes in the body. A serum concentration of 25(OH) D indicates the ideal level of vitamin D in our bodies. The primary function of vitamin D is to regulate calcium and phosphorous absorption. Vitamin D deficiency leads to several diseases. From the therapeutic point of view, vitamin D helps in the treatment of many diseases. Lifestyle has an impact on our bodies' systems of getting sufficient amounts of vitamin D. In this context, due to industrialization and also changes in the environmental factors, any further study or work on vitamin D would be helpful for our society. I believe it offers a more than ample opportunity for me to present this book, *A Critical Evaluation of Vitamin D - Basic Overview*, to the audience.

The book targets physiological, biochemical and immunological aspects of vitamin D, including principles, mechanisms and clinical significance. This book has four sections: 'Vitamin D on Physical and Physiological Activities', 'Vitamin D on Biochemical and Immunological Activities', 'Vitamin D on the Musculoskeletal and Neurological Systems' and 'Vitamin D on the Reproductive System'. Each of these sections is interwoven with the theoretical aspects and experimental techniques of basic and clinical science. This book will be a significant resource for students, scientists, physicians, healthcare professionals and other members of the society who are interested in exploring the role of vitamin D in human life.

In the first section, the authors have revealed the diverse effects of vitamin D on physical performance in athletic and nonathletic populations—the role of vitamin D3 as a potent regulator of oxidative pathways and vitamin D signalling and the multiple effects of vitamin D concerning pathological progress and diseases. In the second section, the authors have disclosed the importance of CYP24A1 mutations in diagnosis; mass spectrometric approaches to quantifying vitamin D metabolites and their distributions in various human diseases; the biological effects of sphingolipids, 1,25(OH)2D3, and its structural analogues in bone and neural disorders; the role of vitamin D pathways in the regulation of the function of both innate and adaptive immune systems and the role of vitamin D in ageing, allergies and conditions like cancer, diabetes mellitus, asthma and various autoimmune diseases. In the third section, we can find information on the multiple signalling pathways of vitamin D and their contribution to myogenesis, the beneficial effects of vitamin D on the skeletal system and other organs, including the cardiovascular system, and the effectiveness of optimization strategies of vitamin D levels following hip fracture and its implication on public health. The last section focuses on the role of vitamin D in female reproduction throughout the different reproductive ages and male fertility and the impact of vitamin D deficiency on the female reproductive system diseases.

I appreciate the support of our higher authorities. I extend my gratitude towards my late mother and late father and my brothers for introducing me to higher education. I am continuously indebted to my wife Anitha for her emotional and technical support throughout this project. The smiles of my daughter, Humsiha, encouraged me to finish this task in an easy way. I must acknowledge the interest and commitment from the Publishing Processing Manager of InTech Mr. Edi Lipovic, whose patience and focus were a fantastic support in this project. Finally, I express deep and sincere appreciation to all the authors for their valuable contributions and scholarly cooperation for the timely completion of this book.

Dr. Sivakumar Gowder
Qassim University, Saudi Arabia

Section 1

Vitamin D on Physical and Physiological Activities

Chapter 1

Vitamin D and Physical Activity

Nikolaos E. Koundourakis and Andrew N. Margioris

Additional information is available at the end of the chapter

http://dx.doi.org/10.5772/65103

Abstract

Vitamin D is synthesized in the skin following exposure to ultraviolet radiation, producing cholecalciferol, while only a small percentage of the circulating vitamin D is of exogenous origin deriving from food. Following two sequential hydroxylations, in the liver and in the kidneys, vitamin D is fully activated. Although its role in bone physiology and calcium homeostasis is well documented, there is emerging evidence that vitamin D exerts a plethora of additional effects on most tissues regulating the musculoskeletal, cardiovascular, and immune systems as well as energy homeostasis. Its deficiency/insufficiency poses a major public health problem observed in all age groups and regardless of latitude and insolation. In muscles, vitamin D deficiency is associated with a decline in neuromuscular function including muscular strength, walking speed, balance, jumping and sprinting performance, and aerobic capacity, although the evidence is still weak regarding its effects in the young and the athletes. Supplementation counteracts the negative effects of vitamin D deficiency on performance although in individuals with adequate levels of vitamin D, additional supplementation does not appear to enhance further physical capabilities. The aim of this chapter is to review our current understanding of diverse effects of vitamin D in physical performance in athletic and nonathletic populations.

Keywords: vitamin D, Physical activity, athletes, musculoskeletal system, cardiovascular system, immune system

1. Introduction

Vitamin D is synthesized in the skin from 7-dehydrocholesterol following exposure to ultraviolet radiation (cholecalciferol or D3), and only a small percentage is exogenous originating from food (ergocalciferol, D2) [1–3]. Following to sequential hydroxylations in the liver and the kidneys, vitamin D is activated (**Figure 1**). Vitamin D plays a crucial role in calcium

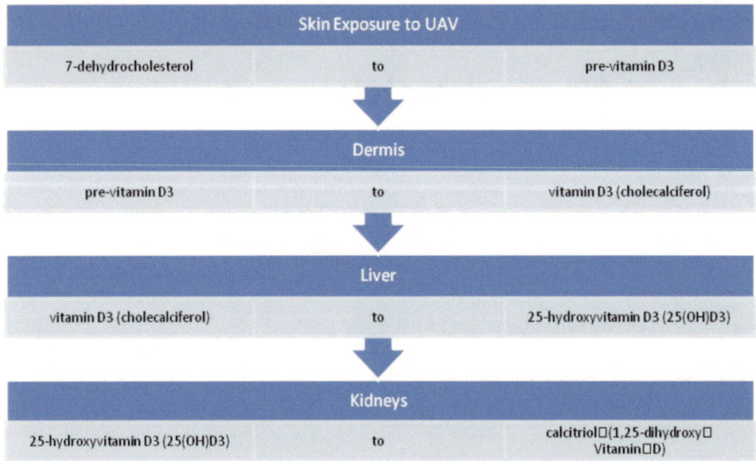

Figure 1 Biochemical pathways of vitamin D production after exposure to UVB.

homeostasis in addition to its multiple effects in a plethora of tissues [3]. Indeed, vitamin D receptors (VDRs) are present in most tissues affecting the transcription of many genes [4–7]. Vitamin D is transported in the blood via vitamin D-binding protein [8, 9]. Vitamin D plays a role in the mineralization of type I collagen matrix in the skeleton [10]. Vitamin D and its nuclear receptor affect the expression of more than one thousand genes acting as a transcription factor [11]. Non-genomic effects of vitamin D have also been described [12–19].

2. Vitamin D deficiency

A surprisingly high prevalence of vitamin D insufficiency has been recently reported worldwide (**Table 1**) regardless of their insolation [20–23]. In Canada, 30–50 % of children and adults are vitamin D deficient and need vitamin D supplementation [24]. Similar data have been reported in Africa, Australia, Brazil, Middle East, Mongolia, and New Zealand documenting a high risk for vitamin D deficiency in adults and children [24–26]. Cross-sectional studies of vitamin D status in adolescents have found deficiency in 17–47 % with an increased risk in black and Hispanic teenagers [24–26]. It should be noted here that there are fewer reports regarding the prevalence of vitamin D in athletes. A high prevalence of vitamin D deficiency has been documented in athletes in both outdoor and indoor sports. For instance, it has been reported that more than 42 % of distance runners in Baton Rouge, Louisiana, were vitamin D insufficient or deficient [21]. Indeed, athletes may be more vulnerable to vitamin D deficiency than age-matched non-athletes even in regions with insolation. A recent meta-analysis pooling 23 studies composed of 2313 athletes found that 56 % of athletes had inadequate vitamin D [23]. In addition, a sizeable number of athletes do not meet the US dietary reference intake for vitamin D in addition to their inadequate endogenous synthesis when they train indoors [27]. In a recent study, Morton et al. [26] examined the prevalence of seasonal variation in vitamin D levels in 20 FA Premier League soccer players residing at a latitude of 53°N. Serum 25-

hydroxyvitamin D (25(OH)D) levels decreased in winter to insufficient levels in 65 % of the soccer players (<50 nmol/L). Koundourakis et al. [28] showed that professional Greek soccer players' training at a latitude of 35.9°N had insufficient vitamin D levels, while nearly identical levels were reported in American national football league players [29], in elite gymnasts in Australia [30], and in young Hawaiian skateboarders [31] and a variety of other athletes worldwide [32–35]. These findings were observed irrespective of sun exposure. In a recent study conducted in Israel in a favorable latitude (31.8°N) for insolation, 73 % of the athletes were vitamin D deficient. In Qatar, 84 % of soccer players were vitamin D insufficient. Furthermore, vitamin D deficiency in regions of high sunlight (high insolation) may be due to clothing, air pollution, and failure to meet the dietary reference intake for vitamin D in addition to their inadequate endogenous synthesis when they train indoors. Indeed, high prevalence of low vitamin D levels has been reported in the general population in the United Arab Emirates, India, and Egypt [36, 37].

Study	Population	Indoor/ outdoor	Season	Country, latitude	Vitamin D levels
Lovell [30]	18 female gymnasts, age = 13.6 ± 1.2 y	Indoor	Spring	Australia, 35.27°S	Insufficient 83.33 %, deficient 33.33 %
Constantini et al. [35]	98 Israeli male and female athletes and dancers, age = 14.7 ± 3.0 y	Outdoor/ indoor	Winter	Israel, 31°N	Insufficient 73 %, insufficiency was higher in dancers (94 %), basketball players (94 %), Tae Kwon Do fighters (67 %), athletes from indoor sports
Kopeć et al. [32]	24 Caucasian male Polish professional soccer players 26.24±3.41 y	Outdoor	Summer Winter	Poland, 51.10°N Cyprus, 30.11°N	Insufficient 37.5 %, deficient 12.5 % Insufficient 45.8 %, deficient 37.5 %
Morton et al. [26]	20 FA Premier League soccer players	Outdoor/ indoor	Winter	United Kingdom, 53°N	Insufficient 65 %
Galan et al. [40]	28 male soccer players, 26.7 ± 3.6 y	Outdoor/ indoor	Winter	Spain, 37°N	Deficient 64 %
Koundourakis et al. [28]	67 Caucasian male professional soccer players, age = 25.6 ± 6.2 y	Outdoor	Spring	Greece, 35.9° N	Insufficient 55.22 %
Halliday et al. [34]	93 male athletes (football, handball, shooting, squash, cycling, body building, martial arts), age = 21.3 ± 6.5 y	Outdoor/ indoor	Spring, summer	Middle East 29°N	Insufficient 91.39 %, deficient 8.60 %
Willis et al. [33]	19 male and female endurance-trained runners, age = 28.3 ± 3.4 y	Outdoor	N/A	United States, 30°N	Insufficient 42.10 %, deficient 10.52 %
Close et al. [211]	61 male athletes of soccer, rugby, jockeys, age = 24.3 ± 4.8 y	Outdoor/ indoor	Winter	United Kingdom, 50°N	Deficient 62 %
Close et al. [214]	30 male soccer and rugby athletes, age = 21.3 ± 1.3 y	Outdoor/ indoor	Winter	United Kingdom, 53°N	Deficient 57 %

Insufficient: vitamin D levels <30 ng/ml [75 nmol/L]; deficient: vitamin D levels < 20 ng/ml [50 nmol/L]; y = years.

Table 1. Vitamin D levels in various athletic populations.

Generally speaking, vitamin D insufficiency is defined (**Table 2**) as levels <30 ng/ml (80 nmol/L), whereas levels below 20 (50 nmol/L) and 10 ng/mL (25 nmol/L) represent deficiency and severe deficiency, respectively [7]. Levels between 40 and 60 ng/mL are the preferred range, while vitamin D intoxication usually does not occur until 25(OH)D3 reaches levels higher that 150 ng/ml [24].

Status	25-hydroxyvitamin D (ng/ml)	25-hydroxyvitamin D (nmol/L)
Severe deficiency	<10	<25
Deficiency	<20	<50
Insufficiency	20–30	50–75
Sufficiency	>30	>75
Toxicity	>150	>375

Table 2. Classification of vitamin D (25-hydroxyvitamin D) levels.

3. Seasonal variation of vitamin D levels

Seasonal variation of vitamin D is a well-documented phenomenon [38]. In the winter months, endogenous vitamin D production is drastically reduced, as a result of reduced exposure to UVB radiation, while in summer exposure to UVB radiation is adequate for vitamin D synthesis by the skin [5]. Similar variations have been reported for athletes. In soccer players, the levels of 25(OH)D3 are normal in 50 % of the athletes, while this number drops to 16.7 % in winter [32]. A study performed in a high latitude (Laramie, WY 41.3°N) revealed vitamin D insufficiency of 63 % in winter compared to only 12 % in autumn and 20 % in spring in both indoor and outdoor athletes [38]. Similar findings were reported in a study conducted at higher latitudes (Ellensburg, WA 46.9°N), using exclusively outdoor athletes. The authors reported that a percentage of 25–30 % of the athletes were vitamin D insufficient from fall to winter [39]. Morton et al. [26] also reported a significant drop of serum levels of 25(OH)D3 in a group of professional soccer players of the English Premier League at the latitude of 53°N between summer and winter. In agreement with these data were findings in professional soccer players in Spain that have been training at a latitude of 37°N showing a statistically significant reduction in serum levels of 25(OH)D3 between October and February [40].

Seasonal variation of athletic performance has been found to parallel that of the seasonal fluctuation of vitamin D levels. Indeed, athletic performance peaks when vitamin D levels are the highest, i.e., during summer [41]. Koch and Raschka [42] reviewed the German literature on the seasonality of physical performance. They reported that early studies indicated that strength and maximal oxygen uptake peaks (VO_{2max}) in late summer. Comparable findings have been reported by Erikssen and Rodahl [43]. More specifically, in a population of 1835 Norwegian men, physical performance exhibited seasonal variability peaking in summer. Similar data were reported for the Swedish national track and field teams where VO_{2max} was observed in summer [44].

Most authors explain the enhanced athletic performance in summer to the higher vitamin D levels. However, other authors disagree [41]. According to Moran et al. [26], the higher physical performance in summer is a consequence of outdoor physical activity in the warmer weather which also results in exposure of athletes to UAV, i.e., these authors see the two phenomena as parallel but independent. Athletes participating in athletic events such as track and field competitions have to be in their best conditioning in the summer since most major competitions take place at this time period. As a consequence, enhanced training and competition in summer would result in both better athletic performance and higher vitamin D levels [1, 45]. In contrast, participants in other sports such as soccer, American football, or basketball attain their best performance in winter while in summer recuperate resulting in a reduction and/or cessation of their training and to a deterioration of exercise performance and capacity [28]. Nevertheless, even when controlling for seasonal fluctuations in time spent exercising, variation in wrist flexor strength and muscle trainability still correlated with seasonal variations of vitamin 25(OH)D levels [41].

4. Effects of vitamin D in calcium absorption and bone physiology

Vitamin D plays an essential role in calcium homeostasis and the mineralization of bones [46, 47]. In the small intestine, vitamin D augments calcium uptake via an energy-dependent transcellular pathway and by a passive paracellular pathway through tight junctions [24, 48].

Insufficient levels of circulating vitamin D cause an increase of serum parathyroid hormone (PTH) due to secondary hyperparathyroidism, i.e., in a desperate effort of the body to keep circulating levels of calcium normal borrowing them from bones via an increase of bone turnover and increased bone resorption [26]. In chronic vitamin D-deficient states, individuals are susceptible to osteopenia, osteoporosis, and increased incidence of bone fractures [49, 50]. Administration of vitamin D is recommended as a first-line strategy reducing the incidence of osteoporotic fractures [51]. Indeed, low 25(OH)D levels in women could be viewed as a general marker of impaired health as the Malmö study has documented correlating with increased overall mortality [52].

Furthermore, low serum vitamin D is associated with cortical bone porosity indicative of increased bone fragility [53]. Vitamin D also enhances the activity of insulin-like growth factor 1 (IGF-1) via an induction the expression of its receptor, an affect crucial in bone formation both in vitro and in vivo [54].

4.1. Implications of these data on physical activity and exercise performance

Low vitamin D levels are related with osteoporosis and the risk of fractures in elderly individuals. Data from cross-sectional and longitudinal studies report that vitamin D insufficiency impairs bone health [4, 7]. As a result of this deterioration in bone health, the increased incident of falls in the elderly can easily result to bone fractures. However, recent evidence indicates that this is also evident in young physically active individuals. A study on Finish military recruits reported that those with insufficient vitamin D levels (<30 ng/mL) were susceptible with a significant risk of developing stress fracture [55]. Similar findings have been reported in a number of other studies [56] including athletes [57, 58]. Deterioration in bone strength could be a problem apart from non-athletes, also in athletic populations. Hypothetically, athletes with low bone mineral density are predisposed to skeletal injury especially in contact individual and/or team sports [4, 7]. The latter could have several implications for the athletic performance especially in professional athletes. Notably the association between the incident of fractures and low vitamin D levels in athletes has not yet been determined, and there are also a number of studies in nonathletic populations that failed to support the hypothesis that high adequate vitamin D levels protect against stress fractures [4].

5. Effects of vitamin D on the skeletal muscle

The potential association between vitamin D and muscle function was also based on early clinical descriptions that myopathy was strongly associated with severe vitamin D deficiency [59]. Currently, vitamin D has been shown to be a potent modulator of skeletal muscle physiology [5]. Although the expression of the vitamin D receptors in the skeletal muscle tissue was questioned [60], recent data strongly indicate that they are also expressed in the skeletal muscle [61]. As we mentioned earlier, $1,25(OH)_2D$ by binding to its nuclear receptor augments the transcription of key muscular proteins [5–7]. It should be noted here that both genomic and non-genomic effects of vitamin D are crucial for muscular performance. Indeed, vitamin D affects muscular calcium and phosphate transport across cell membranes, phospholipid metabolism, and muscle cell proliferation and differentiation [62]. In addition, vitamin D downregulates the expression of myostatin (a negative regulator of muscle mass) and upregulates the expression of follistatin and insulin-like growth factor 2 (IGF-2) [5–7, 63]. Exposure of skeletal muscles to 1,25-dihydroxyvitamin D3 induces the expression of several myogenic markers and transcription factors [63]. A 10-day incubation of muscle cells with 1,25-dihydroxyvitamin D3 induces myosin heavy chain (MHC) type II, a late myogenic marker, while the mean diameter and width of muscular fibers increase compared to parallel controls. Indeed, vitamin D increases the size of MHC type II-positive myotubes [61, 64] and the cross-sectional area (CSA) of skeletal muscle fibers [65]. Vitamin D signaling has been also reported to alter the expression of C2C12 myotube size [5, 61], indicating a direct positive effect on the contractile filaments and thus muscle strength. In addition, it affects the diameter and number of type II muscle fibers [5–7] and in particular type IIa fibers [61]. In severe vitamin D deficiency, a proximal myopathy is observed characterized by type II (fast twitch) muscle fiber atrophy [5, 61]. A recent vitamin D supplementation study showed that in young male populations, a significant increase in the percentage of type IIa fibers was evident in the vitamin D group compared to the placebo group, demonstrating a positive effect on the muscle fiber type morphology, calcium transport, and regulation of intracellular calcium [61].

5.1. Implications of these data on physical activity and exercise performance

The data presented above suggest that vitamin D exerts beneficial effects on skeletal muscles and thus may improve physical performance capacity in both physically nonathletic individuals and athletes. The documented vitamin D-mediated induction of muscle protein synthesis and myogenesis could result in muscles of higher quality and quantity. This could be in turn translated into increased muscle strength since it is well documented that there is a strong linear associations between muscle mass and strength [66]. In addition the association between vitamin D levels and muscle anabolism could also be related with another positive effect, which is particularly important for athletic populations in that it could accelerate muscle recovery from the stress of intense exercise [26]. This hypothetically could be of major importance in periods of continues competitions since it could enable them to maintain their performance capacity at a better level. In the elderly on the other hand, low vitamin D levels could cause a reduced muscular anabolism, sarcopenia, and thus negatively altered muscle mass which in turn leads to a negative effects in the sense of balance, strength, and stability. In addition, in vitamin D-insufficient states, active individuals may not have adequate energy provisions to sustain a moderate- to high-intensity muscular effort. Lastly, hypertrophy of type II muscle fibers could result in enhanced neuromuscular performance [4, 66]. These types of fibers are a major determinant of the explosive type of human movements resulting in high-power output. It is well documented that type II muscle fibers display a markedly faster muscle contraction velocity and thus higher force than type I muscle fibers [67]. Therefore the anaerobic maximal-intensity short-burst activities, such as jumping, sprinting, acceleration, deceleration, and change of direction, which are of crucial importance for the majority of the athletic events, are highly related with the type II muscle fibers. Interestingly, apart from athletes the importance of type II muscle fibers is highlighted also on the elderly. Reversal of type II fiber atrophy as a result of vitamin D supplementation is thought to account for approximately 20 % lower risk of falling [68]. Vitamin D deficiency apart from the atrophy of type II fibers in these individuals enables fat infiltration and consequently fibrosis (fibrosis is observed in different tissues including the muscle due to excess fat accumulation) which are important factors underpinning muscle quality and also may be a predictor of muscle function in older adult [69]. In closing, based on the available literature, we propose that vitamin D is beneficial for individuals since it increases muscle protein synthesis, adenosine triphosphate (ATP) concentration, strength, jump height, jumping and sprinting velocity and power, and aerobic and anaerobic exercise capacity. Physical performance could be significantly enhanced and/or preserved with adequate vitamin D. Vitamin D also prevents muscular degeneration and reverses myalgia [68, 70].

6. Effects of vitamin D on the cardiovascular system

Vitamin D receptors are expressed in human myocardium as well as in vascular smooth muscles and the endothelium [5, 6]. The activated form of vitamin D, i.e., 1,25(OH)2D, participates in the structural remodeling of cardiac muscles and vascular tissues [71] and results in improvements in flow-mediated dilation and blood pressure [72]. Improved cardiac muscle function has been reported in patients with severe vitamin D deficiency following treatment [71]. In addition, animal studies have demonstrated that $1,25(OH)_2D$ directly alters myocyte contractility with accelerated relaxation, which is crucial in the normal cardiac diastolic function [6]. Vitamin D has been also found to regulate the function of calcium channels in cardiac myocytes, providing a rapid influx of calcium into cells promoting myocyte contractility [73]. Data from both animal and human studies show that vitamin D can act as a negative regulator of the renin-angiotensin-aldosterone system (RAAS) [6]. It is well documented that the RAAS maintains vascular resistance since it is a strong vasoconstrictor agent, due to angiotensin II synthesis, blood pressure, electrolyte, and intravascular fluid volume homeostasis. Vitamin D has the ability to decrease RAA activity by suppressing renin gene expression resulting to the downregulation of the RAA system [6, 45]. This is a vital mechanism that is regulated by vitamin D since it has been observed that the elevation in renin synthesis is highly related with left ventricular hypertrophy and the development of hypertension [45]. Another mechanism via which vitamin D insufficiency may induce left ventricular hypertrophy is by affecting and in particular elevating PTH levels. Observational studies have suggested that apart from hypertension, elevated PHT levels also result to left ventricular hypertrophy [74]. Vitamin D insufficiency has been found to be related with increased arterial stiffness and endothelial dysfunction in the conductance and resistance blood vessels in humans [75, 76]. Vitamin D has also some antiatherogenic functions by promoting HDL transport and inhibiting cholesterol uptake by the macrophages and the formation of foam cells [77], while its deficiency stimulates systemic and vascular inflammation, enabling atherogenesis [78]. In contrast, vitamin D has been shown to suppress inflammation via several pathways, such as inhibition of prostaglandin and cyclooxygenase pathways, upregulation of anti-inflammatory cytokines, decrease of cytokine-induced expression of adhesion molecules, reduction of matrix metalloproteinase, and downregulation of the RAA [45].

The presence of vitamin D receptors in the vascular wall suggests that it plays a role in vascular physiology and pathophysiology [74]. Among individuals with peripheral arterial disease, low vitamin D status has been associated with a faster decline of functional performance [79]. Severe vitamin D deficiency results in a disrupted adaptive immune response and an inflammatory milieu, promoting vascular dysfunction, insulin resistance, and arteriosclerosis [45, 80].

In addition, recent evidences indicate that severe vitamin D deficiency is strongly associated with sudden cardiac death [81, 82]. Furthermore, Vitamin D has been found to be a useful biomarker for the prediction of mortality when obtained at admission for chest pain [83]. The mechanisms via which vitamin D exert beneficial effects on human myocardium are as follows: vitamin D inhibits the proliferation of cardiomyoblasts by promoting cell-cycle arrest and enhances the formation of cardiomyotubes without inducing apoptosis. Vitamin D also attenuates left ventricular dysfunction in animal models and humans [84]. However, a group of researchers were unable to find any association between vitamin levels and risk of cardiovascular disease [85]. It should be noted though that analyzed data from more than 13,000 adults in the Third National Health and Nutrition Examination Survey (NHANES III) found a strong association between low vitamin D levels and key cardiovascular risk factors (i.e., hypertension, diabetes, overweight, and hypertriglyceridemia, but not hypercholesterolemia) after adjustment for multiple variables [86].

6.1. Implications of these data on physical activity and exercise performance

As per the available literature presented above, vitamin D plays an important role in the cardiovascular system [75]. Vitamin D affects several aspects of vascular health, including arterial stiffness and endothelial function which are crucial components of aerobic and anaerobic exercise performance and even the ability to perform daily activities. It should be noted that there is a linear association between vascular health and arterial stiffness to endurance capacity [80]. In addition, it is well documented that the most accurate measurement of aerobic capacity, i.e., that of VO_{2max}, is regulated by cardiac output, arterial oxygen content, shunting of blood to active muscle, and extraction of oxygen by muscles. Since low serum vitamin D levels may result in pathological myocardial hypertrophy, increased blood pressure, and endothelial dysfunction [87], these converging lines of evidence could support the idea that inadequate vitamin D levels could negatively influence cardiorespiratory fitness and the ability to perform efficiently during exercise. Thus, vitamin D affects aerobic capacity and VO_{2max}, cardiac output, vascular tone, and supply of oxygen and nutrients to the exercising muscles.

7. Effects of vitamin D on the nervous system

Vitamin D affects the central and peripheral nervous systems. Vitamin D receptors are present throughout the brain including the primary motor cortex, the region coordinating movement [88–91]. Specifically, VDRs have been identified in neuronal and glial cells in several brain areas including the cortex, deep gray matter, cerebellum, brainstem nuclei, spinal cord, and the ventricular system [90]. Furthermore, the enzyme 1α-hydroxylase, the activator of vitamin D precursors, is also present in the brain [89–93]. Vitamin D levels are associated with the conductance velocity of motor neurons and neurotransmission mediated by dopamine, serotonin, acetylcholine, GABA, and the catecholamines [92, 93]. Vitamin D also affects neuronal differentiation, maturation, and growth, its levels correlating to the levels of several neurotrophic factors including nerve growth factors (NGF) and that of neurotrophins, which play crucial roles in the maintenance and growth of neurons [94, 95]. In addition, vitamin D exerts direct neuroprotective effects via the synthesis of proteins binding calcium (Ca^{2+}) ions which are important in neuronal function including transmission. Proper levels of neuronal calcium are critical since their excess may result in the formation of reactive oxygen species (ROS) which lead to neuronal damage. Indeed, vitamin D levels are inversely associated with oxidative stress which damages the brain leading to neuronal apoptosis or necrosis [96]. Vitamin D also affects neuroplasticity, a process via which neural synapses and pathways are adapted to the needs of environmental and behavioral demands adjusting the brain to noxious stimuli diseases or environmental cues [90, 97, 98]. The VDRs in glial cells play in the uptake and release of neurotransmitters, including that of GABA neurotransmission within the motor cortex [99]. GABAergic function is the principal "brake" within the brain affecting muscle relaxation via the corticospinal neurons [100].

7.1. Implications of these data on physical activity and exercise performance

The data described above suggest that vitamin D exerts major effects on several aspects of brain function. The effects of vitamin D on GABAergic tone and on serotonin and dopamine are crucial for muscular coordination and avoidance of central fatigue, a condition associated with the synaptic concentration of several neurotransmitters [101]. A high ratio of serotonin to dopamine affects exercise performance because of its effect on the general feeling of tiredness and the perceptions of effort [102].

Another mechanism via which vitamin D affects the brain and athletic performance may involve the nocireceptors or nociceptors, i.e., the sensory nerve cell that responds to noxious stimuli by sending signals to the spinal cord and brain. The nociceptors are replete in VDRs and 1α-hydroxylase [103, 104]. When these receptors transfer pain signals to the brain, an inhibitory physical response takes place [104]. The relevance of this mechanism and vitamin D and physical activity/performance is based on the recent findings in animal studies that vitamin D depletion could result in nociceptive hyperinnervation and hypersensitivity within deep muscle tissue and a loss of balance without affecting muscle strength or the cutaneous nociceptive response [103]. Based on this finding, we could speculate that nociceptive hyperinnervation and hypersensitivity within deep muscle tissue could result to a false onset of myalgia during physical activity which in vitamin D-deficient individuals could result to a reduction in performance. The beneficial effects of vitamin D-mediated effects of the nervous system on physical activity and exercise performance are further supported by the findings of Dhesi et al. [105]. Reaction times deteriorate with age [105] and play an important role in neuroprotective responses [106]. In cross-sectional work, we identified a correlation between vitamin D status and reaction times, suggesting a neuroprotective role of this vitamin [105]. However, it is still unknown whether the latter mechanisms could be attributed also to young healthy individuals or athletic population.

8. Effects of vitamin D on pulmonary function

Vitamin D affects pulmonary function and health. Vitamin D insufficiency has been associated with impaired pulmonary function, asthma, and chronic obstructive pulmonary disease (COPD) [107, 108]. It should be noted that vitamin D deficiency may harm the lungs not only via its well-known immune effects. The Third National Health and Nutrition Examination Survey (NHANES III) reported a strong positive association between serum 25(OH)D and forced expiratory volume (FEV1) and forced vital capacity (FVC). These effects of vitamin D may be mediated through surfactant, a substance maintaining alveolar structural integrity [109], lung compliance, and oxygen exchange. It should be stressed that lung compliance, FEV1, FVC, and cardiopulmonary oxygen transfer are linearly related with physical performance [110]. Indeed, vitamin D induces the synthesis of alveolar surfactant and plays a critical role in epithelial-mesenchymal interactions during lung growth [111]. Vitamin D plays a role in COPD and more specifically in smooth muscle proliferation in the airways [112, 113]. Zosky et al. [114] have provided evidence supporting a direct role of vitamin D on lung growth in vivo, while vitamin D deficiency results in lung volume deficits [115]. Finally, vitamin D insufficiency correlates with multiple indices of compromised lung function and increased airway reactivity [115].

8.1. Implication of these data on physical activity and exercise performance

The data described above provide evidence implicating vitamin D in alveolar structural integrity, lung compliance, vital capacity, cardiopulmonary oxygen transfer, airway smooth muscle proliferation, and lung remodeling. Regarding athletes, exercise performance and aerobic capacity (VO_{2max}) depend on all these lung functions [28]. Adequate VO_{2max} levels are needed in all sporting disciplines. Vitamin D appears to affect all components of VO_{2max} adequacy.

9. Effects of vitamin D on immunity

Vitamin D affects the innate and adaptive immunity via VDRs [116–118]. Vitamin D enhances monocyte/macrophage activity. Vitamin D directly activates the transcription of several antimicrobial substances including defensin β2 (DEFB) and cathelicidin [118, 119]. Human cathelicidin (hCAP18) causes destabilization of microbial membranes [120]. In college athletes vitamin D insufficiency is associated with higher frequency of illnesses including common colds, influenza, and gastroenteritis [27]. Furthermore, vitamin D affects both T and B cells [121]. Under resting conditions the expressions of VDRs are low in both T and B cells, but they are upregulated in infectious diseases, suggesting a crucial role in adaptive immunity [122–125]. Indeed, vitamin D affects B-cell function, including inhibition of memory- and plasma-cell generation and apoptosis.

It is also known for years that immune cells can activate vitamin D to $1,25(OH)_2D$ via 1α-hydroxylase, an effect resulting in several beneficial paracrine effects including modulation of T-cell function [126]. Vitamin D upregulates the expression of several antimicrobial peptides (AMP) while at the same time downregulates the expression of inflammatory cytokines including that of tumor necrosis factor α (TNFα) and interleukin (IL) 6 [126]. Low levels of vitamin D in the general population and in athletes (especially following intense exercise) result in IL6 and TNFα increase. Vitamin D ameliorates this inflammatory response.

Several studies have also documented a negative association between vitamin D levels and upper respiratory tract infections (URTIs) in young and elderly adults [127–129]. In athletes, the incidence of respiratory illnesses is higher (especially at the elite level) because of the high dements on the lungs during prolonged exercise [130]. This could lead to increased exposure to viral and bacterial pathogens from the external environment, increasing the risk of URTIs. Li and Gleeson [131] have suggested that low vitamin D levels may exaggerate the vulnerability of athletes to URTIs, while individuals with higher vitamin D levels exhibit a lower propensity to URTI [132, 133]. It has been suggested that vitamin D prevents the development of URTIs by affecting the activation of toll-like receptors (TLRs) [134]. Vitamin D interferes with the activation of TLR signaling by microbial antigens via cathelicidin and β-defensin

[134]. Cathelicidin enhances the microbicidal capability of monocytes/macrophages by increasing the oxidative burst of these phagocytic cells, an effect also depended on vitamin D [134].

9.1. Implications of these data on physical activity and exercise performance

It is reasonable to suggest that the well-documented association between optimal vitamin D levels and the well-being and the proper function of the immune system combined with the anti-inflammatory effects of this vitamin D may improve athletic performance. The observation that low vitamin D levels may result in increased pro-inflammatory cytokine levels, such as TNFα and IL6, following periods of intense exercise training may suggest that this pro-inflammatory period may be ameliorated by vitamin D [68]. This effect may accelerate the recovery phase from the stress of intense exercise. High IL6 and TNFα levels may partially explain age-related skeletal and muscular failings in vitamin D-deficient older individuals. Reduction of TNFα levels following vitamin D supplementation may be crucial to combat the "overtraining syndrome" in athletes [135]. Furthermore, the cytokines IL-10 and IL-13 have been shown to promote muscle regeneration and prevent skeletal muscle damage [136, 137]. Regarding the effects of vitamin D on the URTIs, its negative association with URTIs has been addressed by He et al. [138] who reported a higher proportion of subjects who experienced one or more URTI episodes in the vitamin D-deficient state during the 4-month study period than the optimal vitamin D group [138]. In closing, since vitamin D has been shown to affect the inflammatory response and prevent URTIs, its use may be of practical significance in the athletes.

10. Effects of vitamin D on the adipose tissue

Vitamin D plays a crucial role in the physiology of adipocytes. VDRs and the 25-hydroxyvitamin D 1α-hydroxylase (CYP27B1) genes are all expressed in human adipocytes [139]. Vitamin D levels are inversely related to obesity including body fat percentage (BFP), body mass index (BMI), and waist circumference [140, 141]. Vitamin D supplementation for 16 weeks reduces visceral fat in overweight and obese adults [142]. Drincic et al. [143] identified body weight as the single strongest predictor of 25OHD levels followed by fat mass. The loss of weight results in an elevation of vitamin D levels [144]. Epidemiologically, vitamin D levels are negatively associated with BFP in non-athletes [19, 145, 146]. Interestingly, a high BFP in winter months has been found correlating with indoor athletic activities and low vitamin D levels [146].

10.1. Implications for physical activity/exercise performance

The close association between vitamin D levels and adiposity could affect athletic performance. It should be noted that neuromuscular performance capacity including sprinting ability, jumping performance, agility, acceleration, aerobic capacity, and deceleration is inversely associated with BFP [147]. These data suggest that high BFP in conjunction with low vitamin D levels exerts a detrimental effect on exercise performance in athletes as well as in the general population. It should be noted that increased BFP is associated with the development of a systemic chronic low-grade inflammation and insulin resistance, a characteristic of obesity, which also negatively affect exercise performance [148].

11. Effects of vitamin D on the electron transport system

Vitamin D has been also suggested to be related with the production of adenosine triphosphate (ATP) via the aerobic pathway. In particular, this speculation has been based on the observations that vitamin D appears to affect electron transport. Vitamin D receptors are present within mitochondria. Silencing of vitamin D receptors in several cell lines results in growth inhibition accompanied by an increase in mitochondrial membrane potential sensitizing cells to oxidative stress [149]. Vitamin D plays a role on mitochondrial respiratory chain activity acting as a facilitator of the diversion of acetyl-CoA from the energy-producing TCA cycle toward biosynthetic pathways that are essential for cellular proliferation. These effects of vitamin D on aerobic energy production during exercise are needless to say crucial for exercise performance.

12. Insolation and vitamin D on athletic performance

Early studies on collegiate athletes and students showed that cardiovascular fitness, muscle endurance, and speed were enhanced following exposure to ultraviolet radiation [150–152]. In particular, in 1938 Russian authors [151] reported that a course of ultraviolet irradiation improved speed in the 100 m dash in four students compared with matched controls. The authors observed that the experimental group (i.e., under UAV treatment) had a 7.4 % increase in 100 m dash times compared to only 1.7 % of the control group. Notably both groups were undergoing the same training schedule throughout the study. Supportive evidence came from a study by German researchers in 1944 [152]. A treatment under UV irradiation twice a week for a period of 6 weeks resulted in a 13 % improvement in performance on a bike ergometer. No alteration was evident in the control group. Enhanced performance parameters were also reported by Allen and Cureton [150] as a result to UAV treatment in college students. The authors observed that a period of 10 weeks under UV treatment resulted in improvements in both cardiovascular fitness and muscular endurance. UV increased cardiovascular fitness by 19.2 % compared to only 1.5 % improvement in the controls. In 1956, Sigmund [153] examined the effects of UV radiation on reaction times on adolescents and adult individuals. UV treatment improved by 17 % the reaction time compared to controls. In another study in the late 1960s, a single dose of UV irradiation beneficially affected the strength, speed, and endurance of college women [154]. Interestingly, the beneficial effect was evident in white participants but not in African American females, suggesting a different skin response to UV. A follow-up study from the same team [155] observed that the same amount of UV managed to improve aerobic performance in the same population. Enhanced performance was also observed by another study in a 30-yard dash of 15 college women after 6-minute exposure in UV light [156].

13. Vitamin D deficiency in athletes versus non-athletes

It is increasingly apparent that athletes need higher levels of vitamin D because of the higher demands of the daily strenuous training exercise which they undergo [157]. Indeed, experimental findings suggest a major role of vitamin D in muscle mass, strength, and function. Vitamin D affects significantly the athletic ergometrics [158]. Furthermore, athletic performance appears to show a seasonal variability peaking when vitamin D levels peak and declining as they decline reaching nadir when vitamin D levels are at their lowest. It should be noted that vitamin D administration improves athletic performance in vitamin D-deficient athletes [41].

14. Vitamin D in non-athletes

Overwhelming evidence demonstrates the association between vitamin D levels and the ability of non-athletes to perform their daily physical activities and/or exercise (**Table 3**). Early on, myopathy and muscle weakness were attributed to the most obvious consequence of vitamin D deficiency, i.e., that of rickets and osteomalacia [159]. To date, several cross-sectional observations and longitudinal studies have reported a close association between vitamin D and several parameters of physical performance. The majority of the studies have been performed on elderly subjects although similar data have been reported in adults below the age of 65 years [160] and in young individuals [161, 162].

Author	Participants	Physical performance parameters tested	Results
Ward et al. [161]	99 adolescent girls, 12–14 y	Jump power, countermovement jump(CMJ) height, and velocity, Esslinger Fitness Index, force	Positive relationship between vitamin D levels and jump velocity, jump height, power, Esslinger Fitness Index, and force
Foo et al. [162]	301 healthy Chinese adolescent girls, age 15.0 ± 04 y	Handgrip strength, physical activity levels	Adequate vitamin D status had greater handgrip muscle strength than deficient
Gerdhem et al. 155	1044 elderly women, 75 y old (range 75.0–75.9 y old)	30 m gait, knee extension, knee flexion, Romberg balance, self-estimated activity level	Positive relationship between vitamin D levels and gait speed, Romberg balance

http://dx.doi.org/10.5772/65103

Author	Participants	Physical performance parameters tested	Results
			test, self-estimated activity level, and thigh muscle strength
Houston et al. [166]	988 elderly males and females, age 85.2 ± 3.6 y	SPPB test (standing balance, repeated chair stands, 3 m walk [gait speed]), grip and knee extensor strength	Significantly poorer SPPB in deficient individuals versus those in sufficient state. No relationship between vitamin D levels and absolute grip and knee extensor strength. Older adults with deficient 25(OH)D levels had significantly lower muscle strength indicating poorer muscle quality
Visser et al. [167]	1260 elderly males and females >65 y old	Physical activity (30, 30–59, or 60 min/day), timed walking test, and a repeated chair stand test	Vitamin D-deficient and vitamin D-insufficient individuals had poorer physical performance in all tests
Bischoff-Ferrari et al. [168]	4100 elderly males and females >60 y old	8-foot walk test and sit-to-stand test	Vitamin D levels between 40 and 94 nmol/L were associated with better musculoskeletal function in the lower extremities than concentrations of 40 nmol/L. Significant positive association between vitamin D levels and the 8-foot walk test, the sit-to-stand test
Tieland et al. [169]	127 pre-frail and frail elderly people, ≥65 y old	Short physical performance battery (SPPB) test, maximum strength (one repetition maximum)	Significant positive associations were observed between vitamin D levels and gait speed and chair rise
Houston et al. [174]	1155 elderly individuals, aged 65–102 y old	Short physical performance battery (SPPB), handgrip strength	Significant positive associations were observed between vitamin D levels and SPPB score in men and handgrip strength in men and women
Wicherts et al. [177]	1234 elderly men and women, age 79.8 ± 5.9 y	Time taken to walk 3 m, turn 180°, and walk back (walking test); time taken to rise five times from a kitchen chair with arms folded in front of the chest (chair stands); the ability to stand with the heel of the one foot directly in front of, and touching the toes of, the other foot for at least 10 s (tandem stand)	Vitamin D levels were associated with physical performance which was poorer in participants with levels < 10 ng/ml and levels of 10–20 ng/ml
Mastaglia et al. [178]	44 postmenopausal women, age 71 ± 4 y	Walking-speed test, standing balance, and sit-to-stand tests.	Women with vitamin D levels ≥ 20 ng/ml scored higher on the muscle function

Author	Participants	Physical performance parameters tested	Results
		Lower extremity muscle strength was determined using a manual dynamometer	tests and had stronger knee extensor and hip abductor muscles
Ceglia et al. [181]	1219 males, age 4.9 ± 12.8 y	Timed walking test, chair stand test, grip strength	No association was evident between any physical performance test and vitamin D levels
Matheï et al. [182]	367 elderly males and females, age 84.7 ± 3.6 y old	Balance, grip strength, and gait speed	No significant relation between vitamin D levels and balance, gait speed and grip strength, and serum

y, years

Table 3. Vitamin D levels and physical performance in non-athletes.

In the elderly vitamin D deficiency has been associated with myopathy, reduced muscle mass and strength, low exercise performance, and an increased risk of falling [163]. The rise of vitamin D levels has been associated to improved balance [164], reduced frailty and disability [165, 166], independence [167], and better mobility [168, 169]. Improved balance and mobility independent of muscle function may partly explain the association of vitamin D with the reduced risk of falls in older adults. It is well documented that the maintenance of muscle strength and power in this population is also related with the ability to perform all types of physical activities and therefore well-being and functional performance. In the elderly, mobility limitations have been defined as the self-reported inability to walk a mile, climb stairs, or perform heavy housework [170]. This is dependent on multiple components such as muscles, bones, and the cardiovascular system, and vitamin D has been found to play a crucial role in the functionality of these tissues. The impairment in any of these tissues could lead to reduced mobility that in turn could cause accelerated functional decline and disability. Given that both muscle strength and power are related to the functional activities in older adults [161, 171], it could be argued that assessing the potential parameters that could affect muscle strength or power and even muscle quality would identify ways to avoid impaired mobility and functional performance. In addition, proper cardiovascular function is essential to perform adequately all types of activities. In this context, it could be suggested that vitamin D is involved in the regulation of strength and power (as a product of strength) and affects the proper cardiovascular function and also the quality of the muscles (fat infiltration in vitamin D-deficient state). The latter is related with vitamin D since in the deficient state, it has been reported that there is increased fat infiltration in the muscle. This physiological mechanism is associated with lower levels of muscle strength and physical performance, independent of muscle mass [172]. To date, a number of organizations recognize that vitamin D is directly related with the falls in the elderly, including the International Osteoporosis Foundation, the Endocrine Society, and the US Preventive Services Task Force [173]. In accordance are the observations of Gerdhem et al. [164]. The authors evaluated the association between 25-

http://dx.doi.org/10.5772/65103

hydroxyvitamin D levels (25OHD), fall-associated variables (including tests of functional performance), and fracture in ambulatory women. A low 25OHD was associated with lower physical activity, gait speed, and balance. Vitamin D levels below 20 ng/ml have been associated with an increased risk of fractures. It has been suggested that the minimum level of 25OHD required to maximize the reduction of falls and fractures appears to be in the order of 60 nmol/L. Any increases above this do not appear to have any additional beneficial effect in lowering the risk of falling [173].

Regarding physical performance per se, several cross-sectional studies indicate a positive association between physical performance and serum vitamin D in the elderly. In the In-CHIANTI study, a representative sample of 976 persons aged 65 years or older at study baseline was examined [174]. Physical performance was assessed. A significant association between low levels of vitamin D and poor physical performance as assessed by the handgrip strength test and a short physical performance battery (SPPB) test (which included ability to stand from a chair and ability to maintain balance in progressively more challenging positions) was documented. It was found that individuals with serum vitamin D levels < 25 nmol/L performed worse than those with levels above 25 nmol/L. Muscle strength using a handgrip test was also significantly greater in subjects with vitamin D levels higher than 50 nmol/L than those with lower levels [174]. The use of grip strength as a tool is relatively straightforward to stratify adults at risk of impaired mobility. It has been shown in elderly individuals that higher levels of vitamin D correlated with better lower extremity muscle performance compared to age-matched individuals with lower levels [169]. Similar data have been shown in the Rancho Bernardo Study cohort [175]. Likewise in a longitudinal survey of community-dwelling Japanese, older women showed that those with higher levels of vitamin D responded better in a 3-month exercise program than those with lower vitamin D levels [176].

Wicherts et al. [177] examined the association of serum 25-hydroxyvitamin D levels to physical performance and its decline with age. The authors reported that after adjustment for age, gender, chronic diseases, degree of urbanization, body mass index, and alcohol consumption, a relationship was evident between vitamin D levels and physical performance indices. Physical performance was inferior in individuals with low vitamin D compared to age-matched individuals with serum 25OHD levels higher than 30 ng/ml. Thus 25OHD levels below 20 ng/ml appear to be associated with poorer physical performance and a greater decline in physical performance in older men and women.

In another study on elderly females aged over 65 years, the authors examined the association between vitamin D levels and muscle function and strength. Performance was evaluated by a walking-speed test, standing balance, and a sit-to-stand test [178]. It was observed that vitamin D levels above 50 nmol/L were related with higher levels of muscle strength. The authors also reported that the individuals with the lowest 25OHD levels (i.e., below 20 nmol/L) had the poorest values in strength. Stockton et al. [179], based on the observations from their meta-analysis, concluded that vitamin D had no significant effect on lower extremity muscle strength except in individuals with starting serum 25OHD levels <25 nmol/L. In support to this suggestion are the findings of Pramyothin et al. [180]. In a study performed in older Hawaiian women of Japanese ancestry, which are known for its very low rate of falls, a high dietary intake in vitamin D, and a large exposure to sunlight, their mean vitamin D level was 80 nmol/L, while no individuals were in a vitamin D-deficient state. The authors failed to observe any association between vitamin D levels and all the measured performance parameters, i.e., physical strength test (except for the quadriceps), falls, and daily activities. It was concluded that the absence of a relationship was due to the very high level of vitamin D that these women had at baseline. Mastaglia et al. [178] suggest that there is a limit below which vitamin D insufficiency may affect physical performance. In agreement to this hypothesis are the findings of a recent study [181]. The authors reported that in a population-based sample of adult men with a broad age range (mean age is 47 years), there was no association between serum 25(OH)D concentration and lean body mass, muscle strength, and physical function after controlling for multiple lifestyle factors. However, only 20 % of the subjects had a vitamin D level below 50 nmol/L. This observation is giving further support to the cause-effect hypotheses of "sufficient" or "insufficient" vitamin D levels on several indices on human physiology including physical function and athletic performance. However confusing findings have been reported by Matheï et al. [182] in elderly individuals (>80 years) with an 80 % prevalence of vitamin D insufficiency where no correlation was evident between vitamin D levels and physical performance, as assessed by gait speed, handgrip test, and a static balance test. The authors suggest that this absence of association could be at least partly due to age-related downregulation of the VDR in muscles [182]. On the other hand, Ward et al. [161] reported that in vitamin D-insufficient young adolescent girls, there was a correlation between vitamin D levels and muscle power, force, velocity, and jump height. Similarly, Foo et al. [162] found a positive association between vitamin D serum levels and handgrip strength in a population of 301 vitamin D-sufficient adolescent girls. However, studies in young women failed to verify any association between vitamin D status and muscle performance, handgrip strength [183], or any other physical performance parameter [184]. These discrepancies may be partially explained by differences in the experimental protocols, accuracy of vitamin D serum measurements, age of the participants, muscle fitness, health state, gender, and other variables. In summary, most published data suggest an association of vitamin D levels greater than 50 nmol/L with muscle fitness. Vitamin D-associated improvement of balance and mobility are independent variables form muscle function and may partially explain the beneficial effect of vitamin D supplementation on the rate of falls in older individuals [185].

15. Vitamin supplementation in non-athletes

Vitamin D levels in the blood are increasingly recognized as an important factor in muscular well-being and functional performance. As a consequence, it was speculated that vitamin D3 supplementation may exert beneficial effects on physical activity. Indeed, several major trials have demonstrated that vitamin D supplementation lower the risk of falling (**Table 4**). Prince et al. [186] in a randomized, controlled trial in Australia evaluated women with at least one fall in a 12-month period and with a plasma 25-hydroxyvitamin D level <24.0 ng/mL. The authors suggested that elderly with a history of falling and vitamin D insufficiency living in sunny climates benefit from ergocalciferol supplementation resulting in a 19 % reduction in the relative risk of falling, mostly in winter. Pfeifer et al. [187] demonstrated a reduction in falls of 27–39 % in community-dwelling seniors supplemented with 800 IU vitamin D and calcium daily versus calcium alone. This drop in falls was correlated with an improvement in quadriceps strength and in the timed up and go (TUG) test. Similarly, Zhu et al. [188] reported an enhanced muscular strength and TUG test in the individuals within the lowest vitamin D quartile. These results are consistent with another study that showed a 49 % reduction of falls in elderly women from a geriatric ward supplemented with 800 IU per day of vitamin D [189]. Bischoff-Ferrari et al. [189] evaluated several studies involving men and women with an average age of 70 that vitamin D supplementation resulted in a significant 22 % decrease in fall risk.

Study	Quality	Participants	Intervention	Parameters measured	Results
Prince et al. [186]	Double-blind, randomized controlled trial	302 community -dwelling ambulant older women, aged 70–90 y old	1000 mg/day of calcium for 1 year combined with 1000 IU/day of ergocalciferol or identical	Risk of falling	Ergocalciferol supplementation resulted in a 19 % reduction in the relative risk of falling
Zhu et al. [188]	Randomized controlled trial	302 community -dwelling ambulant elderly women; mean age: 76.9 ± 4.5 y	1000 IU/day of vitamin D2 or identical placebo; calcium citrate (1 g calcium/day) in both groups for 1 y	Timed up and go (TUG) test, ankle dorsiflexion, knee flexor, knee extensor, hip abductor, hip flexor, hip extensor, and hip adductor strength were assessed	Significant ↑ in knee flexor strength and all hip muscle strength and TUG in both groups. Ankle dorsiflexion strength significantly↓ in both groups, no change in knee extensor strength after 12 months
Bunout et al. [191]	Randomized double-blind controlled trial	96 healthy elderly subjects aged > 70 y	Participants were randomized to a resistance training or control group. Trained and control groups were further randomized to receive vitamin D 400 IU plus 800 mg of calcium per day or calcium alone for 9 months	Handgrip strength, Isometric quadriceps maximum voluntary strength, endurance, general physical fitness, measuring the timed up and go (TUG), short physical performance battery (SPPB)	Timed up and go ↑ more in trained subjects supplemented with vitamin D. At the end of the follow-up, gait speed was ↑ among subjects supplemented with vitamin D (whether trained or not)
Bischoff et al.	Randomized double-blind controlled	122 elderly women, mean age = 85.15 ±	Calcium 1200 mg and 800 IU/day vitamin D	TUG test, grip a strength, knee flexor strength, knee	↓Risk of falls and in 62 women with complete for all strength test,

Study	Quality	Participants	Intervention	Parameters measured	Results
[192]	trial	6.8 y	versus calcium 1200 mg/day for 12 weeks	extensor strength, risk of falls	there was significant ↑ in all measured parameters
Moreira-Pfrimer et al. [193]	Randomized double-blind controlled trial	46 individuals (11 males and 35 females), age = 77.6 8 ± 8.2 y	6-month period: daily calcium + monthly placebo or daily calcium + oral cholecalciferol (150,000 IU once/month the first 2 months, 90,000 IU once/month until the end)	Maximum isometric strength of hip flexors (SHF) and knee extensors (SKE)	SHF ↑ in the calcium/ vitamin D group by 16.4 % and SKE ↑ by 24.6 %
Pfeifer et al. [194]	Randomized double-blind controlled trial	242 community -dwelling elderly women and men, mean age 77 ± 4 y	1000 mg of Ca or 1000 mg of Ca plus 800 IU of vitamin D per day for 12 months, followed by an 8-month treatment-free observation period	Timed up and go (TUG) test, maximum isometric leg extensor strength	Vitamin D resulted in a significant ↓ falls of 27 % at month 12 and 39 % at month 20 and in significant ↑ and ↓in quadriceps strength and time to perform the TUG test, respectively
Janssen et al. [196]	Randomized double-blind placebo-controlled trial	70 female geriatric patients >65 y old	400 IU/day of cholecalciferol + 500 mg/day of Ca or a placebo +500 mg/day of Ca for 6 months	Handgrip strength, leg extension power, get up and go (GUT) test, modified Cooper test	No significant improvements in strength and physical activity measures
Brunner et al. [197]	Randomized, double-blind, placebo-controlled trial	33,067 women aged 50–79 y old	1000 mg calcium carbonate plus 400 IU vitamin D-3 or matching placebo in a regimen of two pills per day	Grip strength, chair stand test, timed walk test, physical activity status, physical function (self-report), physical activity (self-report)	No evidence of treatment effects on measures of any kind of physical functioning, activity, and strength were observed
Gupta et al. [200]	Randomized, double-blind, placebo-controlled trial	40 healthy males and females, age 31.5 ± 5.0 y	60,000 IU of vitamin D3/w for 8 weeks followed by 60,000 IU/m for 4 months with 1 g of Ca daily or dual placebos for 6 months	Handgrip and gastro-soleus dynamometry, pinch-grip strength, respiratory pressures, 6 min walk test	All measured physical performance parameters ↑ significantly in the treatment group

Study	Quality	Participants	Intervention	Parameters measured	Results
Owens et al. [201]	Placebo-controlled trial	29 male participants, mean age = 22.7 ± 3 y	Oral dose of 10,000 IU/day vitamin D3 or a visually identical placebo for 3 months	Lower limb muscle function, isokinetic torque, percutaneous isometric electromyostimulation	No significant changes in any of the measured performance indices

y, years; ↑, increase; ↓, decrease

Table 4. Vitamin supplementation in nonathletic population.

Recent studies on elderly populations support the data regarding the beneficial effect of vitamin D supplementation on neuromuscular performance improving strength [9] and force production [190]. Furthermore, a significant linear association between vitamin D levels and muscular strength, body sway, and physical performance is evident [106, 191, 192]. In addition, Moreira-Pfrimer et al. [193] reported that vitamin D supplementation results in improved isometric muscle strength following a 6-month supplementation of daily cholecalciferol resulting in an increase of strength of hip flexors (SHF) and strength of knee extensors (SKE).

In addition to the strength of lower extremities, upper extremity strength also benefits from vitamin D supplementation. However, regarding grip strength, a tool that is relatively straightforward to stratify adults at risk of impaired mobility, the results are confusing. In a recent review of the literature, it was reported that grip strength does not seem to respond to vitamin D supplementation in males [194]. The failure to observe any association between grip strength in the male population is in agreement with others. Effects of vitamin D treatment on handgrip strength have been evaluated in several studies, none of them showing significant effects in contrast to women [191, 194–197]. This could be related with the gender of the participants in the studies. A recent research suggests that there are gender differences in the vitamin D that affect functional performance [198].

In an attempt to examine the effects of vitamin D supplementation concomitantly with resistance training, Agergaard et al. [61] performed a study in young and elderly untrained males. The participants of this study were randomized to a daily supplementation of vitamin D plus calcium or just calcium for a period of 16 weeks. The study took place during a period and at latitude of low sunlight (December–April, 56°N). All individuals followed a progressive resistance training program of the quadriceps muscle during the last 12 weeks of the study. Measurements of muscle hypertrophy (defined as changes in CSA) and quadriceps isometric strength levels were performed. Muscle biopsies were analyzed for fiber type morphology, expression of VDR, and Myostatin [61]. The authors failed to observe any difference between the groups in quadriceps CSA and isometric strength despite the observed increase in strength that was evident in both groups compared to baseline. Vitamin D intake and strength training increased both strength and CSA in elderly individuals compared to the young group. In the young vitamin D group, the authors observed that the experimental period resulted in an increase in fiber type IIa percentage and a lower Myostatin mRNA expression compared to the control group. No additive effect of vitamin D intake during 12 weeks of resistance training could be detected on either whole muscle hypertrophy or muscle strength. The authors attributed this lack of response to vitamin D supplementation to the fact that the participants in their study were vitamin D sufficient, compared to the findings of studies that have observed enhanced performance to vitamin D supplementation that have employed individuals in a deficient state [191–193, 196, 199, 200]. They suggest that the muscular performance benefits may be relevant only for individuals with vitamin D deficiency. However, a recent study using insufficient young males failed to report improvements to performance via vitamin D treatment [201]. The authors observed that despite the elevation in vitamin D levels (>120 nmol/L), no effect was observed in performance parameters (muscle strength evaluated by isokinetic dynamometry and percutaneous isometric electromyostimulation). The latter study suggests that other parameters apart from vitamin D levels are responsible for the effect on performance such as age, type of vitamin D that was given, period of supplementation, and amount of supplementation.

A recent review of the literature which included controlled and randomized controlled trials that have measured muscle strength and serum vitamin concentration in 18–40-year-old participants reported that the outcome of the available evidence is that vitamin D supplementation increases upper and lower limb strength in young and healthy individuals [200, 200, 202]. However, this increase seems to be dependent on the supplementation treatment employed. According to Tomlinson et al. [202], studies that have reported individual significant increases in muscle strength employed a total dosage of between 60,000 and 14,000 IU vitamin D/week over 6–4 months. Another study prescribed a daily dose of 2000 IU vitamin D as per previously published systematic reviews which suggested that daily dosages are more beneficial than larger, staggered doses. However, adding to the contradiction, Gupta et al. [200], who also reported significant results, used a large, weekly dose of vitamin D, 60,000 IU D3/week for the first 8 weeks followed by 60,000 IU/month for 4 months. The findings of these studies contradict previous evidence that suggested larger weekly, monthly, or one-off doses of vitamin D are not as effective in producing an improvement in muscular strength. According to Tomlinson et al. [202], the inconsistencies in the literature are partly the result of different dosages of vitamin D and different treatment regimens.

16. Vitamin D levels, Athletes, and exercise performance

The first evidence regarding the possible association between vitamin D and performance comes from early studies [41]. It have been reported that cardiovascular fitness, muscle endurance, and speed were enhanced after exposure to ultraviolet radiation [41].

To date there is a growing number of evidence (**Table 5**) to support the association between this secosteroid and exercise performance indices in athletic and physically active populations. A recent study from our laboratory [28] reported a linear relationship between vitamin D levels and muscle strength as evaluated by squat jump (SJ) and countermovement jump (CMJ), sprinting ability (10 m and 20 m), and VO_{2max} in non-supplemented professional soccer players. Our results are comparable with others showing that vitamin D levels are related with neuromuscular performance capacity and aerobic endurance in young physically active individuals and athletes [202–205]. However, these findings are not universal. Others have failed to find an association between exercise performance and vitamin D levels [206]. Vitamin D status was not associated with grip strength or swimming performance in adolescent swimmers [207] nor with isokinetic peak torque during knee flexion and extension after adjusting for total body mass and lean mass in Qatar soccer players [208]. Hamilton et al. [208] reported that vitamin D levels were not related with lower limb isokinetic muscle function in soccer players. According to the authors, the lack of an association was most probably a result of the different modes of exercise used in their study. They hypothesized that vitamin D could preferentially affect muscle groups that were not evaluated in the study. In this study, soccer players with vitamin D levels <50 nmol/L exhibited significantly lower torque in hamstring and quadriceps muscle groups, suggesting that low vitamin D levels are associated with impaired muscular function during exercise. However, no association was observed between performance and vitamin D levels in hockey players [206]. Indeed, low explosive strength during jumping was not associated with vitamin D levels neither status nor power production during the Wingate anaerobic test [206].

Study	No. of participants	Sport	Study location/period	Parameters measured	Results
Koundourakis et al. [28]	67 male professional players, age 25.6 ± 6.2 y	Soccer	6-week off-season transition period, June–July, 35.9°N latitude	Squat jump (SJ), countermovement jump (CMJ), 10m sprint, 20 m sprint, maximal oxygen consumption (VO_{2max})	Significant associations between vitamin D levels and all the measured performance indices
Fitzgerald et al. [206]	53 junior and collegiate male ice hockey players, age 20.1 ± 1.5 y	Hockey	Off-season period, 44.9°N latitude	Grip strength, Wingate anaerobic test, squat jump (SJ)	Significant associations between vitamin D levels and grip strength, time to half peak force but not Wingate test and SJ
Forney et al. [205]	39 males and females, physically active individuals, age 23.26 ± 0.7 y	Various types of physical activity	July and September, 30.46°N latitude	Maximal oxygen consumption (VO_{2max}), Wingate test, vertical and horizontal jump, sit and reach test, maximum muscle strength of bench press, leg curl, leg extension, upright row, bicep curl, triceps pushdown	Significant positive associations between vitamin D levels and VO_{2max}, but not for any measures of strength and power
Dubnov-Raz et al. [207]	80 competitive adolescent swimmers from both sexes, age 14 ± 1.6 y	Swimming	Israel, located between latitudes 32.40°N and 31.55°N	Grip strength, balance, and swimming performance at several speeds	No significant associations between vitamin D levels and any measures of performance

y, years

Table 5. Vitamin D levels and athletes.

Regarding VO_{2max} and aerobic capacity, evidence from large epidemiologic cross-sectional investigations indicate that 25(OH)D concentration is associated with cardiovascular fitness. Mowry et al. [204] found a positive correlation between baseline cardiorespiratory fitness VO_{2max} and serum 25(OH)D levels in 16- to 24-year-old healthy women. Similarly, Ardestani et al. [209] reported that there was evident a strong association between 25(OH)D levels and VO_{2max} in men and women (20–73 years old) over a broad range of 25(OH)D levels (10–82 ng/mL). Interestingly, this correlation was also evident in subjects with low levels of physical activity after adjusting for age, gender, BMI, and moderate to vigorous physical activity. Early studies have shown that exposed to sun athletes showed improved aerobic fitness compared to controls [28]. A recent study in professional soccer players reported that VO_{2max} was significantly related to vitamin D status [28]. Similar findings were reported in young physically active individuals [205]. The authors also reported that the individuals who had 25OHD levels above the recommended limit of 35 ng/mL, which was used as a cut-off value indicating vitamin D sufficiently in their study, were found to have significantly higher VO_{2max} values (20 %) than individuals below this cut-off point. On the contrary, it was reported that vitamin D status was not associated with neither the VO_{2peak}, as a measure of cardiorespiratory fitness, nor the end stage completed during a skating treadmill GXT in junior and collegiate male hockey players [206]. The authors suggested that these findings could be the result of the fact that none of their players were vitamin D deficient. According to Wilson et al. [210], worsening in aerobic capacity could result only as in vitamin D-deficient individuals. In the study by Fitzgerald et al. [206], the authors reported that although 37.7 % of their athletes had insufficient 25(OH)D levels, levels at the deficiency range are necessary for a compromise of cardiovascular fitness to ensue.

17. Vitamin D supplementation and athletes

Few published studies have reported that vitamin D supplementation benefits neuromuscular and aerobic performance capacity (**Table 6**). Indeed, a randomized placebo-controlled study examined the effects of vitamin D3 supplementation (5000 IU per day) for 8 weeks on musculoskeletal performance [211] in highly trained male professional soccer players. Approximately 62 % of the athletes on supplementation and 73 % of the controls exhibited serum total 25(OH)D < prior to treatment at 50 nmol/L. Vitamin D supplementation increased serum total 25(OH)D from a mean baseline of 29–103 nmol/L, whereas the placebo group showed no significant change. The authors found that vitamin D supplementation increased 10 m sprint time and vertical jump compared to the placebo group. Similar data were found in another study where athletes living at northerly latitudes (the United Kingdom = 53°N) had low vitamin D levels (<50 nmol/L). Additionally these data suggest that inadequate vitamin D levels are detrimental to musculoskeletal performance in athletes, while supplementation may increase both sprinting capacity and jumping ability of the soccer players. In support to these findings, Wyon et al. [212] reported enhanced neuromuscular performance in a different type of athletic population, i.e., in elite classical ballet dancers in a study which employed a 4-month oral supplementation of vitamin D3. Isometric muscular strength and vertical jump height were measured pre- and post intervention. A significant increase of isometric strength (18.7 %) and vertical jump (7.1 %) was observed. The intervention group also sustained significantly less injuries than the controls during the study period. In agreement with these data are the findings of another study from the same group [213]. The authors examined the acute effects of vitamin D supplementation on muscle function using isokinetic dynamometry. White adult male national-level judoka athletes who were involved in full-time training participated in the study and were randomly allocated to the treatment (150,000 IU vitamin D3) or placebo and given blinded supplements by an independent researcher. Participants were tested twice, 8 days apart, on a Monday morning before the start of judo training and after 2 days of rest. A 5–7 mL of blood sample was collected followed by isokinetic concentric quadriceps and hamstring. The treatment group had a significant increase in serum 25(OH)D3 levels (34 %) and muscle strength (13 %). However, other studies were unable to document any benefit following vitamin D supplementation in athletes with adequate or moderately deficient levels of vitamin D prior to supplementation. Close et al. [214] examined the effects of vitamin D3 supplementation on serum 25[OH]D concentrations and on several exercise performance indices in club-level athletes. Participants were randomized into those receiving placebo, 20,000 or 40,000 IU per week of oral vitamin D3 for a 12-week period. At baseline 57 % of the participants were found to be almost vitamin D deficient (mean 51 nmol/L). In the two supplementation groups, the following 6-week and 12-week period either with 20,000 IU vitamin D levels increased to 79 and 85 nmol/l, respectively, or with 40,000 IU vitamin D3 to 98 and 91 nmol/l, respectively. Despite the different supplementation dosages, all individuals managed to reach levels greater than the suggested deficiency limit of 50 nmol/l. However, despite the observed increase in vitamin D serum levels, neither group exhibited an improvement of their exercise performance compared to controls [214]. The inability of the authors to document any beneficial effect of vitamin D on athletic performance may be due the fact that the participants of their study were not vitamin D deficient. It has been shown that the beneficial effects of vitamin D supplementation take place in individuals with significant vitamin D deficiency [158]. The lack of any beneficial effect of vitamin D supplementation was also reported in another study involving football players [215]. A placebo group and one receiving 5000 IU/day of vitamin D were examined. Both groups followed the same high-intensity training regime and had total work, sprint performance (5, 10, 20, 30 m), and jumping ability (squat jump and countermovement jump) measured. No significant differences were evident at baseline between the two groups for any of the measured parameters. The authors reported that although all examined performance indices, apart from the 30 m sprint value, were significantly higher as a result of the high-intensity interval training regime, the mean change scores (obtained values at the end of the study minus those of baseline) did not differ significantly between the two groups, an outcome suggesting that vitamin D supplementation did not had any beneficial effect. It should be noted here that the baseline vitamin D levels of the majority soccer players in this study were deficient. Therefore the proposed mechanism by Dahlquist et al. [158] was not applicable in this study. The authors suggest that other factors apart of vitamin D levels play a pivotal role including their initial training status. It is true though that the studies evaluating the effects of vitamin D supplementation on aerobic capacity in athletes are few and between. Jastrzebski et al. [216] have examined the effects of vitamin

D supplementation on VO_{2max}. The treatment group was composed of 14 elite lightweight rowers having sufficient 25(OH)D concentrations (>30 ng/mL). They were put under vitamin D supplementation of 6000 IU per day for an 8-week period. The authors found a significant increase in VO_{2max} following vitamin D supplementation. Thus, it was concluded that 8 weeks of vitamin D supplementation during the training cycle of the same period of time resulted in enhanced aerobic metabolism in this type of athletes. However, since this is the only study that has examined vitamin D supplementation in athletes, these results cannot be generalized to all sporting disciplines. Furthermore, the low sample size of the study is another limiting factor regarding these findings.

Study	Quality	Sports	Participants	Intervention	Parameters measured	Results
Close et al. [211]	Placebo-controlled trial	Soccer	11 male soccer players, age = 18 ± 5 y	8 weeks, supplementation of 5000 IU/day of vitamin D3 or a cellulose placebo	1RM bench press and back squat, 10 m and 30m sprint, Illinois agility test, vertical jump	Significant ↑ in vertical jump height, 10m sprint times, and a trend for ↑ bench press and back squat 1RM
Wyon et al. [212]	Controlled trial	Ballet	24 elite male and female classical ballet dancers, age 28 ± 3.85 y	4-month supplementation of vitamin D3 (2000 IU per day)	Isometric muscular strength and vertical jump height	Significant ↑ in isometric muscular strength and vertical jump height
Wyon et al. [213]	Randomized controlled double-blind trial	Judoka	22 white male national-level judoka athletes. 27.5± 4 y	Acute effects, 1 week, one single dose of 150 000 IU vitamin D3 or placebo	Isokinetic concentric quadriceps, hamstring muscle strength	Significant ↑ in all measured strength-related indices
Close et al. [214]	Randomized controlled trial	Various sports	30 club-level athletes	Three groups receiving a placebo (PLB) either 20,000 or 40,000 IU/week of oral vitamin D3 for 12 weeks	RM bench press, leg press, vertical jump height	Neither dose given for 12 weeks improved the examined measures of physical performance
Jastrzebski, [216]	Randomized controlled trial	Rowing	14 elite lightweight rowers	8 weeks of 6000 IU/day of vitamin D3 in sufficient athletes (>30 ng/mL)	Maximal oxygen consumption (VO_{2max})	Significant ↑ in VO_{2max}
Jastrzębska et al. [215]	Controlled trial	Soccer	36 young soccer players, age: 17.5 ± 0.6 y	Treatment: around 5000 IU/day placebo— sunflower oil	Wingate test. 5, 10, 20, and 30 m sprint, squat jump (SJ), countermovement jump (CMJ)	No significant difference in any of the measured exercise performance parameters

y, years; ↑, increase

Table 6. Vitamin D supplementation and athletes.

18. Conclusion

This chapter describes the role of vitamin D in athletic performance. It describes the effect of vitamin D in several systems involved in exercise performance and physical activity in both athletes and non-athletes.

Vitamin D undoubtedly has a key role in the osseous health of both athletes and non-athletes. A growing body of evidence suggests that vitamin D is strongly related with the cardiovascular, immune, and neuromuscular systems. It is now clear that vitamin D plays an important role in the ability to perform efficiently during the normal daily activities, while its deficiency may result in poorer exercise performance. It should be mentioned that the vast majority form the evidence regarding vitamin D and exercise come from non-athletes. A growing number of studies have suggested that vitamin D exerts a beneficial effect on exercise performance, particularly in older adults and those with lower levels of circulating vitamin D. Regarding the optimum levels of vitamin D, the current evidence suggests that value levels above 30 ng/mL are needed. The majority of the vitamin D supplementation studies indicate that it may beneficially affect physical activity, frailty, muscular tone, stability, mobility in the elderly, and performance in physically active individuals or athletes only when pretreatment levels are within the insufficiency range (<30 ng/ml). Specifically for athletes, although the results of performance trials are not yet convincing enough to support vitamin D as a direct performance enhancer, obtaining optimal 25(OH)D levels can reduce the risk of debilitating musculoskeletal injury, and due to the active role in muscle, resolution of vitamin D insufficiency has the potential to impact performance.

Author details

Nikolaos E. Koundourakis* and Andrew N. Margioris

*Address all correspondence to: nkoyndoy@hotmail.com

Department of Clinical Chemistry, School of Medicine, University of Crete, Heraklion, Greece

References

[1] Holick MF: Vitamin D: a millennium perspective. J Cell Biochem. 2003;88:296–307. doi: 10.1002/jcb.10338

[2] Muir SW, Montero-Odasso M: Effect of vitamin D supplementation on muscle strength, gait and balance in older adults: a systematic review and meta-analysis. J Am Geriatr Soc. 2011;59:2291–300.

[3] Holick M: Sunlight and vitamin D. J Gen Intern Med. 2002;17:733–5. doi: 10.1046/j.1525-1497.2002.20731.x

[4] Todd JJ, Pourshahidi LK, McSorley EM, Madigan SM, Magee PJ: Vitamin D: recent advances and implications for athletes. Sports Med. 2015;45:213–29. doi: 10.1007/s40279-014-0266-7.

[5] Pojednic RM, Ceglia L: The emerging biomolecular role of vitamin D in skeletal muscle. Exerc Sport Sci Rev. 2014;42(2):76–81. doi: 10.1249/JES.0000000000000013.

[6] Artaza JN, Mehrotra R, Norris KC: Vitamin D and the cardiovascular system. Clin J Am Soc Nephrol. 2009;4(9):1515–22. doi: 10.2215/CJN.02260409.

[7] Moran DS, McClung JP, Kohen T, Lieberman HR: Vitamin D and physical performance. Sports Med. 2013;43(7):601–11. doi: 10.1007/s40279-013-0036-y.

[8] Holick MF: Vitamin D deficiency. N Engl J Med. 2007;357(3):266–81. doi: 10.1056/NEJMra070553

[9] Hamilton B: Vitamin D and human skeletal muscle. Scand J Med Sci Sports. 2010;20:182–90. doi: 10.1111/j.1600-0838.2009.01016.x

[10] Cashman KD: Calcium and vitamin D. Novartis Found Symp. 2007;282:123–38.

[11] Tavera-Mendoza LE, White JH. Cell defenses and the sunshine vitamin. Sci Am. 2007;297(5):62–5, 68–70, 72.

[12] Udowenko M, Trojian T: Vitamin D: extent of deficiency, effect on muscle function, bone health, performance, and injury prevention. Conn Med. 2010;74(8):477–80.

[13] Baker AR, McDonnell DP, Hughes M, Crisp TM, Mangelsdorf DJ, Haussler MR, Pike JW, Shine J, O'Malley W: Cloning and expression of full-length cDNA encoding human vitamin D receptor. Proc Natl Acad Sci U S A. 1988;85:3294–8.

[14] Brumbaugh PF, Haussler DH, Bressler R, Haussler MR: Radioreceptor assay for 1 alpha, 25-dihydroxyvitamin D3. Science. 1974;183:1089–91.

[15] Brumbaugh PF, Haussler MR: Specific binding of 1a,25-dihydroxycholecalciferol to nuclear components of chick intestine. J Biol Chem. 1975;250:1588–94.

[16] Saccone D, Asani F, Bornman L: Regulation of the vitamin D receptor gene by environment, genetics and epigenetics. Gene. 2015;561(2):171–80. doi: 10.1016/j.gene.2015.02.024.

[17] Christakos S, Hewison M, Gardner DG, Wagner CL, Sergeev IN, Rutten E, Pittas AG, Boland R, Ferrucci L, Bikle DD: Vitamin D: beyond bone. Ann N Y Acad Sci. 2013;1287:45–58. doi: 10.1111/nyas.12129.

[18] Soni M, Kos K, Lang IA, Jones K, Melzer D, Llewellyn DJ: Vitamin D and cognitive function. Scand J Clin Lab Invest Suppl. 2012;243:79–82. doi: 10.3109/00365513.2012.681969.

[19] Halliday TM, Peterson NJ, Thomas JJ, Kleppinger K, Hollis BW, Larson-Meyer DE: Vitamin D status relative to diet, lifestyle, injury, and illness in college athletes. Med Sci Sports Exerc. 2011;43(2):335–43. doi: 10.1249/MSS.0b013e3181eb9d4d.

[20] Holick MF: Sunlight and vitamin D for bone health and prevention of autoimmune diseases, cancers, and cardiovascular disease. Am J Clin Nutr. 2004;80 (6 suppl):1678S–88S.

[21] Meyer C. Scientists probe role of vitamin D: deficiency a significant problem, experts say. JAMA. 2004;292(12):1416–8.

[22] Zittermann A. Vitamin D in preventive medicine: are we ignoring the evidence? Br J Nutr. 2003;89(5):552–72.

[23] Farrokhyar F, Tabasinejad R, Dao D, Peterson D, Ayeni OR, Hadioonzadeh R, Bhandari M: Prevalence of vitamin D inadequacy in athletes: a systematic-review and meta-analysis. Sports Med. 2015;45(3):365–78. doi: 10.1007/s40279-014-0267-6.

[24] Wacker M, Holick MF: Vitamin D—effects on skeletal and extraskeletal health and the need for supplementation. Nutrients. 2013;5(1):111-48. doi: 10.3390/nu5010111.

[25] Golden NH, Abrams SA: Committee on Nutrition. Optimizing bone health in children and adolescents. Pediatrics. 2014;134(4):e1229–43. doi: 10.1542/peds.2014-2173.

[26] Morton JP, Iqbal Z, Drust B, Burgess D, Close GL, Brukner PD: Seasonal variation in vitamin D status in professional soccer players of the English Premier League. Appl Physiol Nutr Metab. 2012;37:798–802. doi: 10.1139/h2012-037.

[27] Larson-Meyer DE, Willis KS: Vitamin D and athletes. Curr Sports Med Rep. 2010;9(4): 220–6. doi: 10.1249/JSR.0b013e3181e7dd45.

[28] Koundourakis NE, Androulakis NE, Malliaraki N, Margioris AN: Vitamin D and exercise performance in professional soccer players. PLoS ONE. 2014;9(7):e101659. doi: 10.1371/journal.pone.0101659.

[29] Angeline ME, Gee AO, Shindle M, Warren RF, Rodeo SA: The effects of vitamin D deficiency in athletes. Am J Sports Med. 2013;41:461–4. doi: 10.1177/0363546513475787.

[30] Lovell G: Vitamin D status of females in an elite gymnastics program. Clin J Sport Med. 2008;18(2):159–61. doi: 10.1097/JSM.0b013e3181650eee.

[31] Thacher TD, Clarke BL: Vitamin D insufficiency. Mayo Clin Proc. 2011;86:50–60. doi: 10.4065/mcp.2010.0567.

[32] Kopeć A, Solarz K, Majda F, Słowińska-Lisowska M, Mędraś M: An evaluation of the levels of vitamin D and bone turnover markers after the summer and winter periods in polish professional soccer players. J Hum Kinet. 2013;38:135–40. doi: 10.2478/hukin-2013-0053.

[33] Willis KS, Smith DT, Broughton KS, Larson-Meyer DE. Vitamin D status and biomarkers of inflammation in runners. Open Access J Sports Med. 2012;3:35–42. doi: 10.2147/OAJSM.S31022.

[34] Halliday TM, Peterson NJ, Thomas JJ, Kleppinger K, Hollis BW, Larson-Meyer DE: Vitamin D status relative to diet, lifestyle, injury, and illness in college athletes. Med Sci Sports Exerc. 2011;43(2):335–43. doi: 10.1249/MSS.0b013e3181eb9d4d.

[35] Constantini NW, Arieli R, Chodick G, Dubnov-Raz G: High prevalence of vitamin D insufficiency in athletes and dancers. Clin J Sport Med. 2010;20(5):368–71. doi: 10.1097/JSM.0b013e3181f207f2.

[36] Elsammak MY, Al-Wosaibi AA, Al-Howeish A, Alsaeed J: Vitamin d deficiency in Saudi Arabs. Horm Metab Res. 2010;42:364–8. doi: 10.1055/s-0030-1248296.

[37] Babu US, Calvo MS: Modern India and the vitamin D dilemma: evidence for the need of a national food fortification program. Mol Nutr Food Res. 2010;54:1134–47. doi: 10.1002/mnfr.200900480.

[38] Maxwell JD. Seasonal variation in vitamin D. Proc Nutr Soc 1994;53:533–43.

[39] Storlie DM, Pritchett K, Pritchett R, Cashman L. 12-Week vitamin D supplementation trial does not significantly influence seasonal 25(OH)D status in male collegiate athletes. Int J Health Nutr.2011;2:8–13.

[40] Galan F, Ribas J, Sánchez-Martinez PM, Calero T, Sánchez AB, Muñoz A. Serum 25-hydroxyvitamin D in early autumn to ensure vitamin D sufficiency in mid-winter in professional football players. Clin Nutr.2012;31:132–6. doi: 10.1016/j.clnu.2011.07.008.

[41] Cannell JJ, Hollis BW, Sorenson MB, Taft TN, Anderson JJ. Athletic performance and vitamin D. Med Sci Sport Exer. 2009;41:1102–10.

[42] Koch H, Raschka C: Circannual period of physical performance analysed by means of standard cosinor analysis: a case report. Rom J Physiol. 2000;37(1–4):51–8.

[43] Erikssen J, Rodahl K. Seasonal variation in work performance and heart rate response to exercise. A study of 1,835 middle-aged men. Eur J Appl Physiol Occup Physiol. 1979;42(2):133–40.

[44] Svedenhag J, Sjodin B. Physiological characteristics of elite male runners in and off-season. Can J Appl Sport Sci. 1985;10(3):127–33.

[45] Mozos I, Marginean O: Links between Vitamin D deficiency and cardiovascular diseases. Biomed Res Int. 2015;2015:109275. doi: 10.1155/2015/109275.

[46] Bonjour JP: Calcium and phosphate: a duet of ions playing for bone health. Am Coll Nutr. 2011;30(5 Suppl 1):438S–48S

[47] Holick, MF: Vitamin D deficiency. N Engl J Med. 2007, 357, 266–81.

[48] Christakos S, Dhawan P, Porta A, Mady LJ, Seth T: Vitamin D and intestinal calcium absorption. Mol Cell Endocrinol. 2011;347:25–29. doi: 10.1016/j.mce.2011.05.038.

[49] Kulie T, Groff A, Redmer J, Hounshell J, Schrager S: Vitamin D: an evidence-based review. J Am Board Fam Med. 2009;22(6):698–706. doi: 10.3122/jabfm.2009.06.090037.

[50] Kota S, Jammula S, Kota S, Meher L, Modi K: Correlation of vitamin D, bone mineral density and parathyroid hormone levels in adults with low bone density. Indian J Orthop. 2013;47(4):402–7. doi: 10.4103/0019-5413.114932.

[51] Ethgen O, Hiligsmann M, Burlet N, Reginster JY: Public health impact and cost-effectiveness of dairy products supplemented with vitamin D in prevention of osteoporotic fractures. Arch Public Health. 2015;14:48. doi: 10.1186/s13690-015-0099-3.

[52] Buchebner D, McGuigan F, Gerdhem P, Ridderstråle M, Akesson K: Association between hypovitaminosis D in elderly women and long- and short-term mortality-results from the osteoporotic prospective risk assessment cohort. J Am Geriatr Soc. 2016;64:990–7. doi: 10.1111/jgs.14087.

[53] Sundh D, Mellström D, Ljunggren Ö, Karlsson MK, Ohlsson C, Nilsson M, Nilsson AG, Lorentzon M: Low serum vitamin D is associated with higher cortical porosity in elderly men. J Intern Med. 2016;15. doi: 10.1111/joim.12514. [Epub ahead of print].

[54] Scino Y, Ishizuka S, Shima M, Tanaka H: Vitamin D in bone formation. Osteoporosis Int. 1993;Suppl. 1:S196–198.

[55] Jones BH, Thacker SB, Gilchrist J, Kimsey CD Jr, Sosin DM: Prevention of lower extremity stress fractures in athletes and soldiers: a systematic review. Epidemiol Rev. 2002;24:228–47.

[56] McClung JP, Karl JP: Vitamin D and stress fracture: the contribution of vitamin D receptor gene polymorphisms. Nutr Rev. 2010;68(6):365–9. doi: 10.1111/j.1753-4887.2010.00295.x.

[57] Friedl KE, Evans RK, Moran DS: Stress fracture and military medical readiness: bridging basic and applied research. Med Sci Sports Exerc. 2008;40 (11 Suppl):S609–22. doi: 10.1249/MSS.0b013e3181892d53.

[58] Willis KS: Vitamin D. Status and immune system biomarkers in athletes. Laramie: University of Wyoming; 2008. p. 85.

[59] Prineas JW, Mason AS, Henson RA: Myopathy in metabolic bone disease. Br Med J. 1965;1:1034–6.

[60] Wang Y, Zhu J, DeLuca HF: Where is the vitamin D receptor? Arch Biochem Biophys. 2012;523:123–33. doi: 10.1016/j.abb.2012.04.001.

[61] Agergaard J, Trøstrup J, Uth J, Iversen JV, Boesen A, Andersen JL, Schjerling P, Langberg H: Does vitamin-D intake during resistance training improve the skeletal muscle

hypertrophic and strength response in young and elderly men?—a randomized controlled trial. Nutr Metab (Lond). 2015;12:32. doi: 10.1186/s12986-015-0029-y.

[62] Garay E, Jankowski P, Lizano P, Marczak S, Garay E, Jankowski P, Lizano P, Marczak S, Maehr H, Adorini L, Uskokovic MR, Studzinski GP: Calcitriol derivatives with two different side-chains at C-20. Part 4: further chain modifications that alter VDR-dependent monocytic differentiation potency in human leukemia cells. Bioorg Med Chem. 2007;15(13):4444–55.

[63] Garcia LA, King KK, Ferrini MG, Norris KC, Artaza JN: 1,25(OH)2vitamin D3 stimulates myogenic differentiation by inhibiting cell proliferation and modulating the expression of promyogenic growth factors and myostatin in C2C12 skeletal muscle cells. Endocrinology. 2011;152(8):2976–86. doi: 10.1210/en.2011-0159.

[64] Lee SJ, Lee YS, Zimmers TA, Soleimani A, Matzuk MM, Tsuchida K, Cohn RD, Barton ER: Regulation of muscle mass by follistatin and activins. Mol Endocrinol. 2010;24:1998–2008. doi: 10.1210/me.2010-0127.

[65] Angeline ME, Gee AO, Shindle M, Warren RF, Rodeo SA: The effects of vitamin D deficiency in athletes. Am J Sports Med. 2013;41:461–4. doi: 10.1177/0363546513475787.

[66] Black PN and, Scragg R: Relationship between serum 25-hydroxyvitamin D and pulmonary function in the third national health and nutrition examination survey. Chest. 2005;128:3792–8.

[67] Deschenes MR, Kraemer WJ: Performance and physiologic adaptations to resistance training. Am J Phys Med Rehabil. 2002;81(11), S3–S16.

[68] Shuler FD, Wingate MK, Moore GH, Giangarra C: Sports health benefits of vitamin d. Sports Health. 2012;4(6):496–501. doi: 10.1177/1941738112461621.

[69] Beavers KM, Beavers DP, Houston DK, Harris TB, Hue TF, Koster A, Newman AB, Simonsick EM, Studenski SA, Nicklas BJ, Kritchevsky SB: Associations between body composition and gait-speed decline: results from the Health, Aging, and Body Composition study. Am J Clin Nutr. 2013;97(3):552–60. doi: 10.3945/ajcn. 112.047860.

[70] Birge SJ, Haddad JG: 25-hydroxycholecalciferol stimulation of muscle metabolism. J Clin Invest. 1975;56(5):1100–7.

[71] Oh J, Weng S, Felton SK, Bhandare S, Riek A, Butler B, Proctor BM, Petty M, Chen Z, Schechtman KB, Bernal-Mizrachi L, Bernal-Mizrachi C: 1,25(OH)2 vitamin d inhibits foam cell formation and suppresses macrophage cholesterol uptake in patients with type 2 diabetes mellitus. Circulation. 2009;120(8):687–98. doi: 10.1161/CIRCULATIONAHA.109.856070.

[72] Rigby WF, Denome S, Fanger MW: Regulation of lymphokine production and human T lymphocyte activation by 1,25-dihydroxyvitamin D3. Specific inhibition at the level of messenger RNA. J Clin Invest. 1987;79(6):1659–64.

[73] Simpson RU, Hershey SH, Nibbelink KA: Characterization of heart size and blood pressure in the vitamin D receptor knockout mouse. J Steroid Biochem Mol Biol. 2007;103:521–4.

[74] Saleh FN, Schirmer H, Sundsfjord J, Jorde E: Parathyroid hormone and left ventricular hypertrophy. Eur Heart J. 2003;24:2054–60.

[75] Al Mheid I , Patel R, Murrow J, Morris A, Rahman A, Fike L, Nino Al Mheid I, Patel R, Murrow J, Morris A, Rahman A, Fike L, Kavtaradze N, Uphoff I, Hooper C, Tangpricha V, Alexander RW, Brigham K, Quyyumi AA. Vitamin D status is associated with arterial stiffness and vascular dysfunction in healthy humans. J Am Coll Cardiol. 2011;58(2): 186–92. doi 10.1016/j.jacc.2011.02.051.

[76] Tomaschitz A, Pilz S, Ritz E, Grammer T, Drechsler C, Boehm BO, März W: Independent association between 1,25-dihydroxyvitamin D, 25-hydroxyvitamin D and the renin-angiotensin system: The Ludwigshafen Risk and Cardiovascular Health (LURIC) study. Clin Chim Acta. 2010;411:1354–60. doi: 10.1016/j.cca.2010.05.037.

[77] Wang TJ, Pencina MJ, Booth SL, Jacques PF, Ingelsson E, Lanier K, Benjamin EJ, D'Agostino RB, Wolf M, Vasan RS: Vitamin D deficiency and risk of cardiovascular disease. Circulation. 2008;117(4):503–11. doi: 10.1161/CIRCULATIONAHA.107.706127.

[78] Lee JH, O'Keefe JH, Bell D, Hensrud DD, Holick MF: Vitamin D deficiency an important, common, and easily treatable cardiovascular risk factor? J Am Coll Cardiol. 2008;52(24): 1949–56. doi: 10.1016/j.jacc.2008.08.050.

[79] McDermott MM, Liu K, Ferrucci L, Tian L, Guralnik J, Kopp P, Van Horn L, Liao Y, Green D, Kibbe M, Sufit R, Zhao L, Criqui MH. Vitamin D status, functional decline, and mortality in peripheral artery disease. Vasc Med. 2014;19(1):18–26. doi: 10.1177/1358863X13518364.

[80] Al Mheid I, Patel RS, Tangpricha V, Quyyumi AA: Vitamin D and cardiovascular disease: is the evidence solid? Eur Heart J. 2013;34(48):3691–8. doi: 10.1093/eurheartj/eht166.

[81] Drechsler C, Pilz S, Obermayer-Pietsch B, Verduijn M, Tomaschitz A, Krane V, Espe K, Dekker F, Brandenburg V, März W, Ritz E, Wanner C: Vitamin D deficiency is associated with sudden cardiac death, combined cardiovascular events, and mortality in haemodialysis patients. Eur Heart J. 2010;31:2253–61. doi: 10.1093/eurheartj/ehq246.

[82] Deo R, Katz R, Shlipak MG, Sotoodehnia N, Psaty BM, Sarnak MJ, Fried LF, Chonchol M, de Boer IH, Enquobahrie D, Siscovick D, Kestenbaum B: Vitamin D, parathyroid hormone, and sudden cardiac death: results from the Cardiovascular Health Study. Hypertension. 2011;58:1021–8. doi: 10.1161/HYPERTENSIONAHA.111.179135.

[83] Naesgaard PA, León De La Fuente RA, Nilsen ST, Woie L, Aarsland T, Brede C, Staines H, Nilsen DW: Serum 25(OH)D is a 2-year predictor of all-cause mortality, cardiac death

and sudden cardiac death in chest pain patients from Northern Argentina. PLoS One. 2012;7:e43228. doi: 10.1371/journal.pone.0043228.

[84] Artaza JN, Mehrotra R, Norris KC: Vitamin D and the cardiovascular system. Clin J Am Soc Nephrol. 2009;4:1515–22. doi: 10.2215/CJN.02260409.

[85] Bajaj A, Stone KL, Peters K, Parimi N, Barrett-Connor E, Bauer D, Cawthon PM, Ensrud KE, Hoffman AR, Orwoll E, Schernhammer ES: Circulating vitamin D, supplement use, and cardiovascular disease risk: the MrOS Sleep Study. J Clin Endocrinol Metab. 2014;99:3256–62. doi: 10.1210/jc.2013-4178.

[86] Martins D, Wolf M, Pan D, Zadshir A, Tareen N, Thadhani R, Felsenfeld A Levine B, Mehrotra R, Norris K: Prevalence of cardiovascular risk factors and the serum levels of 25-hydroxyvitamin D in the United States: data from the Third National Health and Nutrition Examination Survey. Arch Intern Med. 2007;167:1159–65.

[87] Gouni-Berthold I, Krone W, Berthold HK: Vitamin D and cardiovascular disease. Curr Vasc Pharmacol. 2009;7:414–22.

[88] Mpandzou G, Aït Ben Haddou E, Regragui W, Benomar A, Yahyaoui M: Vitamin D deficiency and its role in neurological conditions: a review. Rev Neurol (Paris). 2016;172(2):109–22. doi: 10.1016/j.neurol.2015.11.005.

[89] Wrzosek M, Łukaszkiewicz J, Wrzosek M, Jakubczyk A, Matsumoto H, Piątkiewicz P, Radziwoń-Zaleska M, Wojnar M, Nowicka G: Vitamin D and the central nervous system. Pharmacol Rep. 2013;65(2):271–8.

[90] DeLuca GC, Kimball SM, Kolasinski J, Ramagopalan SV, Ebers GC: Review: the role of vitamin D in nervous system health and disease. Neuropathol Appl Neurobiol. 2013;39(5):458–84. doi: 10.1111/nan.12020.

[91] Mpandzou G, Aït Ben Haddou E, Regragui W, Benomar A, Yahyaoui M: Vitamin D deficiency and its role in neurological conditions: a review. Rev Neurol (Paris). 2016;172:109–22. doi: 10.1016/j.neurol.2015.11.005.

[92] Buell JS, Dawson-Hughes B: Vitamin D and neurocognitive dysfunction: preventing "D"ecline? Mol Aspects Med. 2008;29:415–22. doi: 10.1016/j.mam.2008.05.001.

[93] Zehnder D, Bland R, Williams MC, McNinch RW, Howie AJ,Stewart PM, Hewison M: Extrarenal expression of 25-hydroxyvitamin d(3)-1 alpha-hydroxylase. J Clin Endocrinol Metab. 2001;86:888–94.

[94] Cai Q, Tapper DN, Gilmour RF Jr, deTalamoni N, Aloia RC, Wasserman RH: Modulation of the excitability of avian peripheral nerves by vitamin D: relation to calbindin-D28k, calcium status and lipid composition. Cell Calcium. 1994;15:401–10.

[95] Cantorna MT, Hayes CE, and DeLuca HF: 1,25-Dihydroxyvitamin D3 reversibly blocks the progression of relapsing encephalomyelitis, a model of multiple sclerosis. Proc Natl Acad Sci USA. 1996;93:7861–64.

[96] Radak Z, Taylor AW, Ohno H, Goto S: Adaptation to exercise induced oxidative stress: from muscle to brain. Exerc Immunol Rev. 2001;7:90–107.

[97] Pirotta S, Kidgell DJ, Daly RM: Effects of vitamin D supplementation on neuroplasticity in older adults: a double-blinded, placebo-controlled randomised trial. Osteoporos Int. 2015;26(1):131–40. doi: 10.1007/s00198-014-2855-6.

[98] Taghizadeh M, Talaei SA, Djazayeri A, Salami M : Vitamin D supplementation restores suppressed synaptic plasticity in Alzheimer's disease. Nutr Neurosci. 2014;17(4):172–7. doi: 10.1179/1476830513Y.0000000080.

[99] Nutt D: GABA-A receptors: subtypes, regional distribution, and function. J Clin Sleep Med. 2006;2:S7-S11.

[100] Manini TM, Clark BC: Dynapenia and aging: an update. J Gerontol Ser A: Biol Sci Med Sci. 2012;67A(1):28–40. doi: 10.1093/gerona/glr010.

[101] Acworth I, Nicholass J, Morgan B, Newsholme EA: Effect of sustained exercise on concentrations of plasma aromatic and branched-chain amino acids and brain amines. Biochem Biophys Res Commun. 1986;137(1):149–53.

[102] Meeusen R, Watson P, Hasegawa H, Roelands B, Piacentini MF: Central fatigue: the serotonin hypothesis and beyond. Sports Med. 2006;36(10):881–909.

[103] Tague SE, Smith PG: Vitamin D receptor and enzyme expression in dorsal root ganglia of adult female rats: modulation by ovarian hormones. J Chem Neuroanat. 2011;41(1): 1–12. doi: 10.1016/j.jchemneu.2010.10.001.

[104] Dubin AE, Patapoutian A: Nociceptors: the sensors of the pain pathway. J Clin Invest. 2010;120(11):3760–72. doi: 10.1172/JCI42843.

[105] Dhesi JK, Jackson SH, Bearne LM, Moniz C, Hurley MV, Swift CG, Allain TJ: Vitamin D supplementation improves neuromuscular function in older people who fall. Age Ageing. 2004;33(6):589–95.

[106] Lord SR, Clark RD, Webster IW: Physiological factors associated with falls in an elderly population. J Am Geriatr Soc. 1991;39:1194–200.

[107] Shi YY, Liu TJ, Fu JH, Xu W, Wu LL, Hou AN, Xue XD: Vitamin D/VDR signaling attenuates lipopolysaccharide-induced acute lung injury by maintaining the integrity of the pulmonary epithelial barrier. Mol Med Rep. 2016;13(2):1186–94. doi: 10.3892/mmr.2015.4685.

[108] Rehan VK, Torday JS. Vitamin D and Lung Development in Early Life. In: Litonjua AA, editor. Vitamin D and the Lung: Mechanisms and Disease Associations. New York; Humana Press London. P. 46. doi: 10.1007/978-1-61779-888-7.

[109] Rehan VK, Torday JS, Peleg S, Gennaro L, Vouros P, Padbury J, Rao DS, Reddy GS: 1Alpha,25-dihydroxy-3- epi-vitamin D3, a natural metabolite of 1alpha,25-dihydroxy

vitamin D3: production and biological activity studies in pulmonary alveolar type II cells. Mol Genet Metab. 2002;76(1):46–56.

[110] Vedala SR, Paul N, Mane AB: Difference in pulmonary function test among the athletic and sedentary population. Natl J Physiol Pharm Pharmacol. 2013;3(2):118–23. doi: 10.5455/njppp.2013.3.109-114.

[111] Nguyen M, Trubert CL, Rizk-Rabin M, Rehan VK, Besancon F, Cayre YE, Garabedian M: 1,25-Dihydroxyvitamin D3 and fetal lung maturation: immunogold detection of VDR expression in pneumocytes type II cells and effect on fructose 1,6 bisphosphatase. J Steroid Biochem Mol Biol. 2004;89:93–97.

[112] Janssens W, Bouillon R, Claes B, Carremans C, Lehouck A, Buysschaert I, Coolen J, Mathieu C, Decramer M, Lambrechts D: Vitamin D deficiency is highly prevalent in COPD and correlates with variants in the vitamin D-binding gene. Thorax. 2010;65:215–20. doi: 10.1136/thx.2009.120659.

[113] Damera G, Fogle HW, Lim P, Goncharova EA, Zhao H, Banerjee A, Tliba O, Krymskaya VP, Panettieri RA: Vitamin D inhibits growth of human airway smooth muscle cells through growth factor-induced phosphorylation of retinoblastoma protein and checkpoint kinase 1. Br J Pharmacol. 2009;158:1429–41. doi: 10.1111/j.1476-538 1.2009.00428.x.

[114] Zosky GR, Berry LJ, Elliot JG, James AL, Gorman S, Hart PH: Vitamin D deficiency causes deficits in lung function and alters lung structure. Am J Respir Crit Care Med. 2011;183(10):1336–43. doi: 10.1164/rccm.201010-1596OC.

[115] Litonju A, Weiss S: Is vitamin D deficiency to blame for the asthma epidemic? J Allergy Clin Immunol.2007;120:1031–5.

[116] Trochoutsou AI, Kloukina V, Samitas K, Xanthou G: Vitamin-D in the immune system: genomic and non-genomic actions. Mini Rev Med Chem. 2015;15(11):953–63.

[117] Aranow C: Vitamin D and the Immune System. J Invest Med. 2011;59(6):881–6. doi: 10.231/JIM.0b013e31821b8755.

[118] Prietl B, Treiber G, Pieber TR, Amrein K: Vitamin D and immune function. Nutrients. 2013;5(7):2502–21. doi: 10.3390/nu5072502.

[119] Wang TT, Nestel FP, Bourdeau V, Nagai Y, Wang Q, Liao J, Tavera-Mendoza L, Lin R, Hanrahan JW, Mader S, White JH: Cutting edge: 1,25-dihydroxyvitamin D3 is a direct inducer of antimicrobial peptide gene expression. J Immunol. 2004;173(5):2909–12.

[120] Ramanathan B, Davis EG, Ross CR, Blecha F: Cathelicidins: microbicidal activity, mechanisms of action, and roles in innate immunity. Microbes Infect. 2002;4(3):361–72.

[121] Provvedini DM, Tsoukas CD, Deftos LJ, Manolagas SC: 1,25-Dihydroxyvitamin D3 receptors in human leukocytes. Science. 1983;221:1181–3.

[122] Lemire JM, Adams JS, Sakai R, Jordan SC: 1a,25-dihydroxyvitamin D3 suppresses proliferation and immunoglobulin production by normal human peripheral blood mononuclear cells. J Clin Investig. 1984;74:657–61.

[123] Chen S, Sims GP, Chen XX, Gu YY, Chen S, Lipsky PE: Modulatory effects of 1,25-dihydroxyvitamin D3 on human B cell differentiation. J Immunol. 2007;179:1634–47.

[124] Mahon BD, Wittke A, Weaver V, Cantorna MT: The targets of vitamin D depend on the differentiation and activation status of CD4 positive T cells. J Cell Biochem. 2003;89: 922–32.

[125] Hewison M: Vitamin D and immune function: an overview. Proc Nutr Soc. 2012;71(1): 50–61.

[126] Boonstra A, Barrat FJ, Crain C, Heath VL, Savelkoul HF, O'Garra A: 1alpha,25-Dihydroxyvitamin d3 has a direct effect on naive CD4(+) T cells to enhance the development of Th2 cells. J Immunol. 2001;167(9):4974–80.

[127] Berry DJ, Hesketh K, Power C, Hyppönen E: Vitamin D status has a linear association with seasonal infections and lung function in British adults. Br J Nutr. 2011;106:1433–40. doi: 10.1017/S0007114511001991.

[128] Ginde AA, Mansbach JM, and Camargo CA Jr: Association between serum 25-hydroxyvitamin D level and upper respiratory tract infection in the Third National Health and Nutrition Examination Survey. Arch Intern Med. 2009;169:384–90. doi: 10.1001/archinternmed.2008.560.

[129] Laaksi I, Ruohola JP, Tuohimaa P, Auvinen A, Haataja R, Pihlajamäki H, and Ylikomi T: An association of serum vitamin D concentrations < 40 nmol/L with acute respiratory tract infection in young Finnish men. Am J Clin Nutr. 2007;86(3):714–7.

[130] Gleeson M (Editor). Immune function in sport and exercise. Edinburgh: Elsevier, 2005.

[131] Li TL, Gleeson M: The effect of single and repeated bouts of prolonged cycling and circadian variation on saliva flow rate, immunoglobulin A and alpha-amylase responses. J Sports Sci. 2004;22(11–12):1015–24.

[132] de Sa Del Fiol F, Barberato-Filho S, Lopes LC, de Cassia Bergamaschi C: Vitamin D and respiratory infections. J Infect Dev Ctries. 2015;9(4):355–61. doi: 10.3855/jidc.5711.

[133] Bryson KJ, Nash AA, Norval M: Does vitamin D protect against respiratory viral infections? Epidemiol Infect. 2014;142(9):1789–801. doi: 10.1017/S0950268814000193.

[134] Liu PT, Stenger S, Li H, Wenzel L, Tan BH, Krutzik SR, Ochoa MT, Schauber J, Wu K, Meinken C, Kamen DL, Wagner M, Bals R, Steinmeyer A, Zügel U, Gallo RL, Eisenberg D, Hewison M, Hollis BW, Adams JS, Bloom BR, Modlin RL: Toll-like receptor triggering of a vitamin D-mediated human antimicrobial response. Science. 2006;311:1770–3.

[135] Larson-Meyer DE, Willis KS: Vitamin D and athletes. Curr Sports Med Rep. 2010;9(4): 220–6. doi: 10.1249/JSR.0b013e3181e7dd45.

[136] Villalta SA, Rinaldi C, Deng B, Liu G, Fedor B, Tidball JG: Interleukin-10 reduces the pathology of mdx muscular dystrophy by deactivating M1 macrophages and modulating macrophage phenotype. Hum Mol Genet. 2011;20:790–805. doi: 10.1093/hmg/ddq523.

[137] Takahashi Y, Katayose D, Shindoh C: Interleukin-13 prevents diaphragm muscle deterioration in a septic animal model. Tohoku J Exp Med. 1999;189:191–202.

[138] He C, Handzlik M, Fraser WD, Muhamad A, Preston H, Richardson A, Gleeson M: Influence of vitamin D status on respiratory infection incidence and immune function during 4 months of winter training in endurance sport athletes. Exerc Immunol Rev. 2013;19:86–101

[139] Li J, Byrne ME, Chang E, Jiang Y, Donkin SS, Buhman KK, Burgess JR, Teegarden D: 1α,25-dihydroxyvitamin D hydroxylase in adipocytes. J Steroid Biochem Mol Biol. 2008;112(1–3):122–6. doi: 10.1016/j.jsbmb.2008.09.006.

[140] Koszowska AU, Nowak J, Dittfeld A, Brończyk-Puzoń A, Kulpok A, Zubelewicz-Szkodzińska B: Obesity, adipose tissue function and the role of vitamin D. Cent Eur J Immunol. 2014;39(2):260–4. doi: 10.5114/ceji.2014.43732.

[141] Pannu PK, Zhao Y, Soares MJ : Reductions in body weight and percent fat mass increase the vitamin D status of obese subjects: a systematic review and metaregression analysis. Nutr Res. 2016;36(3):201–13. doi: 10.1016/j.nutres.2015.11.013.

[142] Rosenblum JL, Castro VM, Moore CE, Kaplan LM: Calcium and vitamin D supplementation is associated with decreased abdominal visceral adipose tissue in overweight and obese adults. Am J Clin Nutr. 2012;95(1):101–8. doi: 10.3945/ajcn.111.019489.

[143] Drincic AT, Armas LA, Van Diest EE, Heaney RP: Volumetric dilution, rather than sequestration best explains the low vitamin D status of obesity. Obesity (Silver Spring). 2012;20:1444–48. doi: 10.1038/oby.2011.404.

[144] Earthman CP, Beckman LM, Masodkar K, Sibley SD: The link between obesity and low circulating 25-hydroxyvitamin D concentrations: considerations and implications. Int J Obes. 2012;36(3):387–96. doi: 10.1038/ijo.2011.119.

[145] Datta S, Pal M, De A. The dependency of vitamin D status on anthropometric data. Malay J Med Sci. 2014;21(3):54–61.

[146] Forney LA, Earnest CP, Henagan TM, Johnson LE, Castleberry TJ, Stewart LK: Vitamin D status, body composition, and fitness measures in college-aged students. J Strength Cond Res. 2014;28(3):814–24. doi: 10.1519/JSC.0b013e3182a35ed0.

[147] Koundourakis NE, Androulakis NE, Malliaraki N, Tsatsanis C, Venihaki M, Margioris AN: Discrepancy between exercise performance, body composition, and sex steroid response after a six-week detraining period in professional soccer players. PLoS ONE. 2014;9(2):e87803. doi: 10.1371/journal.pone.0087803.

[148] Lira FS, Rosa JC, Cunha CA, Ribeiro EB, do Nascimento CO, Oyama LM, Mota JF: Supplementing alpha-tocopherol (vitamin E) and vitamin D3 in high fat diet decrease IL-6 production in murine epididymal adipose tissue and 3T3-L1 adipocytes following LPS stimulation. Lipids Health Dis. 2011;10:37. doi: 10.1186/1476-511X-10-37.

[149] Consiglio M, Destefanis M, Morena D, Foglizzo V, Forneris M, Pescarmona G, Silvagno F: The vitamin D receptor inhibits the respiratory chain, contributing to the metabolic switch that is essential for cancer cell proliferation. PLoS One. 2014;9:e115816. doi: 10.1371/journal.pone.0115816.

[150] Allen R, Cureton T: Effects of ultraviolet radiation on physical fitness. Arch Phys Med. 1945;10:641–4.

[151] Gorkin Z, Gorkin MJ, Teslenko NE: [The effect of ultraviolet irradiation upon training for 100m sprint.] Fiziol Zh USSR.1938;25:695–701. [Article in Russian].

[152] Lehmann G, Mueller EA: [Ultraviolet irradiation and altitude fitness.] Luftfahrtmedizin. 1944;9:37–43. [Article in German].

[153] Sigmund R: [Effect of ultraviolet rays on reaction time in man.] Strahlentherapie. 1956;101(4):623–9. [Article in German].

[154] Rosentsweig J: The effect of a single suberythemic biodose of ultraviolet radiation upon the strength of college women. J Assoc Phys Ment Rehabil. 1967;21(4):131–3.

[155] Rosentswieg J: The effect of a single suberythemic biodose of ultraviolet radiation upon the endurance of college women. J Sports Med Phys Fitness. 1969;9(2):104–6.

[156] Cheatum BA: Effects of a single biodose of ultraviolet radiation upon the speed of college women. Res Q. 1968;39(3):482–5.

[157] Macdougall JD, Hicks AL, Macdonald JR, Mckelvie RS, Green HJ, Smith KM: Muscle performance and enzymatic adaptations to sprint interval training. J Appl Physiol. 1998;84:2138–142.

[158] Dahlquist DT, Dieter BP, Koehle MS: Plausible ergogenic effects of vitamin D on athletic performance and recovery. J Int Soc Sports Nutr. 2015;12:33. doi: 10.1186/s12970-015-0093-8.

[159] Bouillon R, Verstuyf A: Vitamin D, mitochondria, and muscle. J Clin Endocrinol Metabol. 2013;98(3):961–3. doi: 10.1210/jc.2013-1352.

[160] Redzic M, Lewis RM, Thomas DT: Relationship between 25-hydroxyvitamin D, muscle strength, and incidence of injury in healthy adults: a systematic review. Nutr Res. 2013;33(4):251–8. doi: 10.1016/j.nutres.2013.02.007.

[161] Ward KA, Das G, Berry JL, Roberts SA, Rawer R, Adams JE, Mughal Z: Vitamin D status and muscle function in post-menarchal adolescent girls. J Clin Endocrinol Metab. 2009;94(2):559–63. doi: 10.1210/jc.2008-1284.

[162] Foo LH, Zhang Q, Zhu K, Ma G, Hu X, Greenfield H, Fraser DR. Low vitamin D status has an adverse influence on bone mass, bone turnover, and muscle strength in Chinese adolescent girls. J Nutr. 2009;139(5):1002–7. doi: 10.3945/jn.108.102053.

[163] Pojednic RM, Ceglia L: The emerging biomolecular role of vitamin D in skeletal muscle. Exerc Sport Sci Rev. 2014;42(2):76–81. doi: 10.1249/JES.0000000000000013.

[164] Gerdhem P, Ringsberg KAM, Obrant KJ, Akesson K: Association between 25-hydroxy vitamin D levels, physical activity, muscle strength and fractures in the prospective population-based OPRA Study of Elderly Women. Osteoporos Int. 2005;16(11):1425–31.

[165] Shardell M, Hicks GE, Miller RR, Kritchevsky S, Andersen D, Bandinelli S, Cherubini A, Ferrucci L. Association of low vitamin D levels with the frailty syndrome in men and women. J Gerontol (A Biol Sci Med Sci). 2009;64(1):69–75. doi: 10.1093/gerona/gln007.

[166] Houston DK, Tooze JA, Davis CC, Chaves PH, Hirsch CH, Robbins JA, Arnold AM, Newman AB, Kritchevsky SB: Serum 25-hydroxyvitamin D and physical function in adults of advanced age: the CHS All Stars. J Am Geriat Soc. 2011;59(10):1793. doi: 10.1111/j.1532-5415.2011.03601.x.

[167] Visser M, Deeg DJH, Puts MTE, Seidell JC, Lips P. Low serum concentrations of 25-hydroxyvitamin D in older persons and the risk of nursing home admission. Am J Clin Nutrit. 2006;84(3):616–22.

[168] Bischoff-Ferrari HA, Dietrich T, Orav EJ, Hu FB, Zhang Y, Karlson EW, et al. Higher 25-hydroxyvitamin D concentrations are associated with better lower-extremity function in both active and inactive persons aged ≥60 years. Am J Clin Nutr. 2004;80(3):752–8.

[169] Tieland M, Brouwer-Brolsma E, Nienaber-Rousseau C, van Loon L, De Groot L: Low vitamin D status is associated with reduced muscle mass and impaired physical performance in frail elderly people. Eur J Clin Nutr. 2013;67(10):1050–5. doi: 10.1038/ejcn.2013.144.

[170] Dufour AB, Hannan MT, Murabito JM, Kiel DP, McLean RR: Sarcopenia definitions considering body size and fat mass are associated with mobility limitations: the Framingham Study. J Gerontol A Biol Sci Med Sci.2013;68:168–74. doi: 10.1093/gerona/gls109.

[171] Cruz-Jentoft AJ, Baeyens JP, Bauer JM, Boirie Y, Cederholm T, Landi F, Martin FC, Michel J-P, Rolland Y, Schneider SM, Topinková E, Vandewoude M, Zamboni M: Sarcopenia: European consensus on definition and diagnosis: report of the European Working Group on Sarcopenia in Older People. Age Ageing. 2010;39:412–23. doi: 10.1093/ageing/afq034.

[172] Gilsanz V, Kremer A, Mo AO, Wren TA, Kremer R: Vitamin D status and its relation to muscle mass and muscle fat in young women. Clin Endocrinol Metab. 2010;95(4):1595–601. doi: 10.1210/jc.2009-2309.

[173] Dawson-Hughes B: 70th Anniversary Conference on Vitamins in early development and healthy aging: impact on infectious and chronic disease' symposium 1: vitamins and cognitive development and performance Serum 25-hydroxyvitamin D and muscle atrophy in the elderly. Proc Nutr Soc. 2012;71:46–9. doi:10.1017/S0029665111003260

[174] Houston DK, Cesari M, Ferrucci L, Cherubini A, Maggio D, Bartali B, Johnson MA, Schwartz GG, Kritchevsky SB: Association between vitamin D status and physical performance: the InCHIANTI study. J Gerontol A Biol Sci Med Sci. 2007;62(4):440–6.

[175] Dam TT, von Mühlen D, Barrett-Connor EL: Sex-specific association of serum vitamin D levels with physical function in older adults. Osteoporos Int. 2009;20(5):751–60.

[176] Okuno J, Tomura S, Yabushita N, Kim MJ, Okura T, Tanaka K, Yanagi H: Effects of serum 25-hydroxyvitamin D(3) levels on physical fitness in community-dwelling frail women. Arch Gerontol Geriatr. 2010;50(2):121–6. doi: 10.1016/j.archger.2009.02.011.

[177] Wicherts IS, van Schoor NM, Boeke AJ, Visser M, Deeg DJ, Smit J, Knol DL, Lips P: Vitamin D status predicts physical performance and its decline in older persons. J Clin Endocrinol Metab. 2007;92(6):2058–65.

[178] Mastaglia SR, Seijo M, Muzio D, Somoza J, Nuñez M, Oliveri B: Effect of vitamin D nutritional status on muscle function and strength in healthy women aged over sixty-five years. Nutr Health Aging. 2011;15(5):349–54.

[179] Stockton KA, Mengersen K, Paratz JD, Kandiah D, Bennell KL: Effect of vitamin D supplementation on muscle strength: a systematic review and meta-analysis. Osteoporos Int. 2011 22(3):859–71. doi: 10.1007/s00198-010-1407-y.

[180] Pramyothin P, Techasurungkul S, Lin J, Wang H, Shah A, Ross PD, Puapong R, Wasnich RD: Vitamin D status and falls, frailty, and fractures among postmenopausal Japanese women living in Hawaii. Osteoporos Int. 2009;20(11):1955–62. doi: 10.1007/s00198-009-0910-5.

[181] Ceglia L, Chiu GR, Harris SS, Araujo AB: Serum 25-hydroxyvitamin D concentration and physical function in adult men. Clin Endocrinol (Oxf). 2011;74(3):370–6. doi: 10.1111/j.1365-2265.2010.03926.x.

[182] Matheï C, Van Pottelbergh G, Vaes B, Adriaensen W, Gruson D, Degryse JM: No relation between vitamin D status and physical performance in the oldest old: results from the Belfrail study. Age Ageing. 2013;42(2):186–90. doi: 10.1093/ageing/afs186.

[183] Allali F, El Aichaoui S, Khazani H, Benyahia B, Saoud B, El Kabbaj S, Bahiri R, Abouqal R, Hajjaj-Hassouni N: High prevalence of hypovitaminosis D in Morocco: relationship to lifestyle, physical performance, bone markers, and bone mineral density. Semin Arthritis Rheum. 2009;38:444–51. doi: 10.1016/j.semarthrit.2008.01.009.

[184] Garnero P, Munoz F, Sornay-Rendu E, Delmas PD: Associations of vitamin D status with bone mineral density, bone turnover, bone loss and fracture risk in healthy postmenopausal women. The OFELY study. Bone.2007;40:716–22.

[185] Ceglia L: Vitamin D and its role in skeletal muscle. Curr Opin Clin Nutr Metab Care. 2009;12(6):628–33. doi: 10.1097/MCO.0b013e328331c707.

[186] Prince RL, Austin N, Devine A, Dick IM, Bruce D, Zhu K: Effects of ergocalciferol added to calcium on the risk of falls in elderly high-risk women. Arch Intern Med. 2008;168(1): 103–8. doi: 10.1001/archinternmed.2007.31.

[187] Pfeifer M, Begerow B, Minne H, Suppan K, Fahrleitner-Pammer A, Dobnig H: Effects of a long-term vitamin D and calcium supplementation on falls and parameters of muscle function in community-dwelling older individuals. Osteoporos Int. 2009;20:315–22.

[188] Zhu K, Austin N, Devine A, Bruce D, Prince R: A randomized controlled trial of the effects of vitamin D on muscle strength and mobility in older women with vitamin D insufficiency. J Am Geriatr Soc. 2010;58(11):2063–8. doi: 10.1111/j.1532-5415.2010.03142.x

[189] Bischoff-Ferrari HA, Dawson-Hughes B, Willett WC, Staehelin HB, Bazemore MG, Zee RY, Wong JB: Effect of Vitamin D on falls: a meta-analysis. JAMA. 2004;291(16):1999–2006.

[190] Tracy BL, Ivey FM, Hurlbut D, Martel GF, Lemmer JT, Siegel EL, Metter EJ, Fozard JL, Fleg JL, Hurley BF: Muscle quality. II. Effects of strength training in 65- to 75-yr-old men and women. J Appl Physiol. 1999;86(1):195–201.

[191] Bunout D, Barrera G, Leiva L, Gattas V, de la Maza MP, Avendaño M, Hirsch S: Effects of vitamin D supplementation and exercise training on physical performance in Chilean vitamin D deficient elderly subjects. Exp Gerontol. 2006:41:746–52.

[192] Bischoff HA, Stähelin HB, Dick W, Akos R, Knecht M, Salis C, Nebiker M, Theiler R, Pfeifer M, Begerow B, Lew RA, Conzelmann M.: Effects of vitamin D and calcium supplementation on falls: a randomized controlled trial. J Bone Miner Res. 2003;18:343–51.

[193] Moreira-Pfrimer LD, Pedrosa MA, Teixeira L, Lazaretti-Castro M: Treatment of vitamin D deficiency increases lower limb muscle strength in institutionalized older people independently of regular physical activity: a randomized double-blind controlled trial. Ann Nutr Metab. 2009;54(4):291–300. doi: 10.1159/000235874.

[194] Pfeifer M, Begerow B, Minne H, Suppan K, Fahrleitner-Pammer A, Dobnig H: Effects of a long-term vitamin D and calcium supplementation on falls and parameters of muscle function in community-dwelling older individuals. Osteoporos Int. 2009;20:315–22.

[195] Rejnmark R: Effects of vitamin D on muscle function and performance: a review of evidence from randomized controlled trials. Ther Adv Chronic Dis. 2011;2(1):25–37. doi: 10.1177/2040622310381934.

[196] Janssen HC, Samson MM, Verhaar HJ: Muscle strength and mobility in vitamin D-insufficient female geriatric patients: a randomized controlled trial on vitamin D and calcium supplementation. Aging Clin Exp Res. 2010;22(1):78–84.

[197] Brunner RL, Cochrane B, Jackson RD, Larson J, Lewis C, Limacher M, Rosal M, Shumaker S, Wallace R; Women's Health Initiative Investigators. Calcium, vitamin D supplementation, and physical function in the Women's Health Initiative. J Am Diet Assoc. 2008;108(9):1472–9. doi: 10.1016/j.jada.2008.06.432.

[198] Fuleihan G, Nabulsi M, Tamim H, Maalouf J, Salamoun M, Khalife H, Choucair M, Arabi A, Vieth R: Effect of vitamin D replacement on musculoskeletal parameters in school children: a randomized controlled trial. J Clin Endocrinol Metab. 2006;91(2):405–12.

[199] Lim JS, Kim KM, Rhee Y, Lim SK. Gender-dependent skeletal effects of vitamin D deficiency in a younger generation. J Clin Endocrinol Metab. 2012 97(6):1995–2004. doi: 10.1210/jc.2011-3098.

[200] Gupta R, Sharma U, Gupta N, Kalaivani M, Singh U, Guleria R, Jagannathan NR, Goswami R: Effect of cholecalciferol and calcium supplementation on muscle strength and energy metabolism in vitamin D-deficient Asian Indians: a randomized, controlled trial. Clin Endocrinol (Oxf). 2010;73:445–51. doi: 10.1111/j.13 65-2265.2010.03816.x.

[201] Owens DJ, Webber D, Impey SC, Tang J, Donovan TF, Fraser WD, Morton JP, Close CL: Vitamin D supplementation does not improve human skeletal muscle contractile properties in insufficient young males. Eur J Appl Physiol. 2014;114(6):1309–20. doi: 10.1007/s00421-014-2865-2.

[202] Tomlinson PB, Joseph C, Angioi M: Effects of vitamin D supplementation on upper and lower body muscle strength levels in healthy individuals. A systematic review with meta-analysis. J Sci Med Sport. 2015;18:575–80. doi: 10.1016/j.jsams. 2014.07.022.

[203] Barker T, Henriksen VT, Martins TB, Hill HR, Kjeldsberg CR, Schneider ED, Dixon BM, Weaver LK: Higher serum 25-hydroxyvitamin D concentrations associate with a faster recovery of skeletal muscle strength after muscular injury. Nutrients. 2013;5(4):1253–75. doi: 10.3390/nu5041253.

[204] Mowry DA, Costello MM, Heelan KA: Association among cardiorespiratory fitness, body fat, and bone marker measurements in healthy young females. J Am Osteopath Assoc. 2009;109(10):534–9.

[205] Forney LA, Earnest CP, Henagan TM, Johnson LE, Castleberry TJ, Stewart LK: Vitamin D status, body composition, and fitness measures in college-aged students. J Strength Cond Res. 2014;28(3):814–24. doi: 10.1519/JSC.0b013e3182a35ed0.

[206] Fitzgerald JS, Peterson BJ, Warpeha JM, Wilson PB, Rhodes GS, Ingraham SJ: Vitamin D status and V[combining dot above]O2peak during a skate treadmill graded exercise

test in competitive ice hockey players. J Strength Cond Res. 2014;28(11):3200–5. doi: 10.1519/JSC.0000000000000523.

[207] Dubnov-Raz G, Livne N, Raz R, Rogel D, Cohen AH, Constantini NW: Vitamin D concentrations and physical performance in competitive adolescent swimmers. Pediatr Exerc Sci. 2014;26:64–70. doi: 10.1123/pes.2013-0034.

[208] Hamilton B, Whiteley R, Farooq A, Chalabi H: Vitamin D concentration in 342 professional football players and association with lower limb isokinetic function. J Sci Med Sport. 2014;17:139–43. doi: 10.1016/j.jsams.2013.03.006.

[209] Ardestani A, Parker B, Mathur S, Clarkson P, Pescatello LS, Hoffman HJ, Polk DM, Thompson PD: Relation of vitamin D level to maximal oxygen uptake in adults. Am J Cardiol. 2011;107(8):1246–9. doi: 10.1016/j.amjcard.2010.12.022.

[210] Wilson PB, Fitzgerald JS, Ingraham SJ: Relationship between serum 25-hydroxyvitamin D status and cardiorespiratory fitness: findings from the National Health and Nutrition Examination Survey. Int J Cardiol. 2013;168:3112–3. doi: 10.1016/j.ijcard.2013.04.069.

[211] Close GL, Russell J, Cobley JN, Owens DJ, Wilson G, Gregson W, Fraser WD, Morton JP: Assessment of vitamin D concentration in non-supplemented professional athletes and healthy adults during the winter months in the UK: implications for skeletal muscle function. J Sports Sci. 2013;31(4):344–53. doi: 10.1080/02640414.2012.733822.

[212] Wyon MA, Koutedakis Y, Wolman R, Nevill AM, Allen N: The influence of winter vitamin D supplementation on muscle function and injury occurrence in elite ballet dancers: a controlled study. J Sci Med Sport. 2014;17(1):8–12. doi: 10.1016/j.jsams.2013.03.007.

[213] Wyon MA, Wolman R, Nevill AM, Cloak R, Metsios GS, Gould D, Ingham A, Koutedakis Y: Acute effects of vitamin D3 supplementation on muscle strength in Judoka athletes: a randomized placebo-controlled, double-blind trial. Clin J Sport Med. 2015. [Epub ahead of print]

[214] Close GL, Leckey J, Patterson M, Bradley W, Owens DJ, Fraser WD, Morton JP: The effects of vitamin D(3) supplementation on serum total 25[OH]D concentration and physical performance: a randomised dose–response study. Br J Sports Med. 2013;47:692–6. doi: 10.1136/bjsports-2012-091735.

[215] Jastrzębski Z: Effect of vitamin D supplementation on the level of physical fitness and blood parameters of rowers during the 8-week high intensity training. Facicula Educ Fiz şi Sport. 2014;2:57–67.

[216] Jastrzębska M, Kaczmarczyk M, Jastrzębski Z: The effect of vitamin d supplementation on training adaptation in well trained soccer players. Strength Cond Res. 2016; [Epub ahead of print].

Chapter 2

Vitamin D in Oxidative Stress and Diseases

Francesca Uberti, Vera Morsanuto and
Claudio Molinari

Additional information is available at the end of the chapter

http://dx.doi.org/10.5772/64506

Abstract

The data described in this chapter consider some new information about the benefits of vitamin D_3 comparing the results obtained by the authors on the effects of vitamin D_3 during oxidative stress with other works available in the literature. In particular, vitamin D_3 can induce a concentration-dependent increase in endothelial NO production through eNOS activation consequential to the phosphorylation of p38, AKT, and ERK. Additional information obtained by the author is about the ability of vitamin D_3 to prevent the endothelial cell death through modulation of interplay between apoptosis and autophagy. This effect is obtained by inhibiting superoxide anion generation, maintaining mitochondria function and cell viability, activating survival kinases (ERK and Akt), and inducing NO production. The results also describe that vitamin D_3 causes human endothelial cell proliferation and migration in a 3-D matrix through NO-dependent mechanisms. These findings support the role of vitamin D_3 in the human angiogenic process, suggesting new applications for vitamin D_3 in tissue repair and wound healing. Finally, that the authors have demonstrated the ability of vitamin D_3 to counteract negative effects of oxidative stress in brain cells. These data suggest the potential therapeutic use of vitamin D to treat or prevent degenerative brain diseases.

Keywords: active vitamin D, extraskeletal function, oxidative stress, endothelial cells, neuronal cells

1. Introduction

Vitamin D, discovered as an essential nutrient for the prevention of rickets, is required for optimal absorption of dietary calcium and phosphate [1]. Vitamin D plays an essential role in calcium and phosphate metabolism and maintains mineral homeostasis to ensure metabolic functions and bone mineralization. The calcium requirement of the organism is firstly satisfied by dietary calcium intake and, when this is not sufficient, by calcium mobilization from bone and renal reabsorption. Classical vitamin D–responsive tissues are bones, intestine, kidney, and parathyroid glands [1–3] and in this case, vitamin D induces skeletal effects. In addition many other organs respond to vitamin D, such as prostate, breast, immune cells, skeletal muscle, cardiac tissue, parathyroid glands, skin, and brain; all these organs express vitamin D receptor (VDR) and the enzyme 1α-hydroxylase and vitamin D exert extraskeletal effects [4]. Vitamin D exerts its activity through two mechanisms: the hormone signaling, in which the biologically active form reaches target cells through the bloodstream; and the autocrine/paracrine signaling, in which locally produced-vitamin D_3 affects the surrounding cells [5].

A variety of research results from recent years have shown that vitamin D in its hormonally active form, 1α,25-dihydroxyvitamin D [1α,25(OH)2D; calcitriol] is not only a regulator of calcium and phosphate homeostasis, but has numerous extraskeletal effects. These include the significant impact of the vitamin D hormone on the cardiovascular system, central nervous system, endocrine system, and immune system as well as on cell differentiation and cell growth [4]. The term "vitamin D" refers to two different forms playing an important role in humans: vitamin D2 (ergocalciferol), which is made in some plants but largely in fungi [6], and vitamin D_3 (cholecalciferol), which is made by human skin when exposed to sunlight via the UV irradiation of 7-dehydrocholesterol [6]. This photo-production is influenced by several factors including ethnicity (skin pigmentation), UV exposure (latitude, season, use of sunscreens, and clothing), and age. All these factors may reduce the ability to synthesize the bioactive form of vitamin D. Moreover, low outdoor activity and unhealthy lifestyle habits are also cause for a scarce vitamin D_3 production. Conversely, excessive exposure to sunlight degrades pre-vitamin D and vitamin D_3 into inactive photoproducts. Many organs participate in the synthesis of the bioactive form of vitamin D [7, 8]. Vitamin D_3 needs two hydroxylation steps in liver and kidney for its activation. The final product, hormonally active 1α,25-dihydroxyvitamin D_3 (calcitriol), arrives via the circulation to its target tissues and acts in a genomic or nongenomic manner [9].

All genomic actions of vitamin D are mediated by the VDR. VDR is a transcription factor and a member of the steroid hormone nuclear receptor family. VDR binding to its vitamin D response elements (VDREs) causes the recruitment of coregulatory complexes required for its genomic activity. These complexes can be both gene and cell specific, enabling the selectivity of vitamin D_3 action from cell type to cell type [8]. The profile of VDR binding sites and genes activated by vitamin D_3 varies from cell to cell with some albeit far from total overlap especially when results with different time durations of vitamin D_3 exposure are compared [10]. Moreover, these VDR binding sites can be anywhere in the genome, often many thousands of base pairs away from the gene to be regulated. Generally, these sites are associated with binding sites for other transcription factors [11, 12].

Vitamin D_3 is also able to activate very rapid nongenomic mechanisms [8], lasting from seconds to 10 minutes [13]. This mechanism involves signal transduction pathways including activation of adenylyl cyclase-cAMP-protein kinase A and phospholipase C-diacylglycerol-inositol

(1,4,5)-trisphosphate-protein kinase C signal transduction pathways [14]. Particularly, the second messengers Raf (rapidly accelerated fibrosarcoma)/MAPK play an important role because they may engage in cross-talk with the nucleus to modulate gene expression [13]. This firstly identified nongenomic activation is associated with the rapid stimulation of intestinal calcium transport called "transcaltachia" [15]. Successively, this effect was identified in the chondrocytes of the bone growth plate [15] and in keratinocytes of the skin [16]. Identification of the receptor for vitamin D_3 has focused on the VDR itself albeit in a different configuration to identify agonists able to induce nongenomic effects [17]. Interaction between vitamin D_3 and membrane-associated rapid response steroid binding protein (MARRS) has been studied as well. These receptors are located in the membrane within caveolae/lipid rafts [18] where they are poised to activate kinases, phosphatases, and ion channels.

The aim of this paper is to illustrate the most recent findings about the role of vitamin D_3 as a potent regulator of oxidative pathways.

2. Relevant extraskeletal applications

In a recent study, Berridge suggests the idea that vitamin D plays a crucial activity in keeping the integrity of cell signaling pathways coining the term of "custodian of phenotypic stability" [19]. The hypothesis supported is that a loss of this integrity could be explained by vitamin D deficiency, and this represents a risk factor for different diseases. This role assigned to vitamin D may depend on the regulation of the expression of nuclear factors involved both in physiological control of ROS and in calcium signaling [19]. The hormonally active form of vitamin D, $1\alpha25(OH)2D_3$, influences the expression of various genes, whose products not only are involved in the control of calcium and phosphate homeostasis, but are also able to interact with a wide range of organs and target tissues different from those involved in calcium metabolism [20]. Most experimental studies and meta-analysis showing a large amount of information concluded that vitamin D supplementation is associated with a decrease in total mortality [21]. In effect, vitamin D_3 is involved in a fine balance among organs and tissues that contributes to the maintenance of the homeostasis of the human organism. For example, an important function of vitamin D_3 is to maintain normal parathyroid status, preventing proliferation of parathyroid gland cells. Patients with renal failure-dependent vitamin D_3 deficiency, in presence of adequate calcium levels, may be affected by parathyroid hypertrophy with consequent secondary hyperparathyroidism [22]. Moreover, vitamin D_3 acts as a negative endocrine regulator of the renin-angiotensin system. It enters the bloodstream and downregulates renin production and stimulates insulin secretion from β-cells of the pancreatic islets [1, 3]. Vitamin D_3 is thought to possess positive effects on muscles, affecting the calcium handling in the cells and promoting de novo protein synthesis. Vitamin D_3 maintains calcium balance in cultured muscle cells initially via inositol triphosphate induced-rapid ion release from the sarcoplasmic reticulum and then through ion channels that allows calcium influx from the extracellular compartment. Recently, vitamin D_3 deficiency has been connected with sarcopenia (age-related loss of muscle mass) mainly of type II skeletal muscle fibers. Vitamin D supplementation reduces incidence of fractures, actually decreasing the number of falling

[4, 7]. Sarcopenia was initially defined as "disease of the elderly" and a degenerative consequence of ageing, but successively it can be also diagnosed at younger age [23]. Except for ageing, risk factors for sarcopenia include female gender, low physical activity, low protein intake, and vitamin D deficiency in both older and younger populations [23, 24]. Vitamin D levels decline with age and cutaneous vitamin D levels are up to four times lower in older than in younger individuals [25, 26]. Low vitamin D levels result in atrophy predominantly of the type 2 muscle fibers and it has been associated with an increase in sarcopenia [27]. Finally, the serum level of calcitriol is different in younger and elder people and it is related to incidence of sarcopenia.

Vitamin D_3 is a potent immunomodulator. For example, after a skin lesion, keratinocytes that compose the mucocutaneous barrier upregulate VDR and 1α-hydroxylase expression in order to improve immune defenses. Monocytes and macrophages act in the same way after a Mycobacterium tuberculosis infection or exposure to lipopolysaccharides. An increased cathelicidin and α-defensin 2 production results in all cases and these two proteins could display their antimicrobial effects, promoting innate immunity. It has been hypothesized that monocytes or macrophages may also release self-produced-vitamin D_3 to act locally on activated T and B lymphocytes, which are able to regulate cytokine and immunoglobulin synthesis, respectively [3, 28]. The large number of effects of vitamin D_3 regulating the immune system plays a very important role in fighting infectious diseases [29]. Vitamin D_3 enhances the innate immunity against various infections [30], especially tuberculosis, influenza, and viral upper respiratory tract infections [29]. Evidence exists about the potential antimicrobial activity of vitamin D and its deficiency has deleterious effects on general well-being and longevity. Vitamin D reduces the risk of infection in different manner: by modulating production of antimicrobial peptides, by regulating local immune and immunoinflammatory response, and by enhancing clearance of invading organisms. In addition, vitamin D constitutes an inexpensive prophylactic option and possibly a therapeutic product either by itself or as a synergic agent to the traditional drugs [31]. Ecological studies have shown that the prevalence of certain autoimmune diseases was associated with latitude, suggesting a potential role of sunlight exposure, and consequently vitamin D_3 production, on the pathogenesis of type 1 diabetes mellitus, multiple sclerosis and Crohn's disease. The increased prevalence at higher latitudes has been shown for multiple sclerosis (MS), inflammatory bowel disease, rheumatoid arthritis, and type 1 diabetes [32–36]. In children with asthma, vitamin D levels also seem to correlate positively with asthma control and lung function, and inversely with corticosteroid use [37–39].

Many experimental data show that calcitriol stimulates apoptosis and differentiation and inhibits angiogenesis and proliferation in tumor cells. Numerous association studies suggest that serum vitamin D levels are inversely associated with the risk of many types of cancer. Furthermore, in some studies of patients with cancer, an association between low vitamin D levels and poor prognosis has been observed [40–42]. In breast, colon, and prostate cancer, vitamin D_3-VDR complex could arrest cell cycle in the G1-G0 transition, inducing p21 and p27 synthesis and/or stabilization, blocking cell growth and promoting cell differentiation. In TGF-α/EGFR-dependent tumor, vitamin D_3 can inhibit the growth inducing the recruitment of the activated receptor into early endosomes, and reducing growth signaling [21]. In monocytes and osteoblast culture, it seems that the biologically active form of vitamin D_3 could enhance the expression of a suppressor of the oncogenic cyclin D1 and C/EBPβ. Moreover, vitamin D_3 is able to regulate apoptosis. It could induce this process, like in breast cancer cells for example, where it may modulate Bcl2 and Bax contents. However, it could also show antiapoptotic effects, essential in normal tissue development and function. For instance, it protects keratinocytes from UV-irradiation or chemotherapy-induced apoptosis and melanocytes from TNF-α-mediated apoptosis. In normal keratinocytes, vitamin D_3 induces cell differentiation and maintains a calcium gradient necessary for the integrity of the permeability barrier of the skin. Vitamin D_3 represses tumor invasion and metastasis reducing matrix metalloproteinase activity (MMP) and enhancing the expression of molecules with adhesive properties such as E-cadherin [43]. However, calcitriol is a multifunctional steroid hormone able to regulate signaling pathways related to cancer development and progression. In preclinical studies, it was shown that vitamin D can promote cell differentiation and inhibit proliferation, angiogenesis, and cell migration. Inconsistent results are found in epidemiological studies and in early trials regarding clinical effects of vitamin D supplementation and cancer in terms of prevention and impact in cancer-related mortality [44]. Vitamin D also exerts its effects on cancer through nongenomic factors, modulating inflammatory cytokine expression. This is a common element for cancer and neurodegeneration.

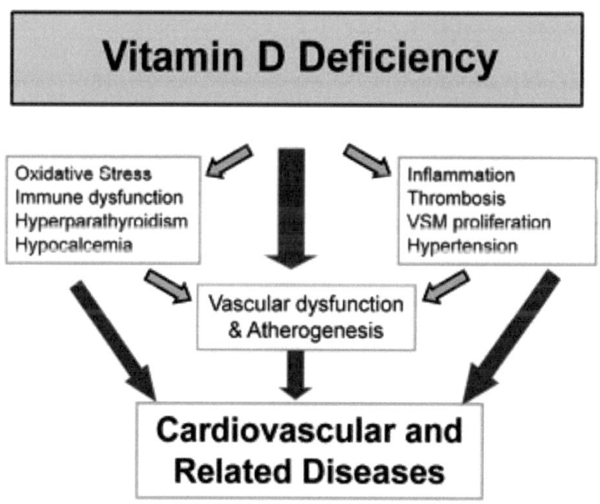

Figure 1. Vitamin D and cardiovascular risk (adapted from [48]).

Another important feature of vitamin D is the mitochondrial detoxification, well known as antioxidant activity. Especially, vitamin D and polyphenols, seem to be promising therapeutic tools for inhibiting radical oxygen species (ROS) formation and arresting cytokine-mediated inflammation [45].

Vitamin D status varies with age [46]. Serum levels of calcitriol are normal from 75 nM, insufficient from 50 to 75 nM, and deficient if less than 50 nM [47]. These serum concentrations depend on sunlight exposure, dietary intake, capacity of the skin, and influence various human diseases (cardiovascular and neurological), as reported in this manuscript (**Figures 1** and **2**).

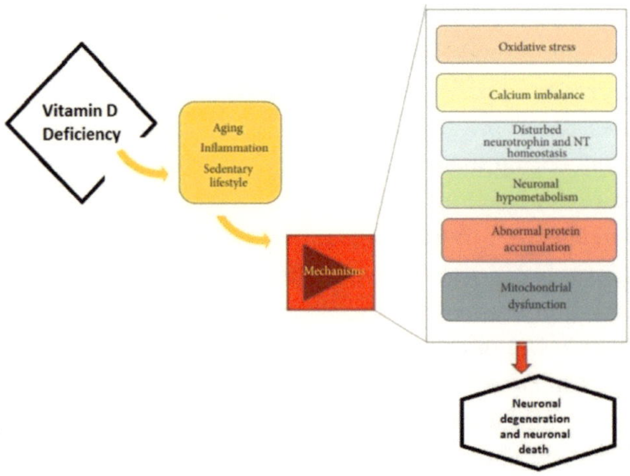

Figure 2. Vitamin D and neurological diseases (adapted from [49]).

2.1. Cardiovascular and cerebral risk

In humans, the relationship among low levels of vitamin D_3, hypercalcemia, osteoporosis, vascular calcification, and cardiovascular diseases has been extensively studied. Vitamin D_3 deficiency is significantly associated with cardiomyopathy and increased risk of cardiovascular disease with consequent mortality in humans [8]. The prospective Intermountain Heart Collaborative Study revealed that vitamin D_3 blood levels below 15 ng/mL compared to vitamin D_3 above 30 ng/mL are associated with significant increases in the prevalence of type 2 diabetes mellitus, hypertension, hyperlipidemia, and peripheral vascular disease, coronary artery disease, myocardial infarction, heart failure, and stroke, as well as with incident death, heart failure, coronary artery disease/myocardial infarction, and stroke [50].

Several data are available from studies in mouse animal model. Heart-VDR null mice show myocardial hypertrophy, while VDR and CYP27B1 null mice develop hypertension as well and display increased production of renin, contributing probably to early atherosclerosis onset [51]. Vitamin D_3 may protect against atherosclerosis inhibiting macrophage cholesterol uptake and foam cell formation, and reducing vascular smooth muscle cell proliferation and expression of adhesion molecules in endothelial cells. Recently, elevated levels of PTH emerged as a possible risk factor for cardiovascular diseases since they could promote development of hypertension; in addition, hyperparathyroidism could cause heart hypercontractility and calcification of the myocardium. Vitamin D_3 could overcome these processes, promoting, for example, synthesis of anti-inflammatory cytokines [4]. Severe vitamin D_3 deficiency in humans is associated with cardiomyopathy [52], and in a number of large epidemiologic studies, the association of increased cardiovascular disease risk with low vitamin D_3 blood levels has been found [53]. Obesity, a condition associated with cardiovascular disease [54], is related with a low vitamin D_3 status due to a sequestration and volumetric dilution of the lipophilic vitamin D_3 in the fat tissue [3, 55, 56]. Many studies suggest a biological beneficial effect of vitamin D_3 on cardiovascular risk factors and cardiovascular health. For example, in a recent study performed on Framingham Offspring Heart Study participants [57], vitamin D_3 deficiency was associated with increased cardiovascular risk, above and beyond established cardiovascular risk factors. In another study, performed on elderly subjects, an association between low serum vitamin D_3 levels and high arterial blood pressure was found [58]. In addition, vitamin D_3 insufficiency/deficiency is associated with myocardial infarction, congestive heart failure, and calcific aortic stenosis, which can lead to the massive vascular calcification seen in chronic kidney disease. Recent clinical studies demonstrated that vitamin D_3 may cause regression of cardiac hypertrophy, reduces cardiovascular morbidity, and mortality in patients who frequently suffer from atherosclerosis [59, 60] and it may improve the cardiac structure and function [61].

The VDR is present in endothelium, vascular smooth muscle, and cardiomyocytes [55, 57] and may protect against atherosclerosis through the inhibition of macrophage cholesterol uptake and foam cell formation, the reduction of vascular smooth muscle cell proliferation, and the decrease of expression of adhesion molecules in endothelial cells [55] and through inhibition of cytokine release from lymphocytes [57]. Several meta-analyses indicate an inverse association between vitamin D_3 status and hypertension, and the antihypertensive effects were associated with a rate in vitamin D_3 levels after dietary supplementation or UVB exposure [62–66]. Mechanistically, this effect could be partly mediated by vitamin D_3's capability to suppress the levels of PTH, which is the cause of arrhythmias, myocardial hypertrophy and increasing blood pressure [67], and the levels of renin [67, 68]. As far as endothelial effects are concerned, it has been demonstrated that vitamin D_3 modulates vascular tone by means of a reduction in calcium influx into the endothelial cells followed by a decrease in endothelium-derived contracting factors production. Vitamin D_3 exerts its physiological effects acting on VDR through both genomic and nongenomic mechanisms, regulating cellular proliferation, differentiation, apoptosis, and angiogenesis in local tissue [69–71]. In particular, many vascular effects, such as increased expression of Ca-ATPase, induction of contractile protein synthesis, and, hence, increased vascular resistance have been reported. Vitamin D_3 is a modulator of vascular wall growth as well [72] and induces a decrease in the expression and/or secretion of proinflammatory and proatherosclerotic factors in the endothelium [73]. The role of endothelium as a target of vitamin D_3 is demonstrated by the study published by Zehnder et al. [74], in which the expression of mRNA and protein for 1α-hydroxylase in human endothelium was shown for the first time. Altogether, these findings demonstrated the direct effects of vitamin D_3 on endothelial function, whose alteration plays an important role in the development of atherosclerosis. Endothelial cells are capable of synthesizing vitamin D_3 as results from the expression of mRNA and protein for the enzyme 25(OH)D_3-1α-hydroxylase [75]. Moreover, studies elsewhere have demonstrated the presence of intracellular vitamin D_3 receptors within endothelial cells (VDR) [76]. Endothelial cells can synthesize vitamin D_3 due to the expression of the key biosynthetic enzyme 25 (OH)D_3-1a-hydroxylase [74]. Because of the presence of both vitamin D_3 and VDR in endothelial cells and the key role of nitric oxide (NO) in the endothelial physiology, it is possible to hypothesize an interaction between vitamin D_3 and NO capable of influencing proliferation and migration of endothelial cells [75]. Although the relationship among vitamin D_3, endothelium, and cardiovascular disease is well established, little is known about the effect of vitamin D_3 on endothelial NO production. In addition, several recent studies have postulated anti-inflammatory, immunomodulatory, and neuroprotective functions for vitamin D_3. A meta-analysis examining the association between vitamin D_3 status and the risk of cerebrovascular events, including more than 1200 stroke cases, found that the pooled relative risk for stroke was 52% higher when comparing 25(OH)D levels ≤12.4 ng/mL with 25(OH)D levels >18.8 ng/mL [77].

As concerned to cerebral effects, the observed widespread distribution of 1α-hydroxylase and the nuclear VDR in both neurons and glial cells suggest that vitamin D_3 may have autocrine and paracrine properties in the brain [78, 79]. VDR is highly expressed in multiple brain regions [80] in the animal [81] and human [82] brain, particularly in the pontine-midbrain area, cerebellum, thalamus, hypothalamus, basal ganglia, hippocampus, olfactory system, and the temporal, orbital, and cingulate areas of brain cortex [83]. Mounting evidence indicates that vitamin D_3 and its receptors play an important role in the brain, ranging from neuroprotection to immunomodulation [84]; cells proliferation and differentiation [85], and plays an important role both in developing [86] and adult brain [80]. Vitamin D_3 can exert these effects since it is able to cross the blood-brain barrier and can bind to VDR within the brain [87, 88]. Upon binding, VDR undergoes a conformational change to form a complex with a retinoid X receptor that controls genic expression [89, 90]. More recent papers demonstrate that gestational vitamin D_3 deficiency induces long-lasting alterations in the brain structure, including changes in volume, cell proliferation and reduction in the expression of nerve growth factors (NGF), glia-derived neurotrophic factor (GDNF) and neurotrophins 3 and 4 [80, 83, 91]. Moreover, vitamin D_3 protects neurons against NMDA, glutamate, 6-hydroxydopamine, and reactive oxygen species [92, 93]. It has been hypothesized that vitamin D_3 exerts its neuroprotective effects via the modulation of neuronal Ca^{2+} homeostasis, in particular through the downregulation of the L-type voltage-sensitive Ca^{2+} channel in hippocampal neurons against excitotoxic insults [80], accompanied by an increase in VDR density. Vitamin D_3 is able to inhibit proinflammatory cytokine and NOS [94] typically increased during various insults or disorders, such as ischemia and reperfusion, Alzheimer's disease, Parkinson's disease, multiple sclerosis, and experimental autoimmune encephalomyelitis. Early vitamin D_3 deficiency may be considered as a risk factor for a number of neurological disorders including schizophrenia, autism [88, 95], multiple sclerosis, Parkinson's disease, and stroke [77, 89]. Vitamin D_3 deficiency is associated with reduced cognitive function, which is an important issue for stroke patients [96]. Moreover, in the rat model of stroke, vitamin D_3 supplementation has been found to reduce brain damage [97] and consequent seizures [77] and oxidative stress [98]. For this reason, it should be very important to study the role of vitamin D_3 in counteracting negative effects of oxidative stress in vitro brain, also in association with stem cell. So, vitamin D_3 may be considered as a potential drug for the treatment of neurodegenerative disorders.

3. Oxidative stress and injury

Xiang et al. [99] showed the ability of vitamin D_3 to stimulate endothelial cell proliferation and to inhibit apoptosis by increasing endothelial nitric oxide synthase (eNOS) expression and nitric oxide (NO) production. NO plays a key role in cardiovascular physiology [100], and its production in the heart by eNOS phosphorylation represents an important regulating factor of both myocardial perfusion after ischemia and myocardial contractility. This has important effects on cardiac cell functions such as oxygen consumption, hypertrophic remodeling, apoptosis, and myocardial regeneration [101–103]. The discovery of NO as a signaling molecule in the cardiovascular system was published in 1998 [104] and it is now recognized as an endogenous vasodilator and an antioxidant factor able to regulate the vascular endothelium supporting its anticoagulant and antithrombogenic capacities, maintaining vascular tone and preventing vascular smooth cell proliferation [105]. NO acts as a signaling molecule by binding to ferrous heme within metalloproteins (such as soluble guanylate cyclase, sGC, cytochrome c oxidase, and hemoglobin). The most important action mechanism of NO is the interaction with heme of the sGC in smooth muscle cells adjacent to the endothelium, catalyzing the conversion of guanosine triphosphate (GTP) to cGMP [106]. cGMP-dependent protein kinases promote the opening of calcium-dependent potassium channel and determine hyperpolarization of the cell membrane of smooth muscle, thus inhibiting cytosolic calcium influx and promoting cell relaxation and vasodilation. Alternatively, NO can determine molecular modifications. Since it is a highly diffusible molecule in tissues and physiological fluids with a very short half-life (few seconds), it rapidly reacts with superoxide anion (O^{2-}) to form peroxynitrite ($ONOO^-$). $ONOO^-$ oxidizes DNA, proteins, lipids, and BH4, uncouples eNOS and limits NO production. $ONOO^-$ directly influences the dilatory capabilities of the arteries and disrupts NO-induced sGC-mediated signal transduction [107]. Nevertheless, NO could be transported by protein carriers or stored locally and cause remote and long-lasting effects in the cardiovascular system [108, 109]. In addition NO is able to enhance endothelial cell survival, proliferation, and migration [110, 111]. Furthermore, NO signaling depends on its own concentration. Thomas and colleagues demonstrated in MCF7 cells (human breast adenocarcinoma cell line) that NO levels ranging from 10 to 30 nM lead to extracellular signal-regulated kinases (ERK) phosphorylation. This effect is due to a cGMP-dependent process, mediating proliferative and protective effects. It has been demonstrated that low NO levels, below 1 nM, are sufficient to obtain this response in endothelial cells. When the NO concentration is between 30 and 60 nM, Akt phosphorylation occurs. This antiapoptotic mechanism depends on Bad and caspase-9 phosphorylation. Hypoxia-inducible factor-1α (HIF-1α) is stabilized when NO is at a high concentration (about 100 nM) and protects against tissue injury, while at 400 nM p53 is posttranslationally modified, resulting in growth arrest or apoptosis. Signal transduction cascades react accordingly to NO not only in terms of concentration but also in terms of duration of exposure with different threshold sensitivities, resulting in distinct phenotypic responses. Indeed, HIF-1α is an immediate but transient responder and protein disappears when the NO concentration falls under the minimum threshold. On the contrary, p53 has a delayed response that takes several hours to occur but is sustained long after NO exposure [112]. Despite the differences in clinical manifestations and neuronal vulnerability, the pathological processes appear similar in neurodegenerative pathways as well.

Cell death and neurodegenerative conditions have been linked to oxidative stress and imbalance between generation of free radicals and antioxidant defenses. Multiple sclerosis, stroke, and neurodegenerative diseases have been associated with the reactive oxygen species and nitric oxide [113]. Recent findings about the production of the reactive oxygen and nitrogen species by NADPH oxidases and nitric oxide synthases and the essential role of glutathione (γ-glutamyl-l-cysteinylglycine) in redox homeostasis demonstrate the importance of these substances in neuronal degeneration. Redox signaling has a profound impact on two transcription factors that modulate microglial fate, nuclear factor kappa-light-chain-enhancer of activated B cells, and nuclear factor (erythroid-derived 2)-like 2, master regulators of the pro-inflammatory and antioxidant responses of microglia, respectively. The relevance of these proteins in the modulation of microglial activity and the interplay between microglia and oxidation has been described. Finally, the relevance of ROS in altering blood-brain barrier permeability is reported. ROS originates in the mitochondrial electron transport chain (mETC) or are produced by NADPH oxidases (NOX) by an incomplete one-electron reduction of oxygen. Three major ROS exist: superoxide anion (O^{2-}), hydrogen peroxide (H_2O_2), and hydroxyl radical (OH). In cellular respiration, most of O_2 generates H_2O_2 through enzymes of the superoxide dismutase (SOD) family, enclosed copper-zinc superoxide dismutase (SOD1 or Cu, Zn-SOD) present in the intermembrane space and in the cytosol, manganese superoxide dismutase (SOD2 or Mn-SOD) present in the mitochondria matrix, and SOD_3 present in the extracellular matrix. Hydrogen peroxide could subsequently be converted to the hydroxyl radical by ferrous, or a final reducing step could convert it to water by catalase, glutathione peroxidase, and peroxyredoxin III activity.

In endothelial cells, ROS into the cytosol could stimulate further ROS production by NOX2 through redox-sensing protein kinase C (PKC) isoforms and Src family kinases activation [114, 115]. Recent examples of the importance of these findings about ROS and NO in the onset or progression of neurodegenerative diseases are also showed [116]. Three NO-synthase isoforms could produce NO: (1) NO-synthase I (or neuronal, nNOS, NOS-1) expressed in neurons and skeletal and smooth muscle; (2) NO-synthase II (or inducible, iNOS, NOS-2) in immune cells; and (3) NO-synthase III (or endothelial, eNOS, NOS-3) in cardiomyocytes and platelets and, predominantly, in endothelial cells [109, 117]. NOS isoforms generate NO at different rates: eNOS and nNOS produce a low concentration of NO, leading to physiological processes regulation, while iNOS generates a high NO concentration in response to inflammatory stimuli, such as in activated macrophages, to establish cytotoxic effects and antipathogen reactions [118]. eNOS is a Ca^{2+}/calmodulin dependent enzyme and contains an N-terminal oxygenase and a C-terminal reductase domain, linked by a calmodulin binding sequence. It works as a homodimer to synthesize NO and L-citrulline from L-arginine and oxygen. In the presence of several cofactors, such as nicotinamide adenine dinucleotide phosphate (NADPH), flavin adenine dinucleotide (FAD), flavin mononucleotide (FMN), heme, tetrahydrobiopterin (BH4), and zinc, it catalyzes a five-electron oxidation of one of the guanidino nitrogens of L-arginine. Since it is a gas, NO then diffuses to the vascular smooth muscle cells and reacts with sGC, leading to cGMP-mediated vasodilatation. In endothelial cells, after enzymatic acylation, myristoylation and reversible palmitoylation, eNOS is predominantly located in the plasma membrane and in the Golgi apparatus associated with caveolin-1 located in caveolae, whose binding inhibits the activity of the enzyme [108, 109]. Two mechanisms leading to eNOS activation and involving G protein coupled receptors and heterotrimeric G protein have been explained: intracellular Ca^{2+} mobilization through phospholipase C (PLC) pathway, and phosphatidylinositol-3-kinase (PI3K)/Akt pathway [119, 120]. An increase in intracellular calcium concentration, indeed, is a critical determinant that causes eNOS dissociation from caveolin-1, leading to the activation of the enzyme. Moreover, serine, threonine, and tyrosine phosphorylation regulate eNOS activity, either enhancing or suppressing it, depending on the residue and the domain that are involved. Several eNOS agonists such as VEGF, bradykinin, and adenosine triphosphate (ATP) promote the activation of phosphorylation. eNOS expression and activation is also sensitive to shear stress caused by blood flow; low shear stress determines lower NO production, increasing atherosclerotic plaque formation [108, 109]. Endothelial dysfunction should be considered as endothelial activation from a quiescent phenotype and it is identified as an initial step in the development of CVD. This is characterized by reduced NO bioavailability that predisposes to vasoconstriction and thrombosis [121]. Several factors can contribute to this process, such as decreased eNOS expression, presence of eNOS antagonist, low L-arginine or BH4 concentration, increased NO degradation or ROS production, exposure to inflammatory cytokines and growth factors [95]. Indeed, the fundamental change is a switch in the signaling from an NO-mediated silencing of cellular processes toward activation by redox signals. The generation of hydrogen peroxide that could react with cysteine groups in proteins alters their function, determining transcriptional factor phosphorylation, induction of nuclear chromatin remodeling, and protease activation. eNOS, normally involved in the maintenance of the quiescent state of the endothelium, could be uncoupled, in the case of substrate L-arginine deficiency. Consequently, superoxide formation occurs if the key cofactor tetrahydrobiopterin is not present, or hydrogen peroxide production is established. The mitochondria are a source of ROS during hypoxia or conditions of increased substrate delivery, such as in obesity-related metabolic disorders or type II diabetes, which are characterized by hyperglycemia and increased circulating free fatty acids. The interaction between ROS and NO results in further endothelial activation and inflammation [122].

4. Protective effects of vitamin D

Although the relationship among vitamin D_3, endothelium, and cardiovascular disease is well established, until recently little was known about the effect of vitamin D_3 on endothelial NO production. NO plays a key role in cardiovascular physiology [100], and its production in the heart by eNOS phosphorylation represents an important regulator of both myocardial perfusion after ischemia and myocardial contractility and has important effects on cardiac cell functions such as oxygen consumption, hypertrophic remodeling, apoptosis, and myocardial regeneration [101–103]. In addition, NO is able to enhance endothelial cells survival, proliferation, and migration [123]. In recent works, cells proliferation has been evaluated in low serum conditions to synchronize cell culture after 24 or 48 h in starvation medium in the presence or absence of vitamin D (1–10–100 nM) for porcine aortic endothelial cells (PAE) [123] and human umbilical vein endothelial cells (HUVEC), respectively [124, 125]. Vitamin D_3 induces a significant dose-dependent increase in cell growth at all concentrations tested. The maximal effect is reached by stimulating PAE cells with 10 nM vitamin D_3, while at the highest concentration tested (100 nM) the vitamin D_3 effect on cell proliferation is less potent ($p < 0.05$). The observed cellular density is almost doubled compared to control samples for PAE cells (690 ± 210 cells/mm^2 vs 354 ± 84 cells/mm^2, $p < 0.01$) and HUVEC. Proliferation assays were also performed in the presence of L-NAME, the arginine analog inhibiting NO synthesis, in order to evaluate NO involvement. In these conditions, vitamin D_3 is no more able to induce cell proliferation, in all cell types. The presence of L-NAME does not alter control cell proliferation, whereas it completely reverts vitamin D-induced cell proliferation. The effective involvement of VDR in these effects has been demonstrated as well. HUVEC were stimulated with the VDR ligand ZK159222. Under these conditions, a significant decrease in vitamin D–induced cell proliferation has been observed, confirming the role of VDR in the effects mediated by vitamin D_3. In addition, an increase in cell mitosis has been observed only in vitamin D_3 treated specimens and not in samples with L-NAME. Cell migration has been evaluated in a three-dimensional model by an anionic hydrogel made of gelatin and polyglutamic acid, which has previously been described as a good substrate for cell growth [126, 127] and was lean on 70% confluent cell monolayers. PAE migration, evaluated by counting the cells migrated in the 3-D matrix for 7 days, increases significantly only in the presence of 100 nM vitamin D_3 ($p < 0.05$). HUVEC migration increases significantly only in the presence of 10 nM and 100 nM vitamin D_3 ($p < 0.05$ and $p < 0.01$ respectively). L-NAME treatment does not affect control cell migration, while it significantly reduces vitamin D-induced hydrogel invasion and ZK159222 reduces the vitamin D-induced migratory effect. Extracellular matrix degradation is one of the main steps in cell migration; for this reason, vitamin D affects on MMP-2 expression in the conditioned medium of cells migrating into the 3-D hydrogel matrix after 7 days has been evaluated by gelatin zymography. Vitamin D_3 addition to PAE culture medium increases MMP-2 production in a dose-dependent manner. In HUVEC, gelatin zymography demonstrates a statistically significant increase in MMP-2 activity in all the vitamin D-treated samples. This effect is more evident in samples stimulated with 100 nM vitamin D_3 ($p < 0.01$). The increase in MMP-2 expression appears to be NO dependent, as L-NAME treatment totally abrogates vitamin D_3 effects on MMP-2 expression, according to the above described results for cell migration. In this context it is very important verify the effect of vitamin D_3 in HUVEC on NO production and the intracellular pathways activated by vitamin D_3 leading to eNOS activation.

Vitamin D_3 is able to stimulate NO production in HUVEC in a dose-dependent manner accompanied by a significant increase in the level of phosphorylation of intracellular kinases. The administration of vitamin D_3 induced the highest production of NO, acutely increased the phosphorylation of eNOS, p38, AKT, and ERK, which are known to be involved in the intracellular signaling leading to NO production [128]. The effects were prevented by L-NAME or specific protein kinase inhibitors such as SB203580, wortmannin, and UO126 related to p38, AKT, and ERK in endothelial cells [129]. Another important finding is the demonstration of the involvement of VDR in vitamin D-induced endothelial NO production [76]. This fact is shown by VDR antagonist ZK159222 and by VDR agonist ZK191784. These data on the role of VDR ligands on endothelial NO production add new information on the possible therapeutic role of these substances. VDR plays an important role in modulating cardiovascular function and early interventional studies in humans demonstrated that VDR analogs therapy seems to be more effective than native vitamin D supplementation in modulating cardiovascular disease risk factors [130]. It is noteworthy that vitamin D_3 response occurs within seconds and, for this reason, it appears to be a nongenomic effect. This fact about the role of the concentration of vitamin D_3, is also important to explain same data reported in the literature about the ability of vitamin D_3 to prevent myocardial ischemia in patients with a decreased serum vitamin D_3 concentration as reported in a case-control study in 1990 on 179 patients [131] in which the odds of having a myocardial infarction increased along. In another study, Giovannucci et al. [132] demonstrates that men with vitamin D_3 deficiency had a higher risk of myocardial infarction compared with those with normal vitamin D_3 concentration.

Recent investigations on cardiac myocytes showed that ROS produced by mitochondrial and oxidative stress can cause multiple changes in the cell structure and the function that are associated with a failing heart [133] and apoptosis [134] or autophagy [135]. On the other hand, ROS, along with NO, show antiapoptotic effects involving several signaling pathways; for example, not only activates proapoptotic signals [135] but also potently induces autophagy [136]. The functional role of autophagy during ischemia/reperfusion is complex, because its pathophysiological functions depend on the severity and duration of ischemia (hypoxia) and the consequent tissue damage during reperfusion in the heart. The level of autophagy may determine whether autophagy itself is protective or detrimental in response to ischemia/reperfusion injury in the heart [137]. Vitamin D_3 is able to induce autophagy in various human cell types. The signaling pathways regulated by vitamin D_3 include, for example, Bcl-2, beclin 1, and mammalian target of rapamycin (mTOR) [138]. The interplay between autophagic and apoptotic pathways is a crucial point to determine the initiation of programmed cell death and whether the members of the beclin 1 and Bcl-2 family were involved. Beclin 1 directly interacts not only with Bcl-2 but also with other antiapoptotic Bcl-2 family proteins such as Bcl-xl, and Bcl-2 was able to inhibit the beclin 1-dependent autophagy [139]. It has been clearly demonstrated that administration of vitamin D_3 to endothelial cells before induction of an oxidative stress can improve cell viability [20]. The mechanisms involved include the prevention of free oxygen radical release and the modulation of the interplay between apoptosis and autophagy. These effects were also accompanied by NO production and preservation of mitochondrial function. HUVEC cultures received an oxidative stress by means of H_2O_2. This method is widely used to reproduce a cellular damage similar to what occurs in myocardial ischemia/reperfusion injury [20]. In light of these data, a cardioprotective role of vitamin D_3 against the ischemic injury can be hypothesized. As observed in NO production and MTT tests, the effect induced by VDR agonist ZK191784 is greater than the one observed after vitamin D_3 alone. Moreover, the combined administration of the two VDR agonists (Vit D and ZK191784) induced an amplified effect. These beneficial effects were observed when VDR agonists were administered before the induction of oxidative stress. This fact supports the hypothesis that vitamin D_3 is able to counteract the negative effects of the oxidant event on endothelial cells, increasing cell viability. In addition, NO release induced by vitamin D_3 during oxidative stress is able to protect cells from death. This result is demonstrated by the observation that the rate of NO production was below 2¼M/s. This threshold prevents the opening of the mitochondrial transition pore and the release of cytochrome C and avoids mitochondrial collapse leading to cell death [140]. The antiapoptotic effects of NO and ROS involved several signaling pathways, and the interplay between autophagic and apoptotic pathways is a crucial point to determine the starting of programmed cell death. Another important finding is that pretreatment with vitamin D_3, alone or in combination with ZK191784, is able to reduce the apoptosis-related gene expression (Bax, caspase-3, caspase-9, caspase-8, and cytochrome C), involving both intrinsic and extrinsic pathways. These findings were confirmed by immunohistochemistry analysis of annexin V and TUNEL assay in which we observed a signal reduction. At the same time, activation of pro-autophagic beclin 1 and the phosphorylation of ERK1/2 and Akt, members of the reperfusion injury salvage kinase pathway, have been shown, indicating a modulation between apoptosis and autophagy. Moreover, vitamin D_3, alone or in combination with ZK191784, administered before the oxidative stress, is able to prevent the loss of mitochondrial potential and the consequent cytochrome C release and caspase activation. In addition, vitamin D_3 alone or in combination with ZK191784 was able to prevent the MPTP opening caused by H_2O_2. These findings depend on changes of mitochondria-trapped calcein intensity and on effects of cyclosporin A, which inhibits MPTP opening, modulating the cyclophilin D activity [141].

Oxidative stress and the consequent mitochondrial fragmentation are involved in several neurodegenerative disorders as well [142–144]. Oxidative stress results from Fenton reaction, which generates ROS and consequently induces irreversible damage to DNA, RNA, proteins, and lipids [145]. Thus, there is compelling evidence that the neuroprotective action of vitamin D_3 is revealed through neuronal calcium regulation, antioxidative pathway, immunomodulation, and detoxification. Vitamin D_3 along with its metabolites influences directly or indirectly almost all metabolic processes such as proliferation, differentiation, apoptosis, inflammatory processes, and mutagenesis. Such multifactorial effects of vitamin D_3 can be a profitable source of new therapeutic solutions for two radically divergent diseases, cancer and neurodegeneration [146]. The role of vitamin D_3 to prevent iron damage in neuroblastoma cells BE(2)M17 has been recently studied [147]. In this research, the mechanisms involved in neurodegeneration, such as cell viability, ROS production, and the most common intracellular pathway have been studied. The BE(2)M17 cell line is an alternative neuronal cell model widely used in neuroscience research [148]. The beneficial effect of vitamin D_3 consists of a significant decrease in iron accumulation and ROS production, two important co-factors of neurodegeneration [149]. The effects of vitamin D_3 have been observed in BE(2)M17 cells through the activation of the VDR receptor and for this reason, the mechanisms underlying the protective effects of vitamin D_3 may be hypothesized to have genomic origin. Moreover, under these conditions, the survival pathway (ERKs) was also activated in presence of vitamin D_3. In addition, in these experiments Fe^{3+}-dependent activation of principal neurodegenerative biomarkers, such as p53 [150], Ki67 [151] and c-Myc [152], has been confirmed. A novel finding is represented by the demonstration that pre-treatment with vitamin D_3 is able to significantly counteract tumoral biomarker activation.

http://dx.doi.org/10.5772/64506

5. Conclusions

The results described herein highlight that vitamin D_3 stimulates endothelial cell proliferation and migration in a 3-D matrix and that these phenomena depend on NO production. Furthermore, this work adds new information to the debate on the benefits of vitamin D_3 supplementation. Indeed, it has been shown for the first time that vitamin D_3 may prevent endothelial cell death through the modulation of the interplay between apoptosis and autophagy. This effect is obtained by inhibiting superoxide anion generation, maintaining mitochondria function and cell viability, activating survival kinases, and inducing NO production. In the recent years, the knowledge about vitamin D_3 has improved and its implications have extended beyond its classical role in bone health in either fields of basic research as well as in human clinical trials, showing the relevance of the vitamin D_3 system. Until now, the available data are significant and confirm its essential role in several physiological and preventive functions. A greater understanding of vitamin D_3 system will shed new light on the use of vitamin D_3 supplementation in very promising fields such as tissue repair, wound healing, and prevention of the human angiogenic process. The former shows that the association between vitamin D_3 status and cardiometabolic outcomes is uncertain and that no clinically significant effect of vitamin D_3 supplementation at the dosages given is found. The latter suggests that vitamin D_3 dietary supplements at moderate to high doses may reduce cardiovascular disease risk. Vitamin D_3 along with its metabolites influences directly or indirectly almost all metabolic processes such as proliferation, differentiation, apoptosis, inflammatory processes, and mutagenesis. Such multifactorial effects of vitamin D_3 can be a profitable source of new therapeutic solutions for two radically divergent diseases as cancer and neurodegeneration [134]. The discussed results could be relevant in the light of the use of vitamin D_3 to promote supplementation or to adjust therapeutic strategies in neurodegenerative disorders.

Author details

Francesca Uberti*, Vera Morsanuto and Claudio Molinari

*Address all correspondence to: francesca.uberti@med.uniupo.it

Physiology Laboratory, Department of Translational Medicine, UPO, Novara, Italy

References

[1] Dusso A.S., Brown A.J., Slatopolsky E. Vitamin D. Am J Physiol Renal Physiol. 2005;289(1):F8–F28.

[2] Christakos S. Recent advances in our understanding of 1,25-dihydroxyvitamin D(3) regulation of intestinal calcium absorption. Arch Biochem Biophys. 2012;523(1):73–76. DOI: 10.1016/j.abb.2011.12.020

[3] Holick M.F., Chen T.C. Vitamin D deficiency: a worldwide problem with health consequences. Am J Clin Nutr. 2008;87(4):1080S–1086S.

[4] Gröber U., Spitz J., Reichrath J., Kisters K., Holick M.F. Vitamin D: update 2013: from rickets prophylaxis to general preventive healthcare. Dermatoendocrinology. 2013;5(3): 331–347. DOI: 10.4161/derm.26738

[5] Lai Y.H., Fang T.C. The pleiotropic effect of vitamin D. ISRN Nephrol. 2013;2013:898125. DOI: 10.5402/2013/898125

[6] Tripkovic L., Lambert H., Hart K., Smith C.P., Bucca G., Penson S. Comparison of vitamin D2 and vitamin D_3 supplementation in raising serum 25-hydroxyvitamin D status: a systematic review and meta-analysis. Am J Clin Nutr. 2012;95(6):1357–1364. DOI: 10.3945/ajcn.111.031070

[7] Girgis C.M., Clifton-Bligh R.J., Hamrick M.W., Holick M.F., Gunton J.E. The roles of vitamin D in skeletal muscle: form, function, and metabolism. Endocr Rev. 2013;34(1): 33–83. DOI: 10.1210/er.2012-1012

[8] Bikle D.D. Vitamin D metabolism, mechanism of action, and clinical applications. Chem Biol. 2014;21(3):319–329. DOI: 10.1016/j.chembiol.2013.12.016

[9] Lehmann B., Meurer M. Vitamin D metabolism. Dermatol Ther. 2010;23(1):2–12. DOI: 10.1111/j.1529-8019.2009.01286.x

[10] Carlberg C., Seuter S., Heikkinen S. The first genome-wide view of vitamin D receptor locations and their mechanistic implications. Anticancer Res. 2012;32:271–282.

[11] Pike J.W., Meyer M.B. The vitamin D receptor: new paradigms for the regulation of gene expression by 1,25-dihydroxyvitamin D(3). Endocrinol Metab Clin North Am. 2010;39:255–569. DOI: 10.1016/j.rdc.2012.03.004

[12] Meyer M.B., Goetsch P.D., Pike J.W. Genome-wide analysis of the VDR/RXR cistrome in osteoblast cells provides new mechanistic insight into the actions of the vitamin D hormone. J Steroid Biochem Mol Biol. 2010;121:136–141. DOI: 10.1016/j.jsbmb. 2010.02.011

[13] Norman A.W. Vitamin D receptor: new assignments for an already busy receptor. Endocrinology. 2006;147(12):5542–5548.

[14] Boland R., Buitrago C., De Boland A.R. Modulation of tyrosine phosphorylation signalling pathways by lα,25(OH)2-vitamin D_3. Trends Endocrinol Metab. 2005;16(6): 280–287.

[15] Norman A.W., Okamura W.H., Hammond M.W., Bishop J.E., Dormanen M.C., Bouillon R., et al. Comparison of 6-s-cis- and 6-s-trans-locked analogs of 1alpha,25-dihydroxy-

vitamin D_3 indicates that the 6-s-cis conformation is preferred for rapid nongenomic biological responses and that neither 6-s-cis- nor 6-s-trans-locked analogs are preferred for genomic biological responses. Mol Endocrinol. 1997;11:1518–1531.

[16] Sequeira V.B., Rybchyn M.S., Tongkao-On W., Gordon-Thomson C., Malloy P.J., Nemere I., et al. The role of the vitamin D receptor and ERp57 in photoprotection by 1α,25-dihydroxyvitamin D_3. Mol Endocrinol. 2012;26(4):574–582. DOI: 10.1210/me.2011-1161

[17] Mizwicki M.T., Norman A.W. The vitamin D sterol-vitamin D receptor ensemble model offers unique insights into both genomic and rapid-response signaling. Sci Signal. 2009;2(75):re4. DOI: 10.1126/scisignal.275re4

[18] Huhtakangas J.A., Olivera C.J., Bishop J.E., Zanello L.P., Norman A.W. The vitamin D receptor is present in caveolae-enriched plasma membranes and binds 1 alpha, 25(OH)2-vitamin D_3 in vivo and in vitro. Mol Endocrinol. 2004;18:2660–2671.

[19] Berridge M.J. Vitamin D cell signalling in health and disease. Biochem Biophys Res Commun. 2015;460(1):53–71. DOI: 10.1016/j.bbrc.2015.01.008

[20] Uberti F., Lattuada D., Morsanuto V., Nava U., Bolis G., Vacca G. Vitamin D protects human endothelial cells from oxidative stress through the autophagic and survival pathways. J Clin Endocrinol Metab. 2014;99(4):1367–1374. DOI: 10.1210/jc.2013-2103

[21] Souberbielle J.C., Body J.J., Lappe J.M., Plebani M., Shoenfeld Y., Wang T.J. Vitamin D and musculoskeletal health, cardiovascular disease, autoimmunity and cancer: recommendations for clinical practice. Autoimmun Rev. 2009;9(11):709–715. DOI: 10.1016/j.autrev.2010.06.009

[22] DeLuca H.F. Overview of general physiologic features and functions of vitamin D. Am J Clin Nutr. 2008;80(6 suppl):1689S–1696S.

[23] Anagnostis P., Dimopoulou C., Karras S., Lambrinoudaki I., Goulis D.G. Sarcopenia in post-menopausal women: is there any role for vitamin D? Maturitas. 2015;82(1):56–64. DOI: 10.1016/j.maturitas.2015.03.014

[24] Girgis C.M., Clifton-Bligh R.J., Turner N., Lau S.L., Gunton J.E. Effects of vitamin D in skeletal muscle: falls, strength, athletic performance and insulin sensitivity. Clin Endocrinol (Oxf). 2014;80:169–181.

[25] MacLaughlin JH. Aging decreases the capacity of human skin to produce vitamin D_3. J Clin Invest. 1985;76:1536–1538.

[26] Burton L.A., Sumukadas D. Optimal management of sarcopenia. Clin Interv Aging. 2010;7(5):217–228.

[27] Visser M., Deeg D.J.H., Lips P. Low vitamin D and high parathyroid hormone levels as determinants of loss of muscle strength and muscle mass (Sarcopenia): the longitudinal aging study. Amsterdam. J Clin Endocrinol Metab. 2003;88:5766–5772.

[28] Bouvard B., Annweiler C., Sallé A., Beauchet O., Chappard D., Audran M. Extraskeletal effects of vitamin D: facts, uncertainties, and controversies. Joint Bone Spine. 2011;87(1): 10–16. DOI: 10.1016/j.jbspin.2010.10.011

[29] Yamshchikov A.V., Desai N.S., Blumberg H.M., Ziegler T.R., Tangpricha V. Vitamin D for treatment and prevention of infectious diseases: a systematic review of randomized controlled trials. Endocr Pract. 2009;15:438–449. DOI: 10.4158/EP09101.ORR

[30] Wacker M., Holick M.F. Vitamin D: effects on skeletal and extraskeletal health and the need for supplementation. Nutrients. 2013;5(1):111–148. DOI: 10.3390/nu5010111

[31] Youssef D.A., Miller C.W., El-Abbassi A.M., Cutchins D.C., Cutchins C., Grant W.B., Peiris A.N. Antimicrobial implications of vitamin D. Dermatoendocrinology. 2011;3(4): 220–229. DOI: 10.4161/derm.3.4.15027

[32] Antico A., Tampoia M., Tozzoli R., Bizzaro N. Can supplementation with vitamin D reduce the risk or modify the course of autoimmune diseases? A systematic review of the literature. Autoimmun Rev. 2012;12:127–136. DOI: 10.1016/j.autrev.2012.07.007

[33] Ponsonby A.L., McMichael A., van der Mei I. Ultraviolet radiation and autoimmune disease: insights from epidemiological research. Toxicology. 2002;181–182:71–78.

[34] Peyrin-Biroulet L., Oussalah A., Bigard M.A. Crohn's disease: the hot hypothesis. Med Hypotheses. 2009;73:94–96. DOI: 10.1016/j.mehy.2009.01.022

[35] Vieira V.M., Hart J.E., Webster T.F., Weinberg J., Puett R., Laden F. Association between residences in U.S. northern latitudes and rheumatoid arthritis: a spatial analysis of the Nurses' Health Study. Environ Health Perspect. 2010;118:957–961. DOI: 10.1289/ehp.0901861

[36] Mohr S.B., Garland C.F., Gorham E.D., Garland F.C. The association between ultraviolet B irradiance, vitamin D status and incidence rates of type 1 diabetes in 51 regions worldwide. Diabetologia. 2008;51:1391–1398. DOI: 10.1007/s00125-008-1061-5

[37] Chinellato I., Piazza M., Sandri M., Peroni D., Piacentini G., Boner A.L. Vitamin D serum levels and markers of asthma control in Italian children. J Pediatr. 2011;158:437–441. DOI: 10.1016/j.jpeds.2010.08.043

[38] Chinellato I., Piazza M., Sandri M., Peroni D.G., Cardinale F., Piacentini G.L. Serum vitamin D levels and exercise-induced bronchoconstriction in children with asthma. Eur Respir J. 2011;37:1366–1370. DOI: 10.1183/09031936.00044710

[39] Searing D.A., Zhang Y., Murphy J.R., Hauk P.J., Goleva E., Leung D.Y. Decreased serum vitamin D levels in children with asthma are associated with increased corticosteroid use. J Allergy Clin Immunol. 2010;125:995–1000. DOI: 10.1016/j.jaci.2010.03.008

[40] Fleet J.C. Molecular actions of vitamin D contributing to cancer prevention. Mol Aspects Med. 2008;29:388–396. DOI: 10.1016/j.mam.2008.07.003

[41] Tretli S., Hernes E., Berg J.P., Hestvik U.E., Robsahm T.E. Association between serum 25 (OH)D and death from prostate cancer. Br J Cancer. 2009;100:450–454. DOI: 10.1038/sj.bjc.6604865

[42] Goodwin P.J., Ennis M., Pritchard K.I., Koo J., Hood N. Prognostic effects of 25-hydroxyvitamin D levels in early breast cancer. J Clin Oncol. 2009;27:3757–3763. DOI: 10.1200/JCO.2008.20.0725

[43] Krishnan A.V., Feldman D. Mechanisms of the anti-cancer and anti-inflammatory actions of vitamin D. Annu Rev Pharmacol Toxicol. 2011;51:311–336. DOI: 10.1146/annurev-pharmtox-010510-100611

[44] Scaranti M., Júnior G.C., Hoff A.O. Vitamin D and cancer: does it really matter? Curr Opin Oncol. 2016;28(3):205–209. DOI:10.9097/CCO.0000000000000282.

[45] Magrone T., Marzulli G., Jirillo E. Immunopathogenesis of neurodegenerative diseases: current therapeutic models of neuroprotection with special reference to natural products. Curr Pharm Des. 2012;18(1):34–42.

[46] Van der Wielen R.P., Lowik M.R., Van dan Berg H., de Groot L.C., Haller J., Moreiras O., Van Staveren W.A. Serum vitamin D concentrations among elderly people in Europe. Lancet. 1995;346(8969):207–210.

[47] M. F. Holick. Vitamin D deficiency. N Engl J Med. 2007;357(3):266–281.

[48] Artaza J.N., Mehrotra R., Norris K.C. Vitamin D and the cardiovascular system. Clin J Am Soc Nephrol. 2009;4(9):1515–1522. DOI: 10.2215/CJN.02260409

[49] Banerjee A., Khemka V.K., Ganguly A., Roy D., Ganguly U., Chakrabarti S. Vitamin D and Alzheimer's Disease: neurocognition to therapeutics. Int J Alzheimers Dis. 2015;2015:192747. DOI: 10.1155/2015/192747

[50] Anderson J.L., May H.T., Horne B.D., Bair T.L., Hall N.L., Carlquist J.F., et al. Relation of vitamin D deficiency to cardiovascular risk factors, disease status, and incident events in a general healthcare population. Am J Cardiol. 2010;106:963–968. DOI: 10.1016/j.amjcard.2010.05.027

[51] Szeto F.L., Reardon C.A., Yoon D., Wang Y., Wong K.E., Chen Y., et al. Vitamin D receptor signaling inhibits atherosclerosis in mice. Mol Endocrinol. 2012;26:1091–1101. DOI: 10.1210/me.2011-1329

[52] Uysal S., Kalayci A.G., Baysal K. Cardiac functions in children with vitamin D deficiency rickets. Pediatr Cardiol. 1999;20:283–286.

[53] Brøndum-Jacobsen P., Benn M., Jensen G.B., Nordestgaard B.G. 25-hydroxyvitamin D levels and risk of ischemic heart disease, myocardial infarction, and early death: population-based study and meta- analyses of 18 and 17 studies. Arterioscler Thromb Vasc Biol. 2012;32:2794–2802. DOI: 10.1161/ATVBAHA.112.248039

[54] Wilson P.W., D'Agostino R.B., Sullivan L., Parise H., Kannel W.B. Overweight and obesity as determinants of cardiovascular risk: the Framingham experience. Arch Intern Med. 2002;162:1867–1872.

[55] Reid I.R., Bolland M.J. Role of vitamin D deficiency in cardiovascular disease. Heart. 2012;98:609–614. DOI: 10.1136/heartjnl-2011-301356

[56] Wortsman J., Matsuoka L.Y., Chen T.C., Lu Z., Holick M.F. Decreased bioavailability of vitamin D in obesity. Am J Clin Nutr. 2000;72:690–693

[57] Wang T.J., Pencina M.J., Booth S.L., Jacques P.F., Ingelsson E., Lanier K., et al. Vitamin D deficiency and risk of cardiovascular disease. Circulation. 2008;117(4):503–511. DOI: 10.1161/CIRCULATIONAHA.107.706127

[58] Martins D., Meng Y.X., Tareen N., Artaza J., Lee J.E., Farodolu C., et al. The effect of short term vitamin D supplementation on the inflammatory and oxidative mediators of arterial stiffness. Health (Irvine Calif). 2014;6(12):1503–1511.

[59] Park C.W., Oh Y.S., Shin Y.S., Kim C.M., Kim Y.S., Kim S.Y., et al. Intravenous calcitriol regresses myocardial hypertrophy in hemodialysis patients with secondary hyperparathyroidism. Am J Kidney Dis. 1999;33:73–81.

[60] Shoji T., Shinohara K., Kimoto E., Emoto M., Tahara H., Koyama H., et al. Lower risk for cardiovascular mortality in oral 1α-hydroxyvitamin D_3 users in a haemodialysis population. Nephrol Dial Transplant. 2004;19:179–184.

[61] Talmor Y., Golan E., Benchetrit S., Bernheim J., Klein O., Green J., et al. Calcitriol blunts the deleterious impact of advanced glycation end products on endothelial cells. Am J Physiol Renal Physiol. 2008;294:F1059–F1064. DOI: 10.1152/ajprenal.00051.2008

[62] Burgaz A., Orsini N., Larsson S.C., Wolk A. Blood 25-hydroxyvitamin D concentration and hypertension: a meta-analysis. J Hypertens. 2011;29:636–645. DOI: 10.1097/HJH.0b013e32834320f9

[63] Witham M.D., Nadir M.A., Struthers A.D. Effect of vitamin D on blood pressure: a systematic review and meta-analysis. J Hypertens. 2009;27:1948–1954. DOI: 10.1097/HJH.0b013e32832f075b

[64] Wu S.H., Ho S.C., Zhong L. Effects of vitamin D supplementation on blood pressure. South Med J. 2010;103:729–737. DOI: 10.1097/SMJ.0b013e3181e6d389

[65] Elamin M.B., Abu Elnour N.O., Elamin K.B., Fatourechi M.M., Alkatib A.A., Almandoz J.P., et al. Vitamin D and cardiovascular outcomes: a systematic review and meta-analysis. J Clin Endocrinol Metab. 2011;96:1931–1942. DOI: 10.1210/jc.2011-0398

[66] Krause R., Bühring M., Hopfenmüller W., Holick M.F., Sharma A.M. Ultraviolet B and blood pressure. Lancet. 1998;352(9129):709–710.

[67] Pilz S., Tomaschitz A., März W., Drechsler C., Ritz E., Zittermann A., et al. Vitamin D, cardiovascular disease and mortality. Clin Endocrinol. 2011;75(5):575–584. DOI: 10.1111/j.1365-2265.2011.04147.x

[68] Li Y.C., Qiao G., Uskokovic M., Xiang W., Zheng W., Kong J. Vitamin D: a negative endocrine regulator of the renin-angiotensin system and blood pressure. J Steroid Biochem Mol Biol. 2004;89–90(1–5):387–392.

[69] Haussler M.R., Whitfield G.K., Haussler C.A., Hsieh J.C., Thompson P.D., Selznick S.H., et al. The nuclear vitamin D receptor: biological and molecular regulatory properties revealed. J Bone Miner Res. 1998;13(3):325–349.

[70] Wu-Wong J.R., Kawai M., Chen Y.W., Nakane M. VS-105: a novel vitamin D receptor modulator with cardiovascular protective effects. Br J Pharmacol. 2011;164(2b):551–560. DOI: 10.1111/j.1476-5381.2011.01473.x

[71] Thompson L., Wang S., Tawfik O., Templeton K., Tancabelic J., Pinson D., et al. Effects of 25-hydroxyvitamin D_3 and 1α,25 dihydroxyvitamin D_3 on differentiation and apoptosis of human osteosarcoma cell lines. J Orthop Res. 2012;30(5):831–844. DOI: 10.1002/jor.21585

[72] Somjen D., Kohen F., Amir-Zaltsman Y., Knoll E., Stern N. Vitamin D analogs modulate the action of gonadal steroids in human vascular cells in vitro. Am J Hypertens. 2000;13(4 Pt 1):396–403.

[73] Talmor Y., Bernheim J., Klein O., Green J., Rashid G. Calcitriol blunts pro-atherosclerotic parameters through NFκB and p38 in vitro. Eur J Clin Invest. 2008;38(8):548–554. DOI: 10.1111/j.1365-2362.2008.01977.x

[74] Zehnder D., Bland R., Chana R.S., Wheeler D.C., Howie A.J., Williams M.C., et al. Synthesis of 1,25-dihydroxyvitamin D_3 by human endothelial cells is regulated by inflammatory cytokines: a novel autocrine determinant of vascular cell adhesion. J Am Soc Nephrol. 2002;13(3):621–629.

[75] Dalan R, Liew H., Tan W.K.A., Chew D.E.K., Leow M.K.S. Vitamin D and the endothelium: basic, translational and clinical research updates. IJC Metab Endocr. 2014;4:4–17. DOI: 10.1016/j.ijcme.2014.06.003

[76] Molinari C., Uberti F., Grossini E., Vacca G., Carda S., Invernizzi M., et al. 1α,25-Dihydroxycholecalciferol induces nitric oxide production in cultured endothelial cells. Cell Physiol Biochem. 2011;27(6):661–668. DOI: 10.1159/000330075

[77] Sun Q., Pan A., Hu F.B., Manson J.E., Rexrode K.M. 25-Hydroxyvitamin D levels and the risk of stroke: a prospective study and meta-analysis. Stroke. 2012;43(6):1470–1477. DOI: 10.1161/STROKEAHA.111.636910

[78] Eyles D.W., Smith S., Kinobe R., Hewison M., McGrath J.J. Distribution of the vitamin D receptor and 1 alpha-hydroxylase in human brain. J Chem Neuroanat. 2005;29(1):21–30.

[79] Prüfer K., Veenstra T.D., Jirikowski G.F., Kumar R. Distribution of 1,25-dihydroxyvitamin D_3 receptor immunoreactivity in the rat brain and spinal cord. J Chem Neuroanat. 1999;16(2):135–145.

[80] Tetich M., Kutner A., Leskiewicz M., Budziszewska B., Lasoń W. Neuroprotective effects of (24R)-1,24-dihydroxycholecalciferol in human neuroblastoma SH-SY5Y cell line. J Steroid Biochem Mol Biol. 2004;89–90(1–5):365–370.

[81] Garcion E., Wion-Barbot N., Montero-Menei C.N., Berger F., Wion D. New clues about vitamin D functions in the nervous system. Trends Endocrinol Metab. 2002;13(3):100–105.

[82] Kalueff A.V., Minasyan A., Keisala T., Kuuslahti M., Miettinen S., Tuohimaa P. The vitamin D neuroendocrine system as a target for novel neurotropic drugs. CNS Neurol Disord Drug Targets. 2006;5(3):363–371.

[83] Eyles D., Brown J., Mackay-Sim A., McGrath J., Feron F. Vitamin D_3 and brain development. Neuroscience. 2003;118(3):641–653.

[84] Aloia J., Bojadzievski T., Yusupov E., Shahzad G., Pollack S., Mikhail M., et al. The relative influence of calcium intake and vitamin D status on serum parathyroid hormone and bone turnover biomarkers in a double-blind, placebo-controlled parallel group, longitudinal factorial design. J Clin Endocrinol Metab. 2010;95(7):3216–3224. DOI: 10.1210/jc.2009-1294

[85] Chopp M., Zhang Z.G., Jiang Q. Neurogenesis, angiogenesis, and MRI indices of functional recovery from stroke. Stroke. 2007;38(2 suppl):827–831.

[86] Stewart A., Wong K., Cachat J., Elegante M., Gilder T., Mohnot S., et al. Neurosteroid vitamin D system as a nontraditional drug target in neuropsychopharmacology. Behav Pharmacol. 2010;21(5–6):420–426. DOI: 10.1097/FBP.0b013e32833c850f

[87] Kienreich K., Grubler M., Tomaschitz A., Schmid J., Verheyen N., Rutters F., et al. Vitamin D, arterial hypertension & cerebrovascular disease. Indian J Med Res. 2013;137(4):669–679.

[88] Chowdhury R., Stevens S., Ward H., Chowdhury S., Sajjad A., Franco O.H. Circulating vitamin D, calcium and risk of cerebrovascular disease: a systematic review and meta-analysis. Eur J Epidemiol. 2012;27(8):581–891. DOI: 10.1007/s10654-012-9729-z

[89] Thompson P.D., Jurutka P.W., Haussler C.A., Whitfield G.K., Haussler M.R. Heterodimeric DNA binding by the vitamin D receptor and retinoid X receptors is enhanced by 1,25-dihydroxyvitamin D_3 and inhibited by 9-cisretinoic acid. Evidence for allosteric receptor interactions. J Biol Chem. 1998;273(14):8483–8491.

[90] Arbelle J.E., Chen H., Gaead M.A., Allegretto E.A., Pike J.W., Adams J.S. Inhibition of vitamin D receptor-retinoid X receptor-vitamin D response element complex formation by nuclear extracts of vitamin D-resistant new world primate cells. Endocrinology. 1996;137(2):786–789.

[91] Cui X., McGrath J.J., Burne T.H., Mackay-Sim A., Eyles D.W. Maternal vitamin D depletion alters neurogenesis in the developing rat brain. Int J Dev Neurosci. 2007;25(4): 227–232.

[92] Brewer L.D., Thibault V., Chen K.C., Langub M.C., Landfield P.W., Porter N.M. Vitamin D hormone confers neuroprotection in parallel with downregulation of l-type calcium channel expression in hippocampal neurons. J Neurosci. 2001;21(1):98–108.

[93] Ibi M., Sawada H., Nakanishi M., Kume.T, Katsuki H., Kaneko S., et al. Protective effects of 1 alpha,25-(OH)(2)D(3) against the neurotoxicity of glutamate and reactive oxygen species in mesencephalic culture. Neuropharmacology. 2001;40(6):761–771.

[94] Garcion E., Nataf S., Berod A., Darcy F., Brachet P. 1,25-DihydroxyVitamin D_3 inhibits the expression of inducible nitric oxide synthase in rat central nervous system during experimental allergic encephalomyelitis. Brain Res Mol Brain Res. 1997;45(2):255–267.

[95] Pilz S., Tomaschitz A., Drechsler C., Zittermann A., Dekker J.M., März W. Vitamin D supplementation: a promising approach for the prevention and treatment of strokes. Curr Drug Targets. 2011;12(1):88–96.

[96] Mena N.P., García-Beltrán O., Lourido F., Urrutia P.J., Mena R., Castro-Castillo V., et al. The novel mitochondrial iron chelator 5-((methylamino)methyl)-8-hydroxyquinoline protects against mitochondrial-induced oxidative damage and neuronal death. Biochem Biophys Res Commun. 2015;463(4):787–792. DOI: 10.1016/j.bbrc.2015.06.014

[97] Wang Y., Chiang Y.H., Su T.P., Hayashi T., Morales M., Hoffer B.J., et al. Vitamin D_3 attenuates cortical infarction induced by middle cerebral arterial ligation in rats. Neuropharmacology. 2000;39(5):873–880.

[98] Soni M., Kos K., Lang I.A., Jones K., Melzer D., Llewellyn D.J. Vitamin D and cognitive function. Scand J Clin Lab Invest Suppl. 2012;243:79–82. DOI: 10.3109/00365513.2012.681969

[99] Xiang W., He X.J., Ma Y.L., Yi Z.W., Cao Y., Zhao S.P., et al. 1,25(OH)2D_3 influences endothelial cell proliferation, apoptosis and endothelial nitric oxide synthase expression of aorta in apolipoprotein E-deficient mice. Zhonghua Er Ke Za Zhi. 2011;49(11):829–833.

[100] Kanno Y., Into T., Lowenstein C.J., Matsushita K. Nitric oxide regulates vascular calcification by interfering with TGF-signalling. Cardiovasc Res. 2008;77(1):221–230.

[101] Massion P.B., Pelat M., Belge C., Balligand J.L. Regulation of the mammalian heart function by nitric oxide. Comp Biochem Physiol A Mol Integr Physiol. 2005;142(2):144–150.

[102] Mount P.F., Kemp B.E., Power D.A. Regulation of endothelial and myocardial NO synthesis by multi-site eNOS phosphorylation. J Mol Cell Cardiol. 2007;42(2): 271–279.

[103] Smith R.S.Jr., Agata J., Xia C.F., Chao L., Chao J. Human endothelial nitric oxide synthase gene delivery protects against cardiac remodeling and reduces oxidative stress after myocardial infarction. Life Sci. 2005;76(21):2457–2571.

[104] Butler A.R. 'The heart less bounding': treating angina pectoris. J R Coll Physicians Edinb. 2006;36(2):185–189.

[105] Brevetti G., Schiano V., Chiariello M. Endothelial dysfunction: a key to the pathophysiology and natural history of peripheral arterial disease?. Atherosclerosis. 2008;197(1): 1–11.

[106] Hill B.G., Dranka B.P., Bailey S.M., Lancaster J.R.Jr., Darley-Usmar V.M. What part of NO don't you understand? Some answers to the cardinal questions in nitric oxide biology. J Biol Chem. 2010;285(26):19699–19704. DOI: 10.1074/jbc.R110.101618

[107] Salisbury D., Bronas U. Reactive oxygen and nitrogen species: impact on endothelial dysfunction. Nurs Res. 2015;64(1):53–66. DOI: 10.1097/NNR.0000000000000068

[108] Feletou M. Focus on endothelium-derived vasoactive mediators. In: The Endothelium: Part 1: Multiple Functions of the Endothelial Cells. San Rafael, CA: Morgan & Claypool Life Sciences. Granger DN and Granger J Eds. 2011.

[109] Shu X., Keller T.C., Begandt D., Butcher J.T., Biwer L., Keller A.S., et al. Endothelial nitric oxide synthase in the microcirculation. Cell Mol Life Sci. 2015;72(23):4561–4575. DOI: 10.1007/s00018-015-2021-0

[110] Cooke J.P., Losordo D.W. Nitric oxide and angiogenesis. Circulation. 2002;105:2133–2135.

[111] Cooke J.P. NO and angiogenesis. Atheroscler Suppl. 2003;4(4):53–60.

[112] Thomas D.D., Espey M.G., Ridnour L.A., Hofseth L.J., Mancardi D., Harris C.C., et al. Hypoxic inducible factor 1alpha, extracellular signal-regulated kinase, and p53 are regulated by distinct threshold concentrations of nitric oxide. Proc Natl Acad Sci U S A. 2004;101(24):8894–8899.

[113] Radi E., Formichi P., Battisti C., Federico A. Apoptosis and oxidative stress in neurodegenerative diseases. J Alzheimers Dis. 2014;42(Suppl 3):S125–152. DOI: 10.3233/JAD-132738

[114] Dan Dunn J., Alvarez L.A., Zhang X., Soldati T. Reactive oxygen species and mitochondria: a nexus of cellular homeostasis. Redox Biol. 2015;6:472–485. DOI: 10.1016/j.redox.2015.09.005

[115] Chen Y., Azad M.B., Gibson S.B. Superoxide is the major reactive oxygen species regulating autophagy. Cell Death Differ. 2009;16(7):1040–1052. DOI: 10.1038/cdd.2009.49

[116] Rojo A.I., McBean G., Cindric M., Egea J., López M.G., Rada P., et al. Redox control of microglial function: molecular mechanisms and functional significance. Antioxid Redox Signal. 2014;21(12):1766–1801. DOI: 10.1089/ars.2013.5745

[117] Stuehr D.J., Santolini J., Wang Z.Q., Wei C.C., Adak S. Update on mechanism and catalytic regulation in the NO synthases. J Biol Chem. 2004;279(35):36167–36170.

[118] Thomas D.D., Ridnour L.A., Isenberg J.S., Flores-Santana W., Switzer C.H., Donzelli S., et al. The chemical biology of nitric oxide: implications in cellular signaling. Free Radic Biol Med. 2008;45(1):18–31. DOI: 10.1016/j.freeradbiomed.2008.03.020

[119] Dudzinski D.M., Michel T. Life history of eNOS: partners and pathways. Cardiovasc Res. 2007;75(2):247–260.

[120] Gonzalez E., Kou R., Michel T. Rac1 modulates sphingosine 1-phosphate-mediated activation of phosphoinositide 3-kinase/Akt signaling pathways in vascular endothelial cells. J Biol Chem. 2006;281(6):3210–3216.

[121] Coutinho T., Rooke T.W., Kullo I.J. Arterial dysfunction and functional performance in patients with peripheral artery disease: a review. Vasc Med. 2011;16(3):203–211. DOI: 10.1177/1358863X11400935

[122] Deanfield J.E., Halcox J.P., Rabelink T.J. Endothelial function and dysfunction: testing and clinical relevance. Circulation. 2007;115(10):1285–1295.

[123] Molinari C., Rizzi M., Squarzanti D.F., Pittarella P., Vacca G., Renò F. 1α,25-Dihydroxycholecalciferol (Vitamin D_3) induces NO-dependent endothelial cell proliferation and migration in a three-dimensional matrix. Cell Physiol Biochem. 2013;31(6):815–822. DOI: 10.1159/000350099

[124] Pittarella P., Squarzanti D.F., Molinari C., Invernizzi M., Uberti F., Renò F. NO-dependent proliferation and migration induced by Vitamin D in HUVEC. J Steroid Biochem Mol Biol. 2015;149:35–42. DOI: 10.1016/j.jsbmb.2014.12.012

[125] Jaffe E.A., Nachman R.L., Becker C.G., Minick C.R. Culture of human endothelial cells derived from umbilical veins. Identification by morphologic and immunologic criteria. J Clin Invest. 1973;52(11):2745–2756.

[126] Layman H., Spiga M.G., Brooks T., Pham S., Webster K.A., Andreopoulos F.M. The effect of the controlled release of basic fibroblast growth factor from ionic gelatin-based hydrogels on angiogenesis in a murine critical limb ischemic model. Biomaterials. 2007;28(16):2646–2654.

[127] Renò F., Rizzi M., Cannas M. Gelatin-based anionic hydrogel as biocompatible substrate for human keratinocyte growth. J Mater Sci Mater Med. 2012;23:565–571. DOI: 10.1007/s10856-011-4519-9

[128] Queen L.R., Ji Y., Xu B., Young L., Yao K., Wyatt A.W. Mechanisms underlying β2-adrenoceptor-mediated nitric oxide generation by human umbilical vein endothelial cells. J Physiol. 2006;576(Pt 2):585–594.

[129] Padmasekar M., Nandigama R., Wartenberg M., Schlüter K.D., Sauer H. The acute phase protein α2-macroglobulin induces rat ventricular cardiomyocyte

hypertrophy via ERK1,2 and PI3-kinase/Akt pathways. Cardiovasc Res. 2007;75(1): 118–128.

[130] Wu-Wong J.R. Potential for vitamin D receptor agonists in the treatment of cardiovascular disease. Br J Pharmacol. 2009;158:395–412. DOI: 10.1111/j.1476-5381.2009.00171.x

[131] Scragg R., Jackson R., Holdaway I.M., Lim T., Beaglehole R. Myocardial infarction is inversely associated with plasma 25-hydroxyvitamin D_3 levels: a community-based study. Int J Epidemiol. 1990;19(3):559–563.

[132] Giovannucci E., Liu Y., Hollis B.W., Rimm E.B. 25-Hydroxyvitamin D and risk of myocardial infarction in men: a prospective study. Arch Intern Med. 2008;168(11):1174–1180. DOI: 10.1001/archinte.168.11.1174

[133] Kwon S.H., Pimentel D.R., Remondino A., Sawyer D.B., Colucci W.S. H_2O_2 regulates cardiac myocytes phenotype via concentration-dependent activation of distinct kinase pathways. J Mol Cell Cardiol. 2003;35(6):615–621.

[134] Han H., Long H., Wang H., Wang J., Zhang Y., Wang Z. Progressive apoptotic cell death triggered by transient oxidative insult in H9c2 rat ventricular cells: a novel pattern of apoptosis and the mechanism. Am J Physiol Heart Circ Physiol. 2004;286(6):H2169–H2182.

[135] Rizzuto R., Pozzan T. Microdomains of intracellular Ca^{2+}: molecular determinants and functional consequences. Physiol Rev. 2006;86(1):369–408.

[136] Høyer-Hansen M., Bastholm L., Szyniarowski P., Campanella M., Szabadkai G., Farkas T., et al. Control of macroautophagy by calcium, calmodulin-dependent kinase kinase-β, and Bcl-2. Mol Cell. 2007;25(2):193–205.

[137] Nishida K., Kyoi S., Yamaguchi O., Sadoshima J., Otsu K. The role of autophagy in the heart. Cell Death Differ. 2009;16(1):31–38. DOI: 10.1038/cdd.2008.163

[138] Wu S., Sun J. Vitamin D, vitamin D receptor, and macroautophagy in inflammation and infection. Discov Med. 2011;11(59):325–335.

[139] Hamacher-Brady A., Brady N.R., Gottlieb R.A. The interplay between pro-death and pro-survival signaling pathways in myocardial ischemia/reperfusion injury: apoptosis meets autophagy. Cardiovasc Drugs Ther. 2006;20(6):445–462.

[140] Razavi H.M., Hamilton J.A., Feng Q. Modulation of apoptosis by nitric oxide: implications in myocardial ischemia and heart failure. Pharmacol Ther. 2005;106(2):147–162.

[141] Pestana C.R., Silva C.H., Uyemura S.A., Santos A.C., Curti C. Impact of adenosine nucleotide translocase (ANT) proline isomerization on Ca^{2+}-induced cysteine relative mobility/mitochondrial permeability transition pore. J Bioenerg Biomembr. 2010;42(4): 329–335. DOI: 10.1007/s10863-010-9297-4

[142] Dev S., Kumari S., Singh N., Kumar Bal S., Seth P., Mukhopadhyay C.K. Role of extracellular hydrogen peroxide in regulation of iron homeostasis genes in neuronal

cells: implication in iron accumulation. Free Radic Biol Med. 2015;86:78–89. DOI: 10.1016/j.freeradbiomed.2015.05.025

[143] Singh N., Haldar S., Tripathi A.K., Horback K., Wong J., Sharma D., et al. Brain iron homeostasis: from molecular mechanisms to clinical significance and therapeutic opportunities. Antioxid Redox Signal. 2014;20(8):1324–1363. DOI: 10.1089/ars.2012.4931

[144] Urrutia P., Aguirre P., Esparza A., Tapia V., Mena N.P., Arredondo M., et al. Inflammation alters the expression of DMT1, FPN1 and hepcidin, and it causes iron accumulation in central nervous system cells. J Neurochem. 2013;126(4):541–549. DOI: 10.1111/jnc.12244

[145] Urrutia P.J., Mena N.P., Núñez M.T. The interplay between iron accumulation, mitochondrial dysfunction, and inflammation during the execution step of neurodegenerative disorders. Front Pharmacol. 2014;5:38. DOI: 10.3389/fphar.2014.00038

[146] Kubis A.M., Piwowar A. The new insight on the regulatory role of the vitamin D_3 in metabolic pathways characteristic for cancerogenesis and neurodegenerative diseases. Ageing Res Rev. 2015;24(Pt B):126–137. DOI: 10.1016/j.arr.2015.07.008

[147] Uberti F., Morsanuto V., Bardelli C., Molinari C. Protective effects of 1α,25-dihydroxyvitamin D_3 on cultured neural cells exposed to catalytic iron. Physiol Rep. 2016;4(11):e12769. DOI:10.14814/PHY2.12769.

[148] Filograna R., Civiero L., Ferrari V., Codolo G., Greggio E., Bubacco L., et al. Analysis of the catecholaminergic phenotype in human SH-SY5Y and BE(2)-M17 neuroblastoma cell lines upon differentiation. PLoS One. 2015;10(8):e0136769. DOI: 10.1371/journal.pone.0136769

[149] Cantu D., Schaack J., Patel M. Oxidative inactivation of mitochondrial aconitase results in iron and H_2O_2-mediated neurotoxicity in rat primary mesencephalic cultures. PLoS One. 2009;4(9):e7095. DOI: 10.1371/journal.pone.0007095

[150] Herold S., Kumar P., Wichert S.P., Kretzschmar B., Bähr M., Rossner M.J. Neurodegeneration in autoimmune optic neuritis is associated with altered APP cleavage in neurons and up-regulation of p53. PLoS One. 2015;10(10):e0138852. DOI: 10.1371/journal.pone.0138852

[151] Wharton S.B., Williams G.H., Stoeber K., Gelsthorpe C.H., Baxter L., Johnson A.L., et al. Expression of Ki67, PCNA and the chromosome replication licensing protein Mcm2 in glial cells of the ageing human hippocampus increases with the burden of Alzheimer-type pathology. Neurosci Lett. 2005;383(1–2):33–38.

[152] Sleiman S.F., Berlin J., Basso M., Karuppagounder S.S., Rohr J., Ratan R.R. Histone deacetylase inhibitors and mithramycin a impact a similar neuroprotective pathway at a crossroad between cancer and neurodegeneration. Pharmaceuticals (Basel). 2011;4(8): 1183–1195.

Chapter 3

The Multiple Roles of Vitamin D Besides Calcium-Phosphorus Metabolism

Jia Guo and Zhangsuo Liu

Additional information is available at the end of the chapter

http://dx.doi.org/10.5772/66643

Abstract

Vitamin D is a kind of steroid hormone and is well known for its important role in regulating the levels of calcium (Ca) and phosphorus (P) as well as in mineralization of bone in body. But the vitamin D signaling also exhibits multiple effects, such as anti-inflammation effects, anticancer effect, and cardiovascular- and kidney-protective effects. From a practical point of view, vitamin D deficiency participates in many pathological progressions and diseases. In some diseases, the administration of vitamin D or vitamin D receptor agonist (VDRA) could rescue the clinical symptoms and improve outcomes. In this review, we briefly deal with these topics, limiting ourselves to comment on some novelty studies about vitamin D signaling, which might help us to understand the multiple effects of vitamin D in some pathological progresses and diseases, which are all worth to be studied further.

Keywords: vitamin D, vitamin D receptor, renal disease, inflammation

1. Introduction

Vitamin D is a kind of steroid hormone and presents naturally in very few foods. However, a number of foods have been fortified and sun exposure produces vitamin D in the skin. The final hormonal form of vitamin D was generated by undergoing two hydroxylation steps, the first to produce 25(OH)D3 in the liver and then second in the kidney to create the hormonal form, $1,25(OH)_2D3$, after consumed or made in the skin.

Vitamin D has been known for its important role in regulating the levels of calcium (Ca) and phosphorus (P) as well as in mineralization of bone in the body. But vitamin D deficiency has become a major health problem worldwide, which was thought to be related with several factors according to many researches, such as gender, age, sunlight exposure, cultures, dietary, and so on. In 2012, the 25(OH)D of 2327 (1744 females and 583 males) healthy Caucasian outpatients were tested in the North-East of Italy. The results showed that the 25(OH)D values had no significant differences between females and males. A nonsignificant variation of 25(OH)D values was also found throughout four age cohorts (<21, 21–40, 41–60, and >60 years), in both genders. In each age group, the values of 25(OH)D did not significantly differ between genders [1].

But in 2016, a single-center analysis of patients from 136 countries showed that 82.5% of the studied patients (57.5% were female and 42.5% were male) have vitamin D deficiency to insufficiency; 26.4% of females and 18.4% of males have extreme deficiency of 25(OH)D. There was higher variability of vitamin D in the group of females than males according to coefficient of variation. The prevalence of hypovitaminosis D is significantly high among population of UAE, Saudi Arabia, and many Middle Eastern countries, especially among women, despite abundant sunshine. This study considered the difference between female and male [2].

Vitamin D deficiency also existed in Europe. According to an alternate suggested definition of vitamin D deficiency (<50 nmol/L), the prevalence was 40.4%. Dark-skinned ethnic subgroups had much higher (3- to 71-fold) prevalence of serum 25(OH)D <30 nmol/L than did white populations [3]. In Sweden, about 56.4% people, 25-OHD, were under 50 nmol/L, which was related to several unfavorable health outcomes [4].

Disorders in mineral metabolism and bone disease are common complications of chronic kidney disease (CKD). In 2009, the board of directors of kidney disease, improving global outcomes (KDIGO), published a new guideline for the treatment of patients of CKD and CKD-related mineral and bone disorders (CKD-MBD). The levels of 25(OH)D (calcidiol) were suggested to be measured and repeated testing determined by baseline values and interventions in patients with CKD stages 3–5 with CKD–MBD or CKD stages 1–5. And vitamin D deficiency and insufficiency should be corrected by using treatment strategies recommended in the general population [5].

However, an increasing amount of data has suggested a possible involvement of vitamin D activity in a great number of different pathophysiological fields not exclusively associated with mineral metabolism, such as modulation of inflammation and immune response, cell proliferation and differentiation, gene expression, and so on [6]. The functions of vitamin D have been widely found to be dependent not only on a widespread ability of different tissues to synthesize the active form of vitamin D but also on the almost ubiquitous distribution of the specific vitamin D receptor (VDR), which then translocates to the nuclei of target cells. In this review, we briefly deal with these topics, limiting ourselves to comment on some novelty studies about vitamin D signaling, which might help us to understand the multiple effects of vitamin D in some pathological progresses and diseases.

2. The effects of vitamin D

2.1. The effects of vitamin D in inflammation and immune response

Inflammation and immune response are the body's immediate responses to damage to its tissues and cells by pathogens, noxious stimuli such as chemicals, or physical injury. Recent

http://dx.doi.org/10.5772/66643

studies have suggested that vitamin D signaling played critical roles in controlling inflammation and immune responses [7–9].

Immune cells such as activated CD8 (highest concentration), CD4 lymphocytes, and macrophages express VDR [10], which indicated that vitamin D signaling takes part in these cells'functional modulation [11]. Vitamin D signaling potently inhibited antigen- and mitogen-induced T-cell proliferation and cytokine production. Several key cytokines in T cells are direct targets of vitamin D signaling, in particular Th1 cytokines, such as IL-2 (interleukin-2) and IFN-g (interferon-gamma). Active vitamin D inhibits IL-2 secretion via impairment of transcription factor NF-AT (nuclear factor of activated T cells) complex formation, because the ligand-bound VDR complex itself binds to the distal NF-AT-binding site of the human IL-2 promoter. IFN-g has been found to be directly inhibited by vitamin D through interaction of the ligand-bound VDR complex with a VDRE (vitamin D-responsive element) in the promoter region of this cytokine [12]. In 2016, it was reported that 1α,25(OH)2 vitamin D3 modulates avian T lymphocyte functions without inducing cytotoxic T lymphocyte (CTL) unresponsiveness [13]. Another study, a cross-sectional study was from the National Health and Nutrition Examination Survey (NHANES) 2003–2006, showed sexually active women with cervicovaginal human papillomavirus (HPV) infection status and serum 25-hydroxyvitamin D (25(OH) D) levels (ng/mL) (n = 2353). Associations between serum 25(OH)D levels (continuous and categorical forms) and cervicovaginal HPV infection (high-risk HPV or vaccine-type HPV) were estimated using weighted logistic regression. The results showed that cervicovaginal HPV prevalence is associated with less-than-optimal levels of serum vitamin D [14].

Besides these, there are still some critical proteins through which vitamin D plays the modulation effect in inflammation and immune responses.

2.1.1. *Toll like receptor 9*

The Toll-like receptor 9 (TLR9) expression could be downregulated by vitamin D in monocytes. In response to decreased TLR9, these cells subsequently secreted less IL-6 as a downstream functional effect. This phenomenon may have significant biological relevance and maybe a factor in the association of vitamin D deficiency with susceptibility to autoimmune disease [15].

2.1.2. *Nuclear factor kappa-B*

The antagonism between VDR and NF-κB is mutual since overexpression of p65 but not p50 subunit of NF-κB inhibited VDRE-mediated transcription in transfected cells. Similar NF-κB-dependent mechanism has also been implicated in the inhibition of IL-8 promoter expression by vitamin D signaling [16].

2.1.3. *Tumor necrosis factor-α*

Tumor necrosis factor-alpha (TNF-α) is a pleiotropic inflammatory cytokine produced by activated immune cells as well as stromal cells. It was also reported that vitamin D inhibits TNF-α in mycobacteria-infected macrophages and peripheral blood mononuclear cells from pulmonary tuberculosis patients. The vitamin D analog cholecalciferol reduces the circulating level of TNF-α in patients with ESRD (end-stage renal disease) [17].

Vitamin D signaling also inhibited the induced production of pro-inflammatory cytokines, IL-1α, IL-6, and IL17 [18]. These activities form the basis of vitamin D-mediated modulation of immune responses and inflammation.

2.2. The effect of vitamin D in cell proliferation and differentiation

Vitamin D has been demonstrated to alter cellular proliferation through multiple mechanisms in highly cell-specific manners. One of the most important mechanisms was the slowing of cell cycle progression induced by vitamin D, typically due to inhibition of advancement from the G1 to the S phase of the cell cycle. Many key regulators, which influence gene transcription and protein stability, including p21waf1, p27kip1, cyclin D1, and others, are taken effect on by vitamin D. The ligand-activated VDR directly or indirectly influences the cell cycle, apoptosis, and/or differentiation by interacting with important transcriptional regulators or cell-signaling systems [19, 20].

Anti-proliferative effects of vitamin D signaling are often, but not always, linked to the promotion of cellular differentiation. Vitamin D and the VDR also have important interactions with other transcriptional regulators and cell-signaling systems. In many cell lines including cancer cells, 1,25$(OH)_2$D3 and vitamin D analogs upregulate the expressions of TGF-β receptor type I protein or androgen receptors, and downregulate the expressions of estrogen receptors or IGF (insulin-like growth factor I). Vitamin D analogs also induce the expression of E-cadherin, promoting translocation of β-catenin from the nucleus to the cell membrane, to control cell growth and differentiation [19]. In our study, we found that VDR could be downregulated by high glucose and take part in the epithelial-epithelial-mesenchymal transition of podocyte in mice [21].

3. Vitamin D and diseases

3.1. Vitamin D and kidney diseases

Kidney is the major site of the synthesis of 1, 25-$(OH)_2$D3, the active form of vitamin D, under physiologic conditions. Additionally, the vitamin D receptor, which binds to, and mediates the activity of 1, 25-$(OH)_2$D3, is widely distributed in the kidney. Thus, the kidney is essential not only for the maintenance of normal calcium and phosphorus homeostasis but also for the activation of vitamin D. There is close relationship among vitamin D, VDR, and the kidney. Any problem of kidney may lead to the vitamin D-signaling deficiency, and vitamin D-signaling deficiency always causes the abnormal function of kidney [22].

3.1.1. Chronic kidney disease

It was reported that when the glomerular filtration rate declines below 60 mL/min, chronic kidney disease (CKD) was always associated with increased cardiovascular events and mortality. One of the most important reasons was that the reduction/absence of kidney α1-hydroxylase, which mediates the final hydroxylation step of 25(OH)D to 1,25-$(OH)_2$D, could lead to the vitamin D deficiency in CKD patients. Vitamin D deficiency causes parathyroid hyperplasia and increased parathyroid hormone; the consequent hyperparathyroidism and hyper-phosphatemia are important risk factors for mortality in CKD patients [23, 24]. Accordingly, vitamin D treatment is associated with a reduced rate of cardiovascular diseases (CVDs) and mortality [25].

It has been shown that calcitriol (the active form of vitamin D) decreases the glomerulosclerosis index and albumin excretion in subtotally nephrectomized rats (SNX rats). Those mice lacking the VDR were more susceptible to hyperglycemia-induced renal injury [26], which may be related to podocyte loss. And the scientists also found that the expression of renin could be inhibited by calcitriol, and a larger decrease in renal glomerulosclerosis in experimental CKD could be found for the combination use of enalapril and paricalcitol (VDR agonist, VDRA), which may be because of the decreased TGF-β expression and macrophage infiltration [27].

In CKD patients treated with paricalcitol, proteinuria decreased after 23 weeks, independently of glomerular filtration rate, blood pressure, or angiotensin-converting enzyme (ACE) inhibitor [28].

3.1.2. *Diabetic nephropathy*

Diabetic nephropathy (DN) is the most common renal complication of diabetes mellitus and a leading cause of end-stage renal disease (ESRD). Intervention of DN remains a medical challenge despite some success.

In recent years, renin-angiotensin system (RAS) inhibitors have been used as the mainstay treatment for DN because the renin-angiotensin system (RAS) was considered as a major mediator of progressive renal injury in DN. But the compensatory renin increase caused by the disruption of renin feedback inhibition becomes one major problem which limit the efficacy use of the RAS inhibitors. These most recent data demonstrate that vitamin D and its analogs have renoprotective and therapeutic potentials in DN through targeting the RAS. Vitamin D negatively regulates the RAS by suppressing renin expression and thus plays a renoprotective role in DN [26, 29]. The molecular mechanism underlying this regulation is that 1,25(OH)2D3 disrupted the cAMP-signaling pathway, a major regulatory pathway involved in renin biosynthesis [30]. It was reported that vitamin D analog therapy could increase long-term survival, reduces heart and kidney weights, and prevents overt renal tissue damage in mouse models. Combination therapy with a RAS inhibitor and a vitamin D analog was found to markedly improve renal injuries in the diabetic vitamin D receptor null-mutant mice which developed more severe renal injuries because of more robust RAS activation [30]. Another study found that combined therapy with losartan and paricalcitol completely reduced proteinuria in a model of experimental diabetic nephropathy, suggesting that the combination of an ACE inhibitor or an ANG II receptor blocker plus a VDRA may be a good therapeutic option [23, 24].

Results from two clinical trials have provided some initial insight into the benefit of vitamin D therapy in diabetic nephropathy patients. Selective vitamin D receptor activation with paricalcitol for the reduction of albuminuria in patients with type 2 diabetes (VITAL study) investigated an active vitamin D analog exclusively in diabetic nephropathy patients. These patients were all the type 2 diabetics with a stable dose of angiotensin-converting enzyme (ACE) inhibitors or angiotensin receptor blockers (ARBs). The 281 patients were randomized into three groups: a single dose of placebo, 1 μg paricalcitol or 2 μg paricalcitol each day in a double-blind fashion. After 4 months of treatment, a significantly greater reduction in 24-h albumin excretion (measured by percent change from baseline) was found in the 2-μg paricalcitol group compared to placebo patients [31]. The second study about the measurement of the impact of vitamin D repletion with cholecalciferol on urinary albumin, MCP-1, and TGF-β1 was recently published by Kim and colleagues measured [32].

3.1.3. Immunoglobulin A nephropathy

The renoprotective effect of vitamin D was also investigated in immunoglobulin A (IgA) nephropathy. An open-label, non-placebo-controlled, randomized study about the effect of vitamin D and urinary protein excretion in 50 IgA nephropathy patients was operated on by using calcitriol. Patients who receive two doses (0.5 mg) of calcitriol per week or no treatment for 48 weeks were randomly assigned (1:1). The urinary protein excretion >0.8 g/d after renin-angiotensin system-inhibitor treatment for at least 3 months was considered as the main criterion for inclusion. The changes of 24-h urinary protein excretion from baseline to last measurement during treatment were compared and considered as the primary end point. The results showed that there was a significant decrease in proteinuria in the calcitriol-treated group compared with the control group (difference between groups, 41%; 95% confidence interval (CI), 5–79%; $P = 0.03$) the primary end point. At least a 15% decrease in proteinuria was set as the secondary end point. And the result were 7 of 24 (29%) controls and 17 of 26 (65%) of those treated with calcitriol ($P = 0.02$). These two groups showed the similar incidence of recorded adverse events, but had no significant differences in the decrease of estimated glomerular filtration rate and change in blood pressure. The conclusion was that the addition of calcitriol to a renin-angiotensin system inhibitor resulted in a safe decrease in proteinuria in patients with IgA nephropathy [33].

3.1.4. Other renal diseases

Vitamin D also play renoprotective role in other renal diseases. For example, cholecalciferol (vitamin D analog) treatment significantly increased serum 25-hydroxy vitamin D and decreased parathyroid hormone levels with no adverse effects in 25-hydroxy vitamin D-deficient renal-transplant patients [34]. Vitamin D receptor agonist doxercalciferol modulates dietary fat-induced renal disease and renal lipid metabolism [35].

Most of these studies have been reviewed recently and support the idea that vitamin D has therapeutic potential in slowing the progression of nephropathy. But these researches have their own limitations. The protection and mechanism of vitamin D need further study.

3.2. Vitamin D and cardiovascular disease

Cardiovascular disease (CVD) is the leading cause of death among patients with chronic kidney disease, with left ventricular hypertrophy (LVH) being a strong, independent risk factor. LVH is also a major risk factor for coronary ischemia, congestive heart failure, and cardiac arrhythmias [24].

There is a growing body of evidence linking vitamin D signaling and CVD in both experimental animals and humans. For instance, animals lacking the VDR or having vitamin D deficiency always show cardiovascular abnormalities, such as hypertension and LVH [36]. Cardiomyocytes isolated from VDR knockout mice developed contractile abnormalities, such as accelerated contraction and relaxation rates [37]. In humans, studies found that vitamin D deficiency is prevalent in neonates with congenital cardiac defects, and lower postoperative 25(OH)D levels are associated with the need for increased inotropic support in neonates undergoing cardiac operations. Low serum 25(OH)D level could also increase the incidence of sudden cardiac death both in healthy people and in hemodialysis patients. Some research showed that vitamin D deficiency may play a role in myocardial injury and postoperative recovery. Furthermore, two studies demonstrated that individuals with lower levels of active vitamin D were at a higher risk of developing hypertension [38, 39]. Vitamin D also was found to modulate the growth, hypertrophy, and differentiation of cardiomyocytes, pointing to a direct role for VDR agonists (VDRA) in cardiac physiology. These findings support that vitamin D deficiency is associated with poor cardiovascular outcomes in experimental animals and humans.

There were some reports about the improved cardiac function, which could be found in experimental animals or patients with administration of vitamin D or VDRAs. For example, after the treatment of paricalcitol, hypertensive rats showed a prevention of LVH and LV dysfunction accompanied with lower levels of brain natriuretic peptide and atrial natriuretic factor. VDRA therapy revealed similar results in another experimental model of LVH, the Cp/ rat model. Moreover, vitamin D is able to modulate contractility of cardiomyocytes in vitro by changing the distribution of the myosin chains and modulating Ca entry into cardiac muscle cells. In patients, the use of VDRA has been associated with improvements in left ventricular function and reductions in LVH [24].

Results from interventional trials, using either nutritional vitamin D or VDR agonists, supported the idea that VDR activation was beneficial for improving the underlying factors of CVD such as hypertension, endothelial dysfunction, atherosclerosis, vascular calcification, and cardiac hypertrophy.

3.3. Vitamin D and inflammatory bowel disease

Immune-mediated diseases such as inflammatory bowel diseases (IBDs) have increased in developed countries over the last 50 years. Scientists did much research to explain the increased incidence of IBD. They found that changing vitamin D status especially in prenatal as well as childhood could affect the development of the resultant immune response and the development of IBD.

There are reasons to believe that vitamin D could be an external factor that may play a role in the development of IBD. Animal models of experimental IBD had been used to explore the relationship between vitamin D signaling and IBD. The results showed that in IL-10 KO mice, the development of IBD symptoms could be accelerated by vitamin D deficiency [40]. Additionally, a fulminating form of IBD resulted in ulceration of the intestine and the premature mortality of the double VDR and IL-10 KO mice within a very short time frame (3–5 weeks of age).

Vitamin D receptor knockout or vitamin D-deficient mice always have T cells deficient that have been implicated in the pathology of IBD. Actually, the active form of vitamin D directly and indirectly suppresses the function of these pathogenic T cells while inducing several regulatory T cells that could suppress IBD progressing [18].

People found that 1, $25(OH)_2D$ and analogs of $1,25(OH)_2D$ could suppress IBD in the IL-10 KO mice which required adequate dietary calcium, suggesting that $1,25(OH)_2D$ may directly and indirectly control immune function through calcium regulation in vivo. Many evidences from multiple different animal models suggested that the increased susceptibility of mice to experimental IBD could be caused by vitamin D and VDR deficiency, which could be suppressed by $1,25(OH)_2D$ [41].

Although at present there has not been a causal relationship between vitamin D status and IBD established in the clinic, there have been some clinic reports that show the influence of vitamin D in IBD. They found most of children and young adult patients in IBD had low vitamin D status (serum 25(OH)D concentration ≤15 ng/mL). Additionally, a randomized, controlled clinical trial from November 2007 to June 2010 at the Clinical and Translational Study Unit of Children's Hospital in Boston showed that most of the 71 patients (age 5–21) in IBD had low vitamin D status in serum (less than 20 ng/mL). A treatment of oral doses of 2000 IU vitamin D3 daily or 50,000 IU vitamin D2 weekly for 6 weeks could raise serum 25(OH)D concentration obviously and the basic conditions of children and adolescents with IBD could be improved. The change in serum PTH concentration did not differ [42].

Overall, the available data do suggest that vitamin D status and analogs would be useful for normalizing T cell function and in the prevention and treatment of human IBD. But there are still many problems that need to be studied, such as which mechanism of the vitamin D mediated occur in human cells and IBD patients.

3.4. Vitamin D and cancer

Numerous studies have suggested that lower vitamin D level is a risk factor for human cancers, and vitamin D deficiency linked with the incidence and mortality of many types of cancers, including breast, colon, and prostate cancers [43, 44].

Beneficial effects of vitamin D have been observed in experimental cancers induced by various chemical carcinogens or genetic mutations, and in studies using human cancer cells implanted in nude mice. The anticancer effects of vitamin D included alterations in cell growth, angiogenesis, tumor invasion and metastatic potential, and immune surveillance. Many studies have demonstrated that vitamin D signaling inhibited the growth of many cancer-derived cell lines in vitro, by inhibiting progression through the cell cycle, inducing apoptosis, and driving the cells to a more differentiated phenotype. VDR also played important role in anticancer effect. Some preliminary data suggested that VDR polymorphisms were more frequently associated with tumor genes which are Fok1, Bsm1, Taq1, Apa1, EcoRV, and Cdx2 [43–45].

And another study showed that vitamin D receptor expression is linked to potential markers of human thyroid papillary carcinoma [46].

Trials in clinic also have shown the potential therapeutic effects of vitamin D in different kinds of cancers. For example, in a pilot study of patients with recurrent prostate cancer, oral calcitriol (starting with 0.5 mg/day and escalating the maximum dose to 2.5 mg/day) for 6–15 months resulted in a significant decrease in the rate of prostate-specific antigen (PSA) rise during therapy (in comparison to PSA increase before therapy) in six out of seven patients [43]. In another research, scientists provided evidence indicating that vitamin D signaling protected the skin from cancer formation by controlling keratinocyte proliferation and differentiation, facilitating DNA repair, and suppressing activation of the hedgehog (Hh) pathway following UVR exposure [47].

Although the anticancer effect of vitamin D signaling has been discovered, there are still problems that need to be discussed both in experimental and in clinical trials. In vitro and in vivo studies have clearly demonstrated the antitumor effects of vitamin D. But the mechanisms of antigrowth effects of vitamin D are quite variable in different cell types, and even in different cell lines derived from the same type of cancer. In clinic, the results of vitamin D and cancer mortality were inconsistent and even oppositive associations. The majority of studies in cancer patients showed that patients with higher vitamin D levels had a decreased risk of mortality. But there were some other voice against this. In conclusion, the relationship of vitamin D status and anticancer effect is still unclear and warrants further studies.

3.5. Other diseases

There are many more reports and studies about vitamin D and its receptor which were found to play crucial role in many fields, such as vitamin D and obstructive sleep apnea in polycystic ovary syndrome (PCOS) patients [44], serum 25-hydroxy vitamin D levels in middle-aged women in relationship to adiposity and height trajectories [48], and vitamin D deficiency impairs skeletal muscle function in a smoking mouse model [49]. In engender, vitamin D deficiency and low ionized calcium are linked with semen quality and sex steroid levels in infertile men [50]. The same result was also found in subfertile women [51]. In cognitive behavior, low vitamin D levels were frequent in depression patients, especially 25-OH D levels <50 nmol/L were associated with cognitive/affective-depressive symptoms, and anhedonia symptoms in particular [52]. Vitamin D supplementation could prevent depression and poor physical function in older adults [53]. Some reports found that the concentration of vitamin D in the plasma of patients with metabolic syndrome was significantly lower than its recommendations [54], and a relationship was detected between vitamin D concentration and exponents of metabolic syndrome [55, 56]. But a randomized controlled trial for 1 year in a Chinese population found that the correction of hypovitaminosis D did not improve the metabolic syndrome risk profile [57].

4. Conclusion

It is now well recognized that vitamin D is involved not only in the control of bone health and mineral metabolism but also in many other physical progresses, such as the control of immune responses, responses to infectious agents, and cell-proliferative mechanisms, particularly in cancer cells. The vitamin D signaling also exhibits multiple effects, such as anti-inflammation effects, anticancer effect, and cardiovascular- and kidney-protective effects. From a practical point of view, vitamin D deficiency participates in many pathological progression and diseases. In some diseases, the administration of vitamin D or VDRA could rescue the clinical symptoms and improve outcomes. In conclusion, vitamin D has many potential therapeutic effects and multiple roles in different diseases, physical and pathological progressions, which are all worth to be studied further.

Author details

Jia Guo[1, 2, 3] and Zhangsuo Liu[1, 2, 3*]

*Address all correspondence to: zhangsuoliu@sina.com

1 Department of Nephrology, The First Affiliated Hospital of Zhengzhou University, Zhengzhou, Henan, China

2 Nephropathy Research Institutes of Zhengzhou University, Zhengzhou, Henan, China

3 Institute of Clinical Medicine, The First Affiliated Hospital of Zhengzhou University, Zhengzhou, Henan, China

References

[1] Lippi G, Montagnana M, Meschi T, Borghi L. Vitamin D concentration and deficiency across different ages and genders. Aging Clin Exp Res. 2012;24(5):548–51.

[2] Haq A, Svobodová J, Imran S, Stanford C, Razzaque MS. Vitamin D deficiency: a single centre analysis of patients from 136 countries. J Steroid Biochem Mol Biol. 2016. pii: S0960-0760(16)30022-X.

[3] Cashman KD, Dowling KG, Škrabáková Z, Gonzalez-Gross M, Valtueña J, et al. Vitamin D deficiency in Europe: pandemic? Am J Clin Nutr. 2016;103(4):1033–44.

[4] Humble MB, Gustafsson S, Bejerot S. Low serum levels of 25-hydroxyvitamin D (25-OHD) among psychiatric out-patients in Sweden: relations with season, age, ethnic origin and psychiatric diagnosis. J Steroid Biochem Mol Biol. 2010;121(1–2):467–70.

[5] KDIGO clinical practice guideline for the diagnosis, evaluation, prevention, and treatment of Chronic Kidney Disease-Mineral and Bone Disorder (CKD-MBD). Kidney Int Suppl. 2009;113:S1–130.

[6] Messa P, Alfieri C, Rastaldi MP. Recent insights into vitamin D and its receptor. J Nephrol. 2011;24 Suppl 18:S30–7.

[7] Shab-Bidar S, Neyestani TR, Djazayery A, Eshraghian MR, Houshiarrad A, Kalayi A, et al. Improvement of vitamin D status resulted in amelioration of biomarkers of systemic inflammation in the subjects with type 2 diabetes. Diabetes Metab Res Rev. 2012;28(5):424–30.

[8] Bikle DD. Vitamin D regulation of immune function. Vitam Horm. 2011;86:1–21.

[9] White JH. Vitamin D metabolism and signaling in the immune system. Rev Endocr Metab Disord. 2012;13(1):21–9.

[10] Yu S, Bruce D, Froicu M, Weaver V, Cantorna MT. Failure of T cell homing, reduced CD4/CD8alphaalpha intraepithelial lymphocytes, and inflammation in the gut of vitamin D receptor KO mice. Proc Natl Acad Sci U S A. 2008;105(52):20834–9.

[11] Ooi JH, Chen J, Cantorna MT. Vitamin D regulation of immune function in the gut: why do T cells have vitamin D receptors? Mol Aspects Med. 2012;33(1):77–82.

[12] Yee YK, Chintalacharuvu SR, Lu J, Nagpal S. Vitamin D receptor modulators for inflammation and cancer. Mini Rev Med Chem. 2005;5(8):761–78.

[13] Boodhoo N, Sharif S, Behboudi S. 1α,25(OH)2 vitamin D3 modulates avian T lymphocyte functions without inducing CTL unresponsiveness. PLoS One. 2016;11(2):e0150134.

[14] Shim J, Pérez A, Symanski E, Nyitray AG. The association between serum 25-hydroxyvitamin D and human papillomavirus cervicovaginal infection in women in the United States. J Infect Dis. 2016. 15;213(12):1886–92.

[15] Dickie LJ, Church LD, Coulthard LR, Mathews RJ, Emery P, McDermott MF. Vitamin D3 down-regulates intracellular Toll-like receptor 9 expression and Toll-like receptor 9-induced IL-6 production in human monocytes. Rheumatology (Oxford). 2010;49(8):1466–71.

[16] Tan X, Wen X, Liu Y. Paricalcitol inhibits renal inflammation by promoting vitamin D receptor-mediated sequestration of NF-kappaB signaling. J Am Soc Nephrol. 2008;19(9):1741–52.

[17] Wu S, Sun J. Vitamin D, vitamin D receptor, and macroautophagy in inflammation and infection. Discov Med. 2011;11(59):325–35.

[18] Bruce D, Yu S, Ooi JH, Cantorna MT. Converging pathways lead to overproduction of IL-17 in the absence of vitamin D signaling. Int Immunol. 2011;23(8):519–28.

[19] Samuel S, Sitrin MD. Vitamin D's role in cell proliferation and differentiation. Nutr Rev. 2008;66(10 Suppl 2):S116–24.

[20] Xiong M, Gong J, Liu Y, Xiang R, Tan X. Loss of vitamin D receptor in chronic kidney disease: a potential mechanism linking inflammation to epithelial-to-mesenchymal transition. Am J Physiol Renal Physiol. 2012.

[21] Guo J, Xia N, Yang L, Zhou S, Zhang Q, Qiao Y, Liu Z. GSK-3β and vitamin D receptor are involved in β-catenin and snail signaling in high glucose-induced epithelial-mesenchymal transition of mouse podocytes. Cell Physiol Biochem. 2014;33(4):1087–96.

[22] Kumar R, Tebben PJ, Thompson JR. Vitamin D and the kidney. Arch Biochem Biophys. 2012;523(1):77–86.

[23] Agarwal R. Vitamin D, proteinuria, diabetic nephropathy, and progression of CKD. Clin J Am Soc Nephrol. 2009;4(9):1523–8.

[24] Valdivielso JM, Cannata-Andia J, Coll B, Fernandez E. A new role for vitamin D receptor activation in chronic kidney disease. Am J Physiol Renal Physiol. 2009;297(6):F1502–9.

[25] Gravellone L, Rizzo MA, Martina V, Mezzina N, Regalia A, Gallieni M. Vitamin D receptor activators and clinical outcomes in chronic kidney disease. Int J Nephrol. 2011;2011:419524.

[26] Zhang Z, Sun L, Wang Y, Ning G, Minto AW, Kong J, et al. Renoprotective role of the vitamin D receptor in diabetic nephropathy. Kidney Int. 2008;73(2):163–71.

[27] Mizobuchi M, Morrissey J, Finch JL, Martin DR, Liapis H, Akizawa T, et al. Combination therapy with an angiotensin-converting enzyme inhibitor and a vitamin D analog suppresses the progression of renal insufficiency in uremic rats. J Am Soc Nephrol. 2007;18(6):1796–806.

[28] Agarwal R, Acharya M, Tian J, Hippensteel RL, Melnick JZ, Qiu P, et al. Antiproteinuric effect of oral paricalcitol in chronic kidney disease. Kidney Int. 2005;68(6):2823–8.

[29] Klaus G. Renoprotection with vitamin D: specific for diabetic nephropathy? Kidney Int. 2008;73(2):141–3.

[30] Li YC. Vitamin D and diabetic nephropathy. Curr Diab Rep. 2008;8(6):464–9.

[31] de Zeeuw D, Agarwal R, Amdahl M, Audhya P, Coyne D, Garimella T, et al. Selective vitamin D receptor activation with paricalcitol for reduction of albuminuria in patients with type 2 diabetes (VITAL study): a randomised controlled trial. Lancet. 2010;376(9752):1543–51.

[32] Kim MJ, Frankel AH, Donaldson M, Darch SJ, Pusey CD, Hill PD, et al. Oral cholecalciferol decreases albuminuria and urinary TGF-beta1 in patients with type 2 diabetic nephropathy on established renin-angiotensin-aldosterone system inhibition. Kidney Int. 2011;80(8):851–60.

[33] Liu LJ, Lv JC, Shi SF, Chen YQ, Zhang H, Wang HY. Oral calcitriol for reduction of proteinuria in patients with IgA nephropathy: a randomized controlled trial. Am J Kidney Dis. 2012;59(1):67–74.

[34] Courbebaisse M, Thervet E, Souberbielle JC, Zuber J, Eladari D, Martinez F, et al. Effects of vitamin D supplementation on the calcium-phosphate balance in renal transplant patients. Kidney Int. 2009;75(6):646–51.

[35] Wang XX, Jiang T, Shen Y, Santamaria H, Solis N, Arbeeny C, et al. Vitamin D receptor agonist doxercalciferol modulates dietary fat-induced renal disease and renal lipid metabolism. Am J Physiol Renal Physiol. 2011;300(3):F801–10.

[36] Xiang W, Kong J, Chen S, Cao LP, Qiao G, Zheng W, et al. Cardiac hypertrophy in vitamin D receptor knockout mice: role of the systemic and cardiac renin-angiotensin systems. Am J Physiol Endocrinol Metab. 2005;288(1):E125–32.

[37] Tishkoff DX, Nibbelink KA, Holmberg KH, Dandu L, Simpson RU. Functional vitamin D receptor (VDR) in the t-tubules of cardiac myocytes: VDR knockout cardiomyocyte contractility. Endocrinology. 2008;149(2):558–64.

[38] Wang TJ, Pencina MJ, Booth SL, Jacques PF, Ingelsson E, Lanier K, et al. Vitamin D deficiency and risk of cardiovascular disease. Circulation. 2008;117(4):503–11.

[39] Fryer RM, Rakestraw PA, Nakane M, Dixon D, Banfor PN, Koch KA, et al. Differential inhibition of renin mRNA expression by paricalcitol and calcitriol in C57/BL6 mice. Nephron Physiol. 2007;106(4):76–81.

[40] Froicu M, Zhu Y, Cantorna MT. Vitamin D receptor is required to control gastrointestinal immunity in IL-10 knockout mice. Immunology. 2006;117(3):310–8.

[41] Cantorna MT. Vitamin D, multiple sclerosis and inflammatory bowel disease. Arch Biochem Biophys. 2012;523(1):103–6.

[42] Pappa HM, Mitchell PD, Jiang H, Kassiff S, Filip-Dhima R, DiFabio D, et al. Treatment of vitamin D insufficiency in children and adolescents with inflammatory bowel disease: a randomized clinical trial comparing three regimens. J Clin Endocrinol Metab. 2012;97(6):2134–42.

[43] Buttigliero C, Monagheddu C, Petroni P, Saini A, Dogliotti L, Ciccone G, et al. Prognostic role of vitamin d status and efficacy of vitamin D supplementation in cancer patients: a systematic review. Oncologist. 2011;16(9):1215–27.

[44] Davis CD, Milner JA. Nutrigenomics, vitamin D and cancer prevention. J Nutrigenet Nutrigenomics. 2011;4(1):1–11.

[45] Welsh J. Cellular and molecular effects of vitamin D on carcinogenesis. Arch Biochem Biophys. 2012;523(1):107–14.

[46] Izkhakov E, Somjen D, Sharon O, Knoll E, Aizic A, Fliss DM, Limor R, Stern N, Steroid J. Vitamin D and obstructive sleep apnea: an unexplored relationship in PCOS patients. Biochem Mol Biol. 2016. 53(2):617–8.

[47] Denzer N, Vogt T, Reichrath J. Vitamin D receptor (VDR) polymorphisms and skin cancer: a systematic review. Dermatoendocrinol. 2011;3(3):205–10.

[48] Lehtinen-Jacks S, Leu Agelii M, Hunsberger M, Zetterberg H, Lissner L. Serum 25-hydroxy vitamin D levels in middle-aged women in relationship to adiposity and height trajectories over three decades. Eur J Clin Nutr. 2016. 70(6):709–14.

[49] Cielen N, Heulens N, Maes K, Carmeliet G, Mathieu C, Janssens W, Gayan-Ramirez G. Vitamin D deficiency impairs skeletal muscle function in a smoking mouse model. J Endocrinol. 2016. 229(2):97–108.

[50] Ozgurhan G, Vehapoglu A, Vermezoglu O, Temiz RN, Guney A, Hacihamdioglu B. Risk assessment of obstructive sleep apnea syndrome in pediatric patients with vitamin D deficiency: A questionnaire-based study. Medicine (Baltimore). 2016;95(39):e4632.

[51] Lehtinen-Jacks S, Leu Agelii M, Hunsberger M, Zetterberg H, Lissner L. Serum 25-hydroxy vitamin D levels in middle-aged women in relationship to adiposity and height trajectories over three decades. Eur J Clin Nutr. 2016;70(6):709–14.

[52] von Känel R, Fardad N, Steurer N, Horak N, Hindermann E, Fischer F, Gessler K. Vitamin D deficiency and depressive symptomatology in psychiatric patients hospitalized witha current depressive episode: a factor analytic study. PLoS One. 2015;10(9):e0138550.

[53] de Koning EJ, van Schoor NM, Penninx BW, Elders PJ, Heijboer AC, Smit JH, Bet PM, van Tulder MW, den Heijer, van Marwijk HW, Lips P. Vitamin D supplementation toprevent depression and poor physical function in older adults: study protocol of the D-Vital study, a randomized placebo-controlled clinical trial. BMC Geriatr. 2015;15:151.

[54] Godala M, Materek-Kuśmierkiewicz I, Moczulski D, Szatko F, Gaszyńska E, TokarskiS, Kowalski J. Estimation of plasma 25(OH)D vitamin level in patients with metabolicsyndrome. Pol Merkur Lekarski. 2015;39(234):364–7.

[55] Bonakdaran S, Fakhraee F, Karimian MS, Mirhafez SR, Rokni H, Mohebati M, Mazidi M, Mousavi M, Ferns GA, Ghayour-Mobarhan M. Association between serum 25-hydroxyvitamin D concentrations and prevalence of metabolic syndrome. Adv Med Sci.2016;61(2):219–223. doi: 10.1016/j.advms.2016.01.002.

[56] Kramkowska M, Grzelak T, Walczak M, Bogdanski P, Pupek-Musialik D, Czyzewska K. Relationship between deficiency of vitamin D and exponents of metabolic syndrome. Eur Rev Med Pharmacol Sci. 2015;19(12):2180–7.

[57] Yin X, Yan L, Lu Y, Jiang Q, Pu Y, Sun Q. Correction of hypovitaminosis D does not improve the metabolic syndrome risk profile in a Chinese population: a randomized controlled trial for 1 year. Asia Pac J Clin Nutr. 2016;25(1):71–7.

Section 2

Vitamin D on Biochemical and Immunological Activities

Chapter 4

Clinical and Biochemical Features of Patients with *CYP24A1* Mutations

Fay Joanne Hill and John A. Sayer

Additional information is available at the end of the chapter

http://dx.doi.org/10.5772/64503

Abstract

The *CYP24A1* gene encodes 1,25-hydroxyvitamin-D_3-24-hydroxylase, a key enzyme responsible for the catabolism of active vitamin D (1,25-dihydroxyvitamin D_3). Loss-of-function mutations in *CYP24A1* lead to increased levels of active vitamin D metabolites. Clinically, two distinct phenotypes have been recognised from this: infants with *CYP24A1* mutations present with infantile idiopathic hypercalcaemia, often precipitated by prophylactic vitamin D supplementation. A separate phenotype of nephrolithiasis, hypercalciuria and nephrocalcinosis often presents in adulthood. *CYP24A1* mutations should be suspected when a classical biochemical profile of high active vitamin D metabolites, high or normal serum calcium, high urine calcium and low parathyroid hormone is detected. Successful treatment with fluconazole, a P450 enzyme inhibitor, has been shown to be effective in individuals with *CYP24A1* mutations. Although *CYP24A1* mutations are rare, early recognition can prompt definitive diagnosis and ensure treatment is commenced.

Keywords: *CYP24A1*, vitamin D, hypercalcaemia, idiopathic infantile hypercalcaemia, nephrolithiasis

1. Introduction

The supplementation of formula milk with vitamin D_3 (cholecalciferol) prompted a rise in infants presenting with symptomatic hypercalcaemia in the United Kingdom during the 1950s [1]. While this public health initiative was proving highly successful in preventing rickets, for the small cohort of infants presenting with failure to thrive, dehydration and nephrocalcinosis, the consequences of their hypercalcaemia were at times fatal. A diagnosis of idiopathic infantile hypercalcaemia was given to many in this cohort. The apparent increased susceptibility of this minority group to vitamin D toxicity prompted research into a genetic predisposition. Fifty-nine years later, *CYP24A1* mutations were identified demonstrating loss-of-function mutations encoding 1,25-hydroxyvitamin D_3 24-hydroxylase, an enzyme with a key role in vitamin D metabolism [2].

More recently, *CYP24A1* mutations have been recognised in an adult population of patients presenting with calcium-containing renal stones. On investigation, these patients typically displayed hypercalciuria, nephrocalcinosis and occasionally chronic kidney impairment. Vitamin D supplementation was not a feature in all cases [3], demonstrating a clinically significant phenotype manifesting from normal dietary vitamin D intake. Importantly, some patients had been symptomatic for many years, undergoing extensive investigations before a diagnosis was made. A continuing focus on preventative medicine, including oral vitamin D supplementation for maintenance of bone health and during pregnancy, is likely to continue to risk triggering manifestations of vitamin D toxicity in individuals carrying biallelic mutations in *CYP24A1*. As diagnostic tests and successful treatments are starting to emerge, it is important to recognise clinical presentations which should prompt screening for *CYP24A1* deficiency [4–6].

2. *CYP24A1* and the vitamin D pathway

The crucial role of vitamin D in calcium and phosphate homeostasis means excessive levels of its active form can precipitate symptomatic hypercalcaemia. The activation of vitamin D takes place in two stages. The first stage takes place in the liver: vitamin D_3 is converted to 25-hydroxyvitamin D_3, a reaction catalysed by 25-hydroxylase (*CYP2R1*). The second stage occurs in the kidney, when 25-hydroxyvitamin D_3 is hydroxylated to 1,25-dihydroxyvitamin D_3, the active form. This stage is catalysed by 1α-hydroxylase, an enzyme encoded by the *CYP27B1* [2].

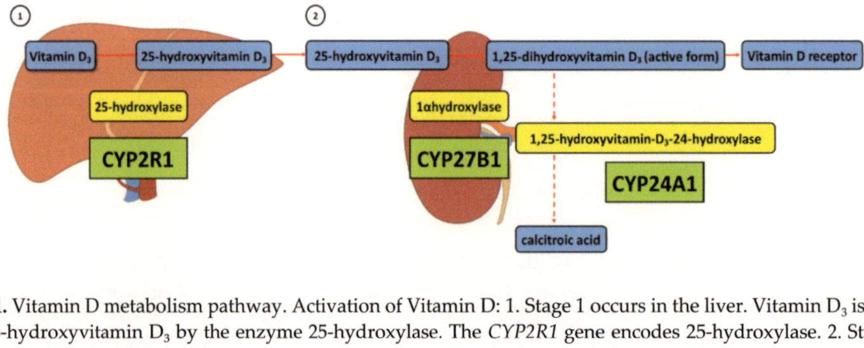

Figure 1. Vitamin D metabolism pathway. Activation of Vitamin D: 1. Stage 1 occurs in the liver. Vitamin D_3 is converted to 25-hydroxyvitamin D_3 by the enzyme 25-hydroxylase. The *CYP2R1* gene encodes 25-hydroxylase. 2. Stage 2 occurs in the kidney. 25-hydroxyvitamin D_3 is converted to 1,25-dihydroxyvitamin D_3 by the enzyme 1α-hydroxylase. The *CYP27B1* gene encodes 1α-hydroxylase. 1,25-dihydroxyvitamin D_3 is the physiologically most active form of vitamin D_3 which binds to the vitamin D receptor. Inactivation of Vitamin D: Several hydroxylation steps occur in the catabolism of 1,25-dihydroxyvitamin D_3 to calcitroic acid. The first of these steps is catalysed by the enzyme 1,25-hydroxyvitamin-D_3-24-hydroxylase, which is encoded by the *CYP24A1* gene.

The inactivation of vitamin D metabolites relies upon two pathways which both include steps catalysed by 1,25-hydroxyvitamin-D_3-24-hydroxylase; *CYP24A1* encodes this mitochondrial enzyme which is part of the cytochrome P450 system [6]. The enzyme is present in vitamin D target cells, predominantly located in the intestine and kidneys (**Figure 1**) [5].

2.1. Phenotypes

2.1.1. Idiopathic infantile hypercalcaemia

The first recognised phenotype of *CYP24A1* mutations was in infants diagnosed with idiopathic infantile hypercalcaemia. These individuals presented with vomiting, dehydration, fevers and failure to thrive. On investigation, a typical biochemical profile of high serum calcium and suppressed parathyroid hormone levels emerged. Renal ultrasound often demonstrated nephrocalcinosis, deposition of calcium salts within the kidney. It was not initially known whether the underlying pathophysiology of idiopathic infantile hypercalcaemia (IIH) was due to excess production of vitamin D metabolites, or an inability to inactivate vitamin D. A candidate gene approach was used to investigate families with typical presentations of idiopathic infantile hypercalcaemia. This research revealed a recessive loss-of-function mutation, in which patients with *CYP24A1* mutations were unable to inactivate vitamin D as they were deficient in the enzyme catalysing this pathway (1,25-hydroxyvitamin-D_3-24-hydroxylase). Affected children presented either after sustained low-dose vitamin D prophylaxis or directly following bolus doses of vitamin D. One sibling in which vitamin D prophylaxis was avoided was proven to carry the same mutation but had remained clinically silent. This supported evidence directly linking exogenous vitamin D supplementation with precipitation of symptomatic hypercalcaemia [2].

2.1.2. Adult nephrolithiasis

Hypercalciuria is the most common cause of calcium-containing kidney stones. The recognition that 40–45% of patients with idiopathic hypercalciuria have at least one relative with nephrolithiasis implicates a genetic predisposition in many cases [4]. *CYP24A1* mutations have now been proven in a cohort of adults presenting with nephrolithiasis, hypercalciuria, nephrocalcinosis and intermittent hypercalcaemia [4]. These patients had undergone extensive investigations before the cause of their nephrolithiasis was known, and multiple stone episodes and nephrocalcinosis may lead to progressive chronic kidney disease (CKD) [7]. This is important in highlighting the potential clinical spectrum of the phenotype, which may manifest without the trigger of vitamin D exposure. A typical biochemistry profile was found within this phenotype group, with normal/high serum calcium levels, suppressed parathyroid hormone, high levels of active vitamin D metabolites (25-hydroxyvitamin D_3 and 1,25-dihydroxyvitamin D_3) and low levels of inactivated vitamin D (24,25-dihydroxyvitamin D_3). A recent study screening patients with known calcium nephrolithiasis for *CYP24A1* mutations did not identify any biallelic variants in a cohort of 166 patients, suggesting *CYP24A1* mutations are a rare cause of idiopathic nephrolithiasis [8]. However, given our increased understanding of this phenotype, it is imperative that recognition of the typical biochemical pattern (suppressed PTH, hypercalcaemia, hypercalciuria) in any patients with nephrolithiasis prompts investigation for *CYP24A1* mutations [4, 6, 8]. Establishing a molecular diagnosis in this small cohort of patients can facilitate correct treatment and lifestyle modification (**Table 1**) [9].

Clinical features	Biochemical profile
Idiopathic infantile hypercalcaemia:	• ↑ 25-hydroxyvitamin D_3
• Vomiting	• ↑ 1,25-dihydroxyvitamin D_3
• Dehydration	• ↓ 24,25-dihydroxyvitamin D_3
• Failure to thrive	• ↑ or high normal serum calcium
• Fever	• ↑ urine calcium
• Adult presentation:	• ↓ parathyroid hormone
• Nephrolithiasis	
• Nephrocalcinosis	

Table 1. Key features of *CYP24A1* mutation phenotypes.

2.1.3. Investigation

In patients with *CYP24A1* mutations, an elevation in total vitamin D levels is typically seen. In particular, 1,25-dihydroxyvitamin D_3 levels are increased, but this assay is not routinely performed in many laboratories. Conversely, serum 24,24-dihydroxyvitamin-D_3 levels are sometimes low or undetectable in patients with *CYP24A1* mutations. A blood test that calculates the ratio between vitamin D metabolites could be utilised in future clinical practice as a screening tool for *CYP24A1* mutations in those patients presenting with a typical biochemical profile. In the first study of this, Molin et al. used liquid chromatography–tandem mass spectrometry to calculate the ratio of active to inactive vitamin D metabolites: Molar ratio (R) of 25-hydroxyvitamin-D_3: 24,25-dihydroxyvitamin D_3. A large increase in the ratio of active to inactive vitamin D metabolites, usually $R > 80$, was demonstrated in subjects who had biallelic mutations resulting in loss of function of *CYP24A1*. Importantly, through use of a ratio calculation, this test can avoid misleading results in patients who might have low 24,24-dihydroxyvitamin-D_3 levels due to vitamin D deficiency [6].

2.2. *CYP24A1* variants

Several different loss-of-function mutations have now been identified within the *CYP24A1* gene. The mutations are reported to be inherited in an autosomal recessive pattern, although it is not yet clear whether partial penetrance or environmental factors may alter manifestation of a recognised phenotype. One study showed individuals with biallelic mutations presented with the clinically recognised phenotype and that heterozygous carriers were not sufficient to manifest clinical disease. However, it was hypothesised that infants with haploinsufficiency/heterozygous variants may be more sensitive to hypercalcaemia during childhood while the kidney is still developing, and this could become relevant in considering additional vitamin D supplementation which might overwhelm the 1,25-hydroxyvitamin-D_3-24-hydroxylase enzyme pathway in this cohort (**Table 2**) [4, 6].

Year mutation reported	Age at presentation	Phenotype	*CYP24A1* mutation	Reference
2011	6 months	IIH	A475fsX490 homozygote	Schlingmann et al. [2]
2011	6 months	IIH	delE143 and E151X	Schlingmann et al. [2]
2011	Asymptomatic	Identified on family screening	delE143 and E151X	Schlingmann et al. [2]
2011	8 months	IIH	L409S and R396W	Schlingmann et al. [2]
2011	Asymptomatic	Identified on family screening	L409S and R396W	Schlingmann et al. [2]
2011	11 months	IIH	delE143 and R159Q	Schlingmann et al. [2]
2011	7 months	IIH	E322K and R396W	Schlingmann et al. [2]
2011	3.5 months	IIH	E322K and R396W	Schlingmann et al. [2]
2011	7 weeks	IIH	R396W homozygote	Schlingmann et al. [2]
2011	5 weeks	IIH	Complex deletion	Schlingmann et al. [2]
2012	10 months	IIH	Homozygous delE143	Dauber et al. [10]
2012	44 years	Intermittent hypercalcaemia, hypercalciuria, nephrolithiasis	2 canonical intron-exon splice junction mutations (IVS5 +1G>A and IVS6 -2A>G)	Tebben et al. [11]
2013	4 months	IIH	Homozygous R396W	Fencl et al. [12]
2013	9 years	Nephrocalcinosis, nephrolithiasis	Homozygous delE143	Dinour et al. [4]
2013	19 years	Nephrolithiasis, nephrocalcinosis, bladder calcification	Compound heterozygous L409S and W268X	Dinour et al. [4]
2013	13 years	Nephrolithiasis, nephrocalcinosis, hypercalcaemia, hypercalciuria	Compound heterozygous L409S and W268X	Dinour et al. [4]
2013	9 years	Nephrocalcinosis, hypercalciuria	Compound heterozygous delE143 and L148P	Nesterova et al. [5]
2013	25 years	Nephrolithiasis, hypercalcaemia, hypercalciuria	Compound heterozygous delE143 and L409S	Nesterova et al. [5]
2013	4.5 months	IIH	Homozygous R396W	Skalova et al. [13]

Year mutation reported	Age at presentation	Phenotype	*CYP24A1* mutation	Reference
2013	3 months	IIH followed by adult presentation with nephrocalcinosis, CKD, hypercalcaemia and hypercalciuria	Homozygous W210R	Meusburger et al. [14]
2014	~20 years	Nephrolithiasis, hypercalcaemia, hypercalciuria	Homozygous delE143	Jacobs et al. [15]
2015	10 years	Nephrolithiasis, hypercalcaemia, hypercalciuria	Homozygous delE143	Sayers et al. [7]
2015	45 years	Nephrocalcinosis, hypercalcaemia, hypercalciuria	Compound heterozygous G469Afs*22 and P21R	Figueres et al. [19]
2015	32 years	Nephrolithiasis, nephrocalcinosis, hypercalcaemia, hypercalciuria	Compound heterozygous L409S and R157W	Figueres et al. [19]
2015	28 days	IIH	Compound heterozygous R157W and M374T	Figueres et al. [19]
2015	2 months	IIH	Compound heterozygous L409S and R396W	Figueres et al. [19]
2015	6 months	IIH	Homozygous L409S	Figueres et al. [19]
2015	2 months	IIH	Compound heterozygous R396W and R396G	Figueres et al. [19]
2015	6 months	IIH	Compound heterozygous delE143 and L409S	Figueres et al. [19]
2015	1 day	Hypercalcaemia, apnoea	Heterozygous M374T	Molin et al. [6]
2015	3 days	Infection, hypercalcaemia, suppressed PTH	Heterozygous M374T	Molin et al. [6]
2015	11 days	Prematurity, hypercalcaemia, suppressed PTH	Heterozygous G322A	Molin et al. [6]
2015	4 days	Prematurity, hypercalcaemia, suppressed PTH	Heterozygous R439C	Molin et al. [6]
2015	13 days	Small for gestational age, hypercalcaemia, suppressed PTH	Heterozygous M374T	Molin et al. [6]

http://dx.doi.org/10.5772/64503

Year mutation reported	Age at presentation	Phenotype	*CYP24A1* mutation	Reference
2015	24 years	Hypercalcaemia, suppressed PTH, nephrocalcinosis, CKD	Homozygous delE143	Jobst-Schwan et al. [3]
2015	Asymptomatic	Identified on family screening	Homozygous delE143	Jobst-Schwan et al. [3]
2015	26 years	Nephrocalcinosis, hypercalcaemia, hypercalciuria	Homozygous delE143	Tray et al. [16]
2015	21 years	Nephrocalcinosis, nephrolithiasis, hypercalcaemia	Heterozygous delE143 and R396W	Tray et al. [16]
2015	5 months	IIH	Compound heterozygous R396W and W134G	Dinour et al. [17]
2015	9 months	IIH	Compound heterozygous G315X and W134G	Dinour et al. [17]
2015	5 months	IIH	Homozygous delE143	Dinour et al. [17]
2015	35 years	Nephrolithiasis, nephrocalcinosis and hypercalcaemia during pregnancy	Homozygous delE143	Dinour et al. [17]

CKD, chronic kidney disease; IIH, idiopathic infantile hypercalcaemia; PTH, parathyroid hormone.

Table 2. Identified mutations in *CYP24A1*.

2.3. Treatment

CYP24A1 mutations lead to calcium stone formation, and conventional treatments for calcium stones are recommended. These would include maintaining a high fluid intake and avoiding excess dietary sodium. Specific measures would include avoiding dietary vitamin D supplements (in foods and drinks) and avoidance of excessive sunlight exposure [7]. Ketoconazole was first demonstrated as an effective treatment for reducing the effects of vitamin D toxicity in patients with *CYP24A1* mutations. As a non-specific P450 enzyme inhibitor ketoconazole inhibits the enzyme catalysing production of 1,25-dihydroxyvitamin D_3, (1α-hydroxylase), thereby decreasing levels of active vitamin D_3. However, *CYP24A1*-deficient individuals require lifelong treatment as they will always lack the enzyme to inactivate vitamin D, and the side-effect profile of ketoconazole, which includes hepatotoxicity, hypogonadism and adrenal insufficiency, makes it unsuitable for this purpose [4]. More recently, low-dose fluconazole, also acting as a P450 enzyme inhibitor, has been shown to reduce serum calcium levels and urinary calcium excretion in a patient with *CYP24A1* mutation. It is likely that this drug, alongside lifestyle modifications such as avoiding excess sun exposure and following a low calcium and oxalate diet, will become the main treatment offered to patients diagnosed with *CYP24A1* mutations (**Figure 2**) [7, 18, 19].

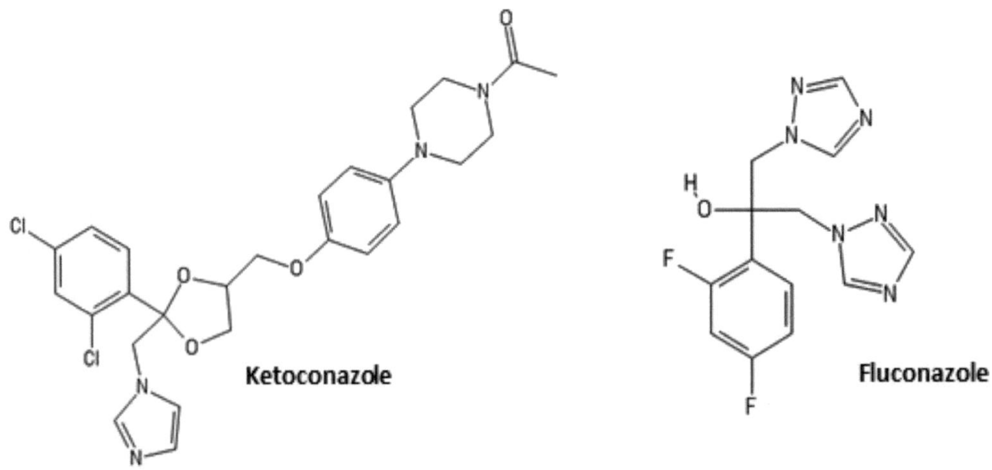

Figure 2. Chemical structures of ketoconazole, an imidazole antifungal agent, and fluconazole, a triazole antifungal agent. Azole agents are cytochrome inhibitors primarily used as antifungal agents. They are heterocyclic ring compounds and are generally classified as either imidazoles (e.g. ketoconazole) or triazoles (e.g. fluconazole), containing two or three nitrogen atoms, respectively, in the azole ring. They exhibit their antifungal action through inhibition of lanosterol 14-α demethylase, a cytochrome P450 enzyme important for the synthesis of a fungal plasma membrane constituent.

2.4. Evidence for genetic heterogeneity of idiopathic infantile hypercalcaemia

Since the discovery of *CYP24A1* mutations underlying idiopathic infantile hypercalcaemia (IIH) in 2011, a cohort of IIH patients has been identified without *CYP24A1* mutations. In 2015, a new loss-of-function mutation in *SLC34A1*, which encodes the renal sodium–phosphate cotransporter 2A (NaPi-IIa), was recognised in this group [20]. These patients presented with a classical IIH phenotype, with symptoms of hypercalcaemia. Importantly, however, their symptoms did not resolve with removal of vitamin D supplementation. Instead, their hypercalcaemia corrected rapidly after commencing phosphate replacement, highlighting the different mechanism driving the hypercalcaemia. In patients with *SLC34A1* mutations, renal phosphate wasting leads of inappropriately high levels of 1,25-dihydroxyvitamin-D_3, which in turn causes hypercalcaemia. It is crucial to distinguish between patients carrying mutations in *CYP24A1* versus *SLC43A1*, as different intervention is required to successfully treat their hypercalcaemia [20]. As *SLC34A1* mutations have also been identified as a cause of nephrolithiasis, there is overlap between *SLC34A1* and *CYP24A1* mutation phenotypes in both paediatric and adult presentations [21].

http://dx.doi.org/10.5772/64503

3. Conclusions

Overall, *CYP24A1* mutations are rare and account for a small proportion of symptomatic hypercalcaemia or nephrolithiasis cases. However, a greater awareness of their phenotypes will increase clinical suspicion in patients presenting with a typical biochemical profile. Testing for mutations in *CYP24A1* can establish a definitive diagnosis, avoiding protracted further investigations and allowing treatment to commence. Alongside dietary and lifestyle advice, aimed at minimising vitamin D intake, fluconazole is proving a promising lifelong treatment to prevent effects of vitamin D toxicity.

Acknowledgements

JAS is supported by the Northern Counties Kidney Research Fund.

Author details

Fay Joanne Hill[1] and John A. Sayer[1,2*]

*Address all correspondence to: john.sayer@ncl.ac.uk

1 Renal Services, Newcastle upon Tyne NHS Foundation Trust Hospitals, Newcastle upon Tyne, United Kingdom

2 Institute of Genetic Medicine, Newcastle University, Newcastle upon Tyne, United Kingdom

References

[1] British Paediatric Association. Hypercalcemia in infants and vitamin D. British Medical Journal 1956; 2: 149.

[2] Schlingmann KP, Kaufmann M, Weber S, Irwin A, Goos C, John U et al. Mutations in CYP24A1 and idiopathic infantile hypercalcemia. The New England Journal of Medicine 2011; 365(5): 410–421.

[3] Jobst-Schwan T, Pannes A, Schlingmann KP, Eckardt K-U, Beck BB, Wiesener MS. Discordant clinical course of vitamin-D-hydroxylase (CYP24A1) associated hypercalcemia in two adult brothers with nephrocalcinosis. Kidney and Blood Pressure Research 2015; 40: 443–451.

[4] Dinour D, Beckerman P, Ganon L, Tordjman K, Eisenstein Z, Holtzman EJ. Loss-of-function mutations of CYP24A1, the vitamin D 24-hydroxlylase gene, cause long-standing hypercalciuric nephrolithiasis and nephrocalcinosis. The Journal of Urology 2013; 190: 552–557.

[5] Nesterova G, Malicdan MC, Yasuda K, Sakaki T, Vilboux T, Ciccone C. 1,25-(OH)2D-24 hydroxylase (CYP24A1) deficiency as a cause of nephrolithiasis. Clinical Journal of the American Society of Nephrology 2013; 8: 649–657.

[6] Molin A, Baudoin R, Kaufmann M, Souberbielle JC, Ryckewaert A, Vantyghem MC. CYP24A1 mutations in a cohort of hypercalcemic patients: evidence for a recessive trait. Journal of Clinical Endocrinology and Metabolism 2015; 100(10): E1343–E1352.

[7] Sayers J, Hynes AM, Srivastava S, Dowen F, Quinton R, Datta HK, Sayer JA. Successful treatment of hypercalcaemia associated with a CYP24A1 mutation with fluconazole. Clinical Kidney Journal 2015; 8(4): 453–455.

[8] Sayers J, Hynes AM, Rice SJ, Hogg P, Sayer JA. Searching for CYP24A1 mutations in cohorts of patients with calcium nephrolithiasis. OA Nephrology 2013; 1(1): 1–6.

[9] Sayer JA. Re: Loss-of-function mutations of CYP24A1, the vitamin D 24-hydroxlylase gene, cause long-standing hypercalciuric nephrolithiasis and nephrocalcinosis. The Journal of Urology 2015; 68: 164–165.

[10] Dauber A, Nguyen TT, Sochett E, Cole DEC, Horst R, Abrams SA et al. Genetic defect in CYP24A1, the vitamin D 24-hydroxylase gene, in a patient with severe infantile hypercalcaemia. The Journal of Clinical Endocrinology and Metabolism 2012; 97(2): E268–E274.

[11] Tebben PJ, Milliner DS, Horst RL, Harris PC, Singh RJ, Wu Y et al. Hypercalcaemia, hypercalciuria, and elevated calcitriol concentrations with autosomal dominant transmission due to CYP24A1 mutations: effects of ketoconazole therapy. The Journal of Clinical Endocrinology and Metabolism 2012; 97(3): E423–E427.

[12] Fencl F, Bláhová K, Schlingmann KP, Konrad M, Seeman T. Severe hypercalcemic crisis in an infant with idiopathic infantile hypercalcaemia caused by mutation in CYP24A1 gene. European Journal of Pediatrics 2013; 172: 45–49.

[13] Skalova S, Cerna L, Bayer M, Kutilek S, Konrad M, Schlingmann KP. Intravenous pamidronate in the treatment of severe idiopathic infantile hypercalcaemia. Iranian Journal of Kidney Diseases 2013; 7(2): 160–164.

[14] Meusburger E, Mündlein A, Zitt E, Obermayer-Pietsch B, Kotzot D, Lhotta K. Medullary nephrocalcinosis in an adult patient with idiopathic infantile hypercalcaemia and a novel CYP24A1 mutation. Clinical Kidney Journal 2013; 6: 211–215.

[15] Jacobs TP, Kaufman M, Jones G, Kumar R, Schlingmann KP, Shapses S. A lifetime of hypercalcaemia and hypercalciuria, finally explained. The Journal of Clinical Endocrinology and Metabolism 2014; 99(3): 708–712.

[16] Tray KA, Laut J, Saidi A. Idiopathic infantile hypercalcaemia, presenting in adulthood-no longer idiopathic nor infantile: two case reports and review. Connecticut Medicine 2015; 79(10): 593–597

[17] Dinour D, Davidovits M, Aviner S, Ganon L, Michael L, Modan-Moses D et al. Maternal and infantile hypercalcaemia caused by vitamin-D-hydroxylase mutations and vitamin D intake. Pediatric Nephrology 2015; 30: 145–152.

[18] Dusso AS, Gomez-Alonso C, Cannata-Andia JB. The hypercalcaemia of CYP24A1 inactivation: new ways to improve diagnosis and treatment. Clinical Kidney Journal 2015; 8(4): 456–458.

[19] Figueres M-L, Linglart A, Bienaime F, Allain-Launay E, Roussey-Kessler G, Ryckewaert A. Kidney function and influence of sunlight exposure in patients with impaired 24-hydroxylation of vitamin D due to CYP24A1 mutations. American Journal of Kidney Disease 2015; 65(1): 122–126.

[20] Schlingmann KP, Ruminska J, Kaufmann M, Dursun I, Patti M, Kranz B et al. Autosomal-recessive mutations in SLC34A1 encoding sodium-phosphate cotransporter 2a cause idiopathic infantile hypercalcemia. Journal of American Society of Nephrology 2015; 27: 604–614.

[21] Oddsson A, Sulem P, Helgason H, Edvardsson VO, Thorlefisson G, Sveinbjörnsson G et al. Common and rare variants associated with kidney stones and biochemical traits. Nature Communications 2015; 6(7975): 1–9

Chapter 5

Synthesis of Low Abundant Vitamin D Metabolites and Assaying Their Distribution in Human Serum by Liquid Chromatography-Tandem Mass Spectrometry (LC-MS/MS) as a New Tool for Diagnosis and Risk Prediction of Vitamin D-Related Diseases

Lars Kattner and Dietrich A. Volmer

Additional information is available at the end of the chapter

http://dx.doi.org/ 10.5772/64518

Abstract

This chapter provides an overview of versatile and efficient chemical syntheses of vitamin D derivatives by application of either linear or convergent synthesis approaches. Synthesis of the most relevant naturally occurring vitamin D metabolites and their deuterated counterparts to use as calibration and reference standards in LC-MS/MS assays is also shown. The chapter then summarizes the most important mass spectrometric approaches to quantify important vitamin D metabolites in human biofluids. In addition, new developments are described that are aimed at the pathobiological interpretation of the measured vitamin D metabolite distributions in various human diseases.

Keywords: vitamin D deficiency, biomarkers, low abundant vitamin D metabolites, assay development, LC-MS/MS, diagnosis, disease risk prediction

1. Introduction

Vitamin D is mostly known for its role in the regulation of calcium and phosphorous homeostasis [1–3]. Consequently, vitamin D deficiency may cause various disorders related to bone mineralization [4]. Drugs based on vitamin D analogs are commonly used to treat bone diseases (osteoporosis, osteomalacia, and rickets) or psoriasis. More recently, it has been suggested that vitamin D deficiency is also connected to a wide range of other diseases beyond bone mineralization, such as diabetes, autoimmune diseases, cardiovascular diseases, and cancer, as various clinical and epidemiological studies have shown [5–9]. However, the development of drugs for treatment of these diseases based on appropriate vitamin D analogs has mostly failed, either due to their rapid metabolic clearance or their calcemic effects. Vitamin D analogs are usually hormonally active compounds with pleiotropic functions and their levels in the body are strictly regulated by the hormonal system. In cases of oversupply, they are enzymatically degraded to avoid harmful effects such as calcemia, leading mainly to inactivation and conversion into water soluble degradation products suitable for renal clearance. Consequently, prevention of vitamin D deficiency rather than therapy of a vitamin D-related disease is a promising approach. Due to low concentration and short half-life of the active metabolite, 1α,25-dihydroxyvitamin D_3 (1,25$(OH)_2D_3$, calcitriol) (**4**) in serum, the biosynthetic precursor 25-hydroxyvitamin D_3 (25(OH)D_3) (**3**) is usually measured as a marker of vitamin D status. Additionally, more than 50 naturally occurring but low abundant vitamin D metabolites of mostly unknown physiological function have been identified. Common assays, particularly ELISA and radio-immunoassays (RIA), are often restricted to 25(OH)D_3 because of sensitivity and specificity limitations [10–12]. Fortunately, recent mass spectrometry advances have permitted reaching deeper into the metabolic cascade of vitamin D, including low abundant species [13–17]. Consequently, a wide variety of naturally occurring metabolites of potential biological activity can be analyzed simultaneously by liquid chromatography-tandem mass spectrometry (LC-MS/MS), which is currently considered the "gold standard" technique. However, this method has suffered in the past from the limited availability of the required vitamin D metabolites needed as calibration and reference standards. Consequently, as a prerequisite, all metabolites of interest have to be available for chemical synthesis, as described in this chapter.

2. Metabolism of vitamin D

Vitamin D_3 (**2**) is mainly synthesized in the skin from 7-dehydrocholesterol (**1**) by UV irradiation and subsequent thermal isomerization of the intermediate product previtamin D_3. Vitamin D_3 is then hydroxylated in the liver to 25(OH)D_3 (**3**), followed by further oxidation in the kidneys to 1,25$(OH)_2D_3$ (calcitriol) (**4**), which is usually considered the biologically active metabolite (**Scheme 1**). Calcitriol **4** is metabolized via CYP24 to other oxidative products, mainly 1α,24(R),25-trihydroxyvitamin D_3 (1,24,25$(OH)_3D_3$) (**5**), followed by subsequent enzymatic oxidation and degradation of the carbon side chain to water soluble calcitroic acid [1, 18] for final clearance from the body. 25(OH)D_3 (**3**) is degraded in an analogous manner via 24(R),25-dihydroxyvitamin D_3 (24,25$(OH)_2D_3$) **6**. Alternatively, 25(OH)D_3 (**3**) and 1,25$(OH)_2D_3$ (**4**) can be metabolized to their corresponding 26,23-lactones (**7,8**) [19]. Vitamin D metabolites also have the potential to be metabolized through a C-3 epimerization pathway, leading to C-3-*epi*-metabolites such as **9-12**, with an inversion of the stereogenic center at position C-3 [20, 21]. These reactions lead to products that are believed to be inactive. Since high concentrations of 3-*epi*-25-hydroxyvitamin D_3 (3-*epi*-25(OH)D_3) (**10**) were found in infants [22], it was initially assumed that this pathway is only a consequence of an immature vitamin D metabolism. It has also been shown, however, that the epimerization pathway is favored if, for any reason, side chain degradation by CYP24 is inhibited [23].

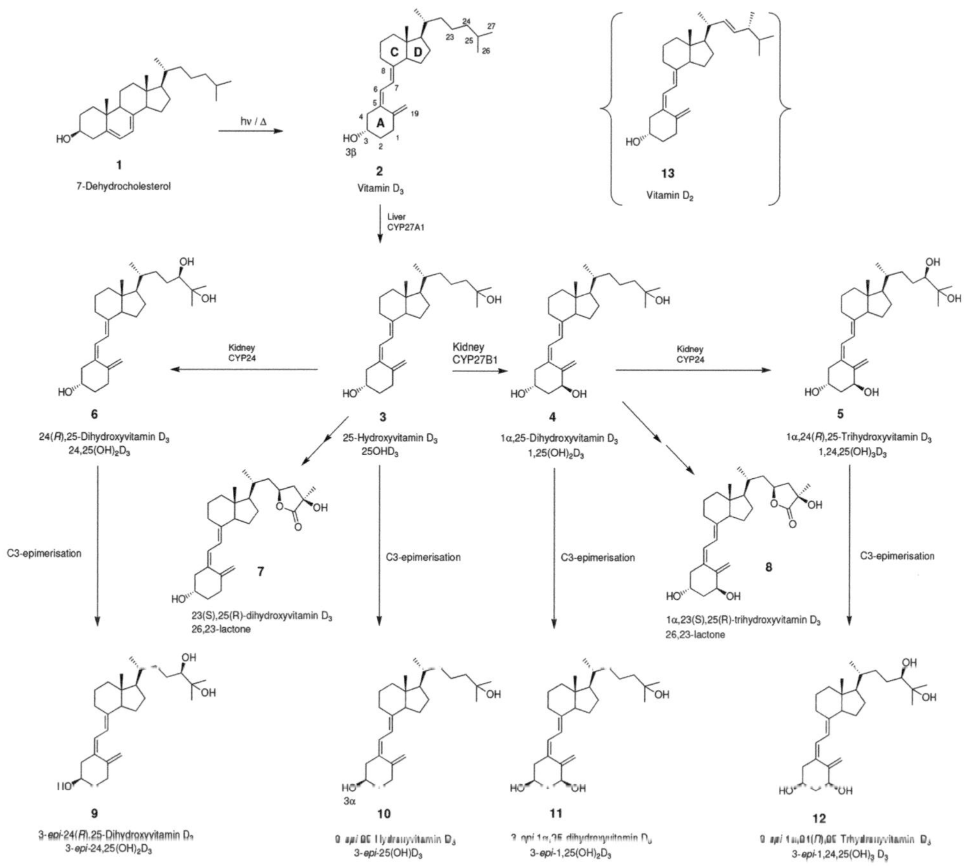

Scheme 1. Metabolic pathways of vitamin D.

Finally, the corresponding metabolites of vitamin D_2 (**13**) also have to be considered [18], because food from plant origin (in particular mushrooms) and food supplements may contain vitamin D_2. Vitamin D_2 metabolites have similar physiological function as compared to their corresponding vitamin D_3 analogs, although their potency is apparently lower [24]. A detailed comparison of the metabolism of vitamin D_2 and D_3 [18] reveals that $25(OH)D_3$ (**3**) and $1,25(OH)_2D_3$ (**4**) are preferentially metabolized at their side chains, particularly at C-23, C-24, and C-26, leading to metabolites susceptible to further oxidation, which is similar for vitamin D_2, even though there are slight differences as a consequence of the C-22/23 carbon double bond and a methyl group at C-24. Thus, C-24 hydroxylation of vitamin D_2 may occur initially, leading to 24-hydroxyvitamin D_2, following by C-1 hydroxylation in the kidneys, giving 1,24-dihydroxyvitamin D_2, which is subsequently oxidized in the side chain. Furthermore, 1,25-

dihydroxyvitamin D_2, formed in an analogous manner to its D_3 counterpart, may be metabolized by side chain oxidation either to C-1,24,25 or C-1,25,26 trihydroxylated vitamin D_2, again followed by further side chain oxidation. Each of these both pathways again leads to calcitroic acid as final metabolite.

3. Synthesis of low abundant vitamin D metabolites

The commercial unavailability of many relevant vitamin D_3 and D_2 species has limited the scope of LC-MS/MS assays in the past. For use as reference standards, vitamin D_3 and D_2 metabolites, as well as their corresponding stable isotopes (labeled with ^2H (D) or ^{13}C), have to be synthesized by application of a versatile and cost-effective methodology. A large variety of chemical syntheses of vitamin D derivatives has been developed in the last few decades by several academic and commercial groups [1, 25–27], mainly with the aim of synthesizing new analogs for drug discovery and development purposes. These efforts have resulted in a large number of more than 3000 synthesized compounds [28]. Critical evaluation of these methods reveals, however, that only a few of them are suitable to reproducibly generate metabolites of interest in gram quantities at reasonable costs within short time frames. In this chapter, we will review some of the more suitable strategies that have been successfully applied and optimized in our laboratory.

The synthesis of deuterated vitamin D_3 and D_2 metabolites is mostly accomplished using the same procedures as those developed for nondeuterated metabolites.

The biosynthetic and technical synthesis (starting from an appropriate steroid precursor (analog to **1**) is not favorable, due to low yields reached in the photochemical ring-opening reaction. It usually fails if a 1-hydroxy substituent is present. By far the most suitable methods for synthesis of vitamin D_3 and D_2 metabolites start with readily available Vitamin D_2 (**13**). Two alternative strategies are applied, either by keeping the vitamin D skeleton intact during the course of the reaction sequence (linear synthesis) (**Scheme 2**), or by first cleaving the molecule to obtain the A-ring and the CD-ring building block, and subsequently reconnecting both molecules after separate appropriate chemical modifications (convergent synthesis) (**Scheme 3**) [1, 26, 27, 29, 30].

3.1. Linear synthesis of vitamin D metabolites

Following a linear synthesis (**Scheme 2**), the *cis*-diene moiety of vitamin D_2 (**13**) is protected by sulfur dioxide after silylation of the hydroxyl group at C-3 as a *tert*-butyldimethylsilyl ether, leading to the SO_2 adduct **14**. Labeling with deuterium at C-6,19,19 can be carried out conveniently by treatment with D_2O or deuterated alcohol (e.g., isopropanol). Finally, thermal removal of SO_2 and photochemical isomerization leads to threefold deuterated vitamin D_2 (d-**13**). This simple and convenient method to obtain C-6,19,19 deuterated metabolites [31] can also be applied to a wide variety of other vitamin D_3 and D_2 derivatives. In order to obtain metabolites of the corresponding C-3-*epi* series of vitamin D_3 or D_2 metabolites, the configuration of the C-3 OH group of vitamin D_2 (**13**) can be epimerized under the so-called "Mitsunobu conditions"

[32], by treatment with an aromatic acid and an azodicarboxylate, resulting in the corresponding ester **15** with concomitant inversion of the stereogenic center at C-3 (orientation of the substituent has changed from β to α). The ester is finally cleaved by saponification with potassium hydroxide or by reduction with lithium aluminum hydride to afford C-3-*epi* vitamin D_2 (**16**), which can be used to synthesize a wide variety of C-3 epimers, including 3-*epi*-25-hydroxyvitamin D_3 (3-*epi*-25(OH)D_3) (**10**, d-**10**) and 3-*epi*-1α,25-dihydroxyvitamin D_3 (3-*epi*-1,25$(OH)_2D_3$) (**11**, d-**11**). Yields are only moderate due to the favored formation of a C-3,4-5,6-7,8 all-*trans* triene system by elimination. To further proceed with the synthesis, the C-22/23 double bond in the side chain of SO_2 adduct **14** is cleaved by ozonolysis, leading to alcohol **17** after reductive workup with sodium boronhydride, which can be used to add a wide variety of modified side chains as appropriate. For instance, **17** can be converted to an iodide, which is coupled in a nickel-zinc-mediated reaction with ethyl acrylate after prior removal of the SO_2 protective group to afford **18**. At this stage, an allylic oxidation by selenium dioxide can be carried out, leading to a 1α-hydroxy series of metabolites, such as **19** as appropriate precursor. Silylation of the newly generated hydroxyl group, followed by Grignard reaction with methylmagnesium bromide, leads to 1α,25-dihydroxy metabolites. If a deuterated Grignard reagent, labeled with 3 deuterium atoms at their methyl groups, is employed, the corresponding sixfold deuterium labeled metabolite is obtained, containing 3 deuterium atoms at C-26 and C-27, respectively. Finally, the C-5/6 double bond has to be isomerized photochemically, and the silyl protective groups have to be removed, ending up with nondeuterated or deuterated 25(OH)D_3 (**3**, d-**3**) or 1,25$(OH)_2D_3$ (**4**, d-**4**) [33], as shown in **Scheme 2**. If 3-*epi*-vitamin D_2 (**16**) is employed in this sequence, 3-*epi*-25$(OH)_2D_3$ (**10**, d-**10**) is obtained. 3-*epi*-1,25$(OH)_2D_3$ (**11**, d-**11**) is obtained, if 1α-hydroxylation by selenium dioxide was carried out, and the resulting C-1-OH-group is epimerized by oxidation and final reduction.

Vitamin D_2 metabolites are synthesized by application of analogous strategies as applied for their corresponding D_3 counterparts, although they are more challenging to synthesize, for two reasons: vitamin D_2 metabolites contain an olefinic double bond in C-22/23-position, which has to be arranged in *trans* geometry during the course of a suitable coupling reaction of a CD-ring precursor and an alternated side chain building block by olefination. Most synthetic olefination methods yield a mixture of *cis* and *trans* double bonds, leading to product mixtures that are difficult to separate due to their similar polarity. If the reaction conditions are too basic, C-20 epimerization can also occur, leading to even more complex product mixtures. Additionally, vitamin D_2 metabolites contain a chiral methyl group in C-24 position that has to be ranged in a defined configuration. When a suitable building block bearing this methyl group is employed in the synthesis, epimerization of the chiral center may occur under basic reaction conditions, along with the risk for epimerization at C-20 in the coupling reaction. Thus, alcohol **17** is converted to its corresponding phenylsulfone, which is coupled by olefination with aldehyde **20**, a method originally invented by Julia [34]. Note that **20** can be synthesized in a few routine steps from commercially available (S)-(+)-3-hydroxyisobutyric acid methyl ester, which is submitted to a Grignard reaction with methyl-magnesium bromide. Analogously to the synthesis of the other sixfold deuterated metabolites labeled at C-26/27, the addition of deuterated Grignard reagent in the course of the reaction leads to sixfold deuterated vitamin

D_2 metabolites. Finally, the double bond at C-5/6 of **21** has to be photochemically isomerized and the silyl protective groups have to be removed, leading to nondeuterated or deuterated 25(OH)D_2 (**22**, d-**22**) [35].

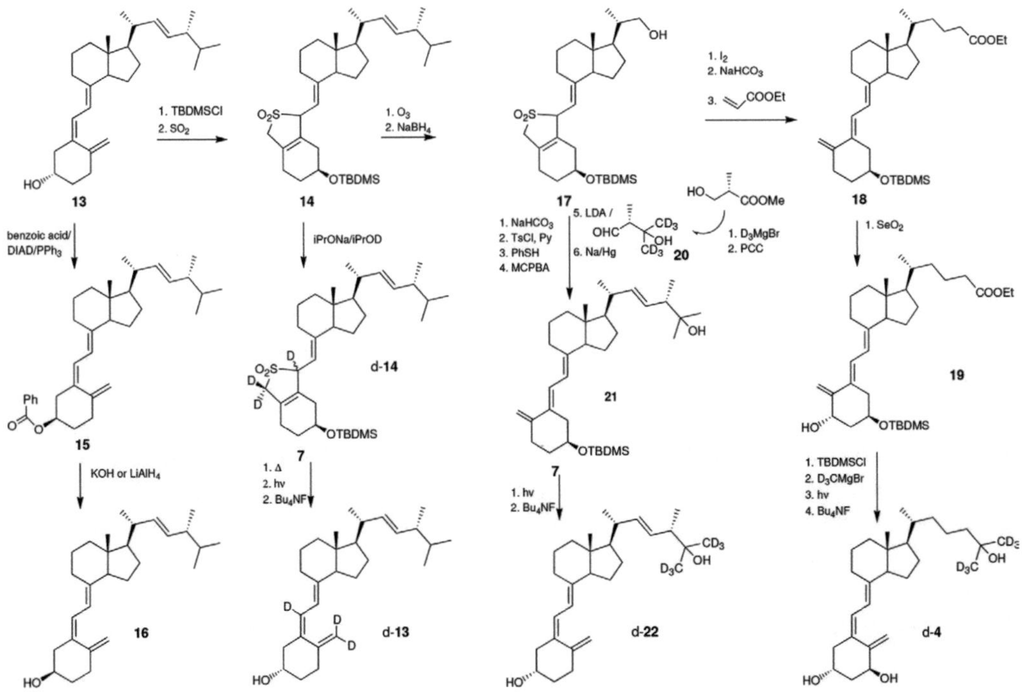

Scheme 2. Linear synthesis of vitamin D metabolites. TBDMSCl: *tert*-butyldimethylsilyl chloride, SO_2: sulfur dioxide, O_3: ozone, $NaBH_4$: sodium borohydride, *i*PrOD: deuterated isopropanol, Bu_4NF: tetra-*n*-butylammonium fluoride, $NaHCO_3$: sodium hydrogen carbonate, TsCl: *p*-toluenesulfonyl chloride, PhSH: thiophenol, MCPBA: 3-chloroperbenzoic acid, I_2: iodine, PCC: pyridinium chlorochromate, D_3CMgBr: deuterated methylmagnesium bromide, SeO_2: selenium dioxide, DIAD: diisopropyl azodicarboxylate, PPh_3: triphenylphosphine, KOH: potassium hydroxide, $LiAlH_4$: lithium aluminium hydride.

3.2. Convergent synthesis of vitamin D metabolites

Alternatively, a convergent approach can be applied (**Scheme 3**), which is more versatile than a linear synthesis and allows for more harsh reaction conditions and wider scope of suitable substrates and reagents. In a classical and widely applied strategy, vitamin D_2 (**13**) is submitted to ozonolysis, leading to Inhoffen-Lythgoe diol **23** and allylic alcohol **24** after reductive workup with sodium boronhydride. After appropriate chemical modification of the CD-ring precursor, this compound is coupled with an A-ring building block. Two strategies can be distinguished. In the most common approach, invented by Lythgoe and further developed by the Hoffmann-La Roche group [36], allylic alcohol **24** is converted via an allylic chloride **25** in diphenylphosphine oxide **26**, which is coupled with a C-8 ketone by a Wittig-Horner olefination. The A-ring diphenylphosphine oxide **26** can be coupled before or after appropriate alteration of the side chain at the CD-ring building block. To employ an A-ring diphenylphosphine oxide such as **26** is favourable, if no further substituents except a C-3 OH group and a C-10 *exo*-methylene group in the A-ring are needed, as it is already present in vitamin D_2 (**13**). As example for an efficient application of this strategy, the synthesis of 24,25$(OH)_2D_3$ (**6**) and its deuterated counterpart (d-**6**), their 3-*epi*-analogs (**9**/d-**9**) and 1α-analogs (**5**/d-**5**), is shown in **Scheme 3** [37]. Here, diol **23** is converted to its corresponding iodide, which is coupled with enone **27** (obtained in a few routine steps from (R)-(-)-lactic acid using a zinc-copper-catalyzed reaction mediated by ultrasound, leading to **28**. By Grignard reaction of **28** with nondeuterated or deuterated methyl-magnesium bromide, **29** or d-**29** are obtained. The 1,2-diol moiety is protected as isopropylidene ether, followed by oxidation with pyridinium chlorochromate to obtain ketone **30**/d-**30**, which can be coupled with diphenylphosphine oxide **26** to obtain d-24,25$(OH)_2D_3$ (d-**6**) after removal of silyl- and isopropylidene protective groups. To access the corresponding C-3-*epi*-analog, vitamin D_2 is first bis-hydroxylated at C-7/8 using potassium permanganate. Inversion of the stereogenic center at C-3 under "Mitsunobu-conditions" using picolinic acid leads to the corresponding C-3-*epi*-triol **31**. The yields of this reaction are much higher as compared to the synthesis of **16** used in the linear synthesis, because an elimination reaction leading to a favored C-3,4-5,6-7,8-all-*trans* system is avoided [38]. Silylation and cleavage of the diol moiety of triol **31** by lead tetraacetate and reductive workup with sodium boronhydride lead to the CD-ring fragment **32** and C-3-*epi* allylic alcohol **33**, which can be used after conversion to the corresponding diphenylphosphine oxide **34**. Note that **34** is coupled to d-**30** to obtain d-3-*epi*-24,25$(OH)_2D_3$ (d-**9**) or its corresponding nondeuterated counterpart **9**. 3-*epi*-diphenylphosphine oxide **34** can be used as a versatile A-ring building block to be employed

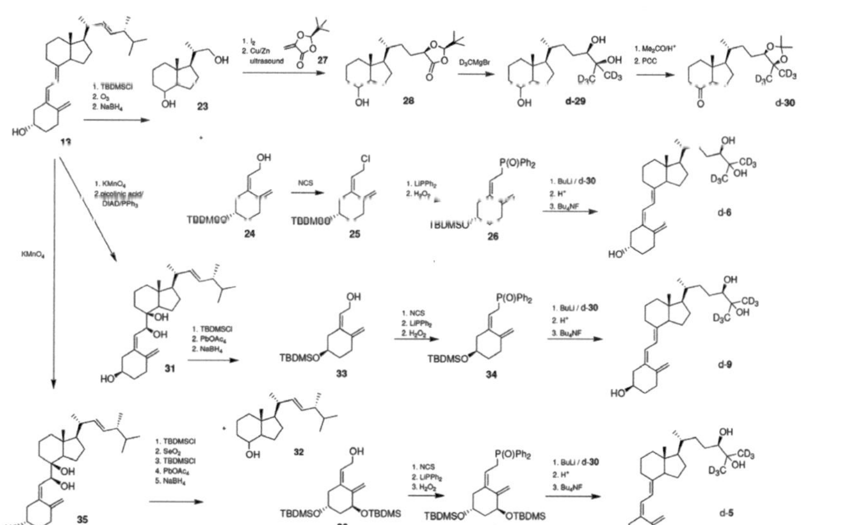

Scheme 3. Convergent synthesis of 24,25$(OH)_xD$ metabolites. TBDMSCl: *tert*-butyldimethylsilyl chloride, O_3: ozone, $NaBH_4$: sodium borohydride, I_2: iodine, Cu: copper, Zn: zinc, D_3CMgBr: deuterated methylmagnesium bromide, Me_2CO: acetone, PCC: pyridinium chlorochromate, $KMnO_4$: potassium permanganate, DIAD: diisopropyl azodicarboxylate, NCS: N-chlorosuccinimide, $LiPPh_2$: lithium diphenylphosphide, H_2O_2: hydrogen peroxide. BuLi: *n*-butyllithium, Bu_4NF: tetra-*n*-butylammonium fluoride, $Pb(OAc)_4$: lead tetraacetate, SeO_2: selenium dioxide.

in a Wittig-Horner reaction for the synthesis of a wide variety of 3-*epi* vitamin D_3 and D_2 metabolites, by connection with appropriate CD-ring building blocks with modified side chain, as shown for the synthesis of 3-*epi*-24,25$(OH)_2D_3$ (**9**, d-**9**) as representative example. Finally, permanganate oxidation of vitamin D_2 (**13**) leads to triol **35**, which is converted in a few routine steps, including a stereoselective 1α-allylic oxidation with SeO_2 via **36** to diphenylphosphine oxide **37**, which is then coupled with d-**30** to afford 1,24,25$(OH)_3D_3$ (**5**/d-**5**).

Two other low abundant metabolites, 23(S),25(R)-dihydroxyvitamin D_3 26,23-lactone **7** and 1α-1,23(S),25(R)-trihydroxyvitamin D_3 26,23-lactone **8**, are synthesized in an analogous manner [29] (**Scheme 4**), by coupling a CD-ring ketone carrying an appropriate side chain with diphenylphosphine oxide **26** or **37**. Here CD-ring-building block **32** obtained from vitamin D_2 **13** is used, which is converted to aldehyde **38**. In order to prepare an appropriate modified side chain, isopentenol **39** is converted using a few routine steps to sulfone **40**, which is then coupled with **38** by a Julia olefination to afford **41**. Removal of the isopropylidene protective group, oxidation to an unsaturated α-hydroxy carboxylic acid and subsequent iodolactonization leads to ketone **44** via **42** and **43**, which is coupled with diphenylphosphine oxide **26** or **37**, respectively, to give **7** and **8**.

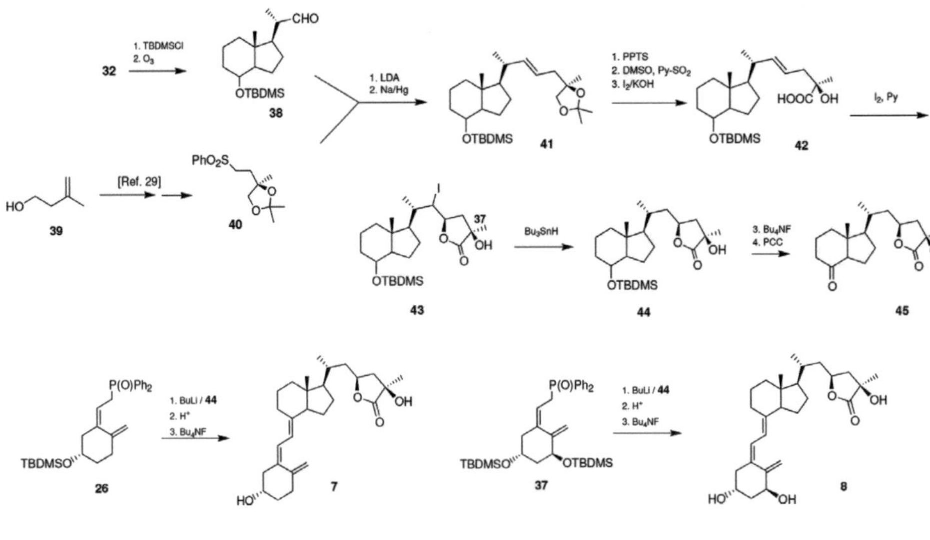

Scheme 4. Convergent synthesis of 23,26$(OH)_x$-lactone metabolites. TBDMSCl: *tert*-butyldimethylsilyl chloride, O_3: ozone, LDA: lithium diisopropylamide, Na: sodium, Hg: mercury, PPTS: pyridinium *p*-toluenesulfonate, DMSO: dimethyl sulfoxide, Py: pyridine, SO_2: sulfur dioxide, I_2: iodine, KOH: potassium hydroxide, Bu_3SnH: tributyltin hydride, Bu_4NF: tetra-*n*-butylammonium fluoride, PCC: pyridinium chlorochromate, BuLi: *n*-butyllithium.

3.3. Convergent Palladium-catalyzed synthesis of vitamin D metabolites

The A-ring building block is preferably synthesized *de novo*, if further substituents apart from a C-3-OH group and a C-10 exo methylene group in the A-ring (already present in vitamin D_2) are needed. This is particularly valuable for synthetic analogs containing substituents at C-2, which were developed in the past as new drugs [30]. However, natural metabolites may also contain additional substituents in the A-ring, such as 4β,25-dihydroxy vitamin D_3 [39]. In these cases, Pd catalyzed tandem reactions can be applied (**Scheme 5**). Initially invented by Trost [40], an acyclic enyne **46** is coupled with a CD-ring vinyl bromide **47** obtained by Peterson olefination from the corresponding ketone. Closure of the A-ring and its connection with CD-ring is preferably carried out in one pot. Alternatively, the corresponding triflate **48** can be employed in a Zn-mediated reaction, which was invented by Negishi [41]. Additionally, CD-ring alkene boranates **49** can be used via an analogous coupling method originally invented by Suzuki [42]. Finally, a cyclic A-ring enynes, such as **50**, can be coupled with CD-ring triflate **50**, a coupling method invented by Sonogashira [43]. In all these cases, the CD-ring vinyl bromide **47**, boranate **49**, or triflate **51** are usually carrying the modified side chain already, before coupling with the acyclic or cyclic A-ring enyne. A major drawback of these methods is the fact that synthesis of acyclic or cyclic enynes or other A-ring building blocks not derived from vitamin D_2 involves many synthesis steps, and in some cases separation of diastereomeric mixtures is also necessary to obtain enantiomerically and diastereomerically pure product. However, the suitable starting materials can usually be obtained from the natural chiral pool (i.e., terpenes such as carvone [44], malic acid [45], quinic acid [43], or carbon hydrates such as D-glucose [46] or D-xylose [47]), which are usually readily available. A wide range of substrates are tolerated. Consequently, these advanced strategies involving Pd catalyzed reactions have widely been applied in recent years and are by now well established.

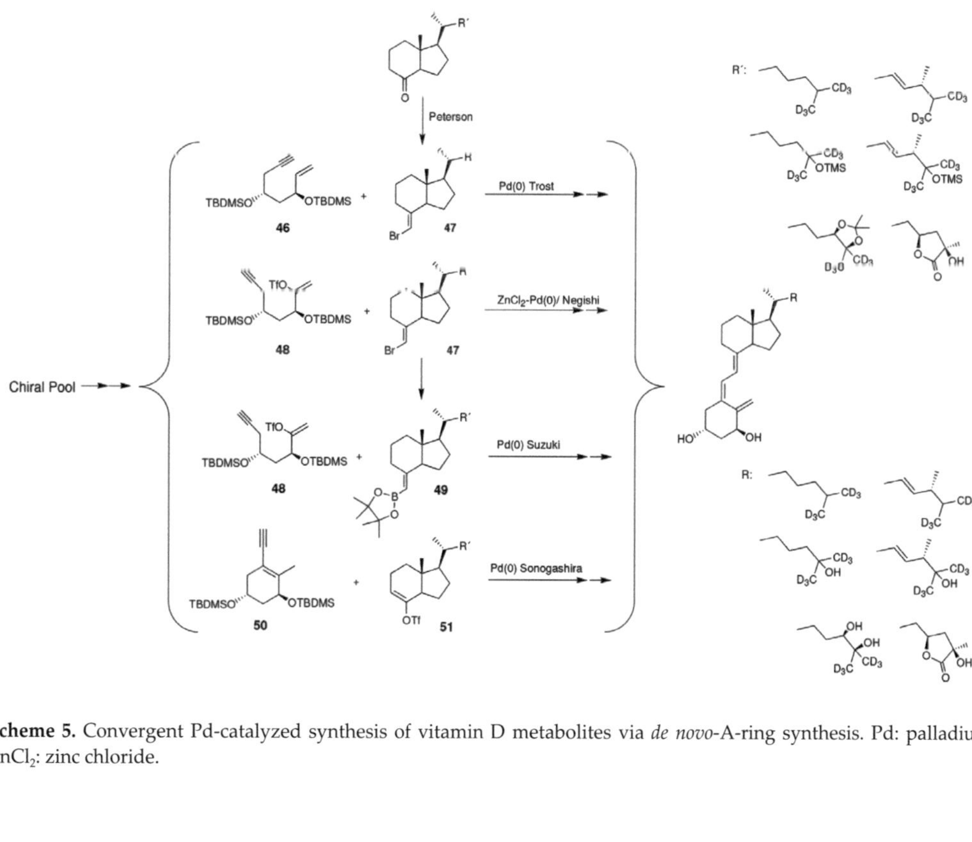

Scheme 5. Convergent Pd-catalyzed synthesis of vitamin D metabolites via *de novo*-A-ring synthesis. Pd: palladium, $ZnCl_2$: zinc chloride.

4. LC-MS/MS assays for vitamin D metabolites

In the second part of this chapter, an overview of the mass spectrometric analysis of vitamin D metabolites in biological samples is presented. Because of the chemical nature of these secosteroidal molecules, liquid chromatography-tandem mass spectrometry (LC-MS/MS) currently provides the optimum analytical platform for analysis of vitamin D metabolites. While mass spectrometry assays are often not as rugged and more expensive than most non-mass spectrometric assays (in particular immunoassays), they provide the ability to capture multiple metabolites simultaneously at very low concentration levels. Immunoassays measure the 25(OH)D_3 and 25(OH)D_2 metabolites only and have limitations with respect to detection sensitivity as well as selectivity and specificity issues.

In most published assays, the metabolites are determined from serum or plasma matrices. Other biological matrices have rarely been used, although vitamin D metabolites have successfully been quantified in saliva [48] and various soft tissues [49]. In the following sections, we discuss the requirements and characteristics of LC-MS/MS of vitamin D metabolites; that is, chromatographic separation, ionization, *m/z* analysis, interferences and accuracy issues, multimetabolite screening applications, and the clinical role of vitamin D fingerprinting methods.

4.1. Liquid chromatographic separation of vitamin D metabolites

Due to the hydrophobic structure, vitamin D metabolites are generally easily separated on reversed-phase liquid chromatography stationary phases (e.g., octadecyl C_{18}) materials utilizing hydrophobic interactions. The vitamin D metabolites generally elute in the order trihydroxylated < dihydroxylated < monohydroxylated metabolites, that is, for vitamin D_3-related molecules, the order of chromatographic retention is 1,25$(OH)_2D_3$ < 25(OH)D_3 < D_3. Between corresponding D_2 and D_3 analogs, the D_2 metabolites elute marginally later than the D_3 versions. For the two important isomers of dihydroxylated vitamin D metabolites, 24,25$(OH)_2D_3$ and 1,25$(OH)_2D_3$, the order of retention is 24,25$(OH)_2D_3$ < 1,25$(OH)_2D_3$.

Possibly more interesting—from a chromatography point of view—are the biochemically formed isomers after stereochemical reversal ($\beta \rightarrow \alpha$) at C-3; for example, the epimers 25(OH)D_3 and 3-*epi*-25(OH)D_3. The subtle diastereomeric differences between these species can be distinguished using specialized columns, such as combined C18/chiral [50], pentafluorophenyl (PFP) [51–53], and cyano (CN) [54–57].

4.2. Ionization of vitamin D metabolites for mass spectrometric analysis

Mass spectrometric determination of vitamin D metabolites using liquid chromatography-mass spectrometry (LC-MS) is not trivial because of the structural limitations that the analytes provide with respect to attaching a charge to the molecules. Gas chromatography-mass spectrometry techniques have been used in the past for qualitative analysis and structure determinations, but these methods have been almost completely replaced by modern LC-MS methods. While analysis of transformation products such as vitamin D sulfates is relatively easy by LC-MS as the metabolites can be simply analyzed as deprotonated molecules, ionizing the relatively nonpolar vitamin D metabolites is not straightforward. For LC-MS, electrospray ionization (ESI) or atmospheric pressure chemical ionization (APCI) are the most common ionization techniques, which usually result in the formation of $[M+H]^+$ ions for most biological molecules. APCI has been successfully applied to steroids, as gas-phase chemical ionization often efficiently transfers a proton to these types of molecules. ESI usually relies on a charging mechanism in the liquid phase and one would therefore expect ESI to be less efficient than APCI for vitamin D metabolites. In reality this is not the case, however, and both techniques have equally been applied to vitamin D analysis, with the analytical figures of merit being quite similar between the two ionization techniques. There appears to be a slight trend toward application of ESI rather than APCI in recent years.

An important, often neglected consideration in the choice of ionization source is the (usually detrimental) impact of coionized sample components that coelute with vitamin D metabolites from the LC column. Here, ionization suppression effects caused by these coeluting molecules may impact the outcome even stronger than the differences seen in ionization efficiencies between ESI and APCI [58, 59]. However, no systematic comparison of ionization efficiency/ ionization suppression effects between ESI and APCI has been performed yet. Furthermore, coionization of coeluting components also leads to the formation of isobaric interferences when sample matrices such as plasma or serum are analyzed, which will be discussed in more detail below.

Another useful ionization technique is atmospheric pressure photoionization (APPI). This technique is very common in fields such as petroleum analysis, but surprisingly, it is virtually unknown in the clinical community. To our knowledge, only one study has systematically compared APPI with APCI and shown that APPI generated significantly higher ion currents for $25(OH)D_3$ than APCI [60]. A second study applied APPI to quantification of $25(OH)D_3$ without comparison to ESI or APCI [61].

Finally, another option to overcome problems of ionization efficiency and resulting problems with detection sensitivity is derivatization of vitamin D metabolites, to convert them into better responding transformation products. Such procedures are now quite common in the vitamin D analytical field; they comprise introduction of a chemical group that is readily ionized or is permanently charged. Cookson-type triazoline-diones and triazoline-dione-related reagents (e.g., 4-phenyl-1,2,4-triazoline-3,5-dione [PTAD]) are most often applied, which utilize the reaction of the reagent's dienophile with the *cis*-diene group at C-5/C-6 and C-10/C-19 of vitamin D (Diels-Alder [4+2] reaction) (**Scheme 6**).

Several promising PTAD assays have been described in the literature [39, 49, 62–64]. The advantage of using the *cis*-diene moiety is that all vitamin D metabolites contained in the sample are converted when the reagent is present at sufficient concentrations (i.e., in large excess quantities) as all relevant vitamin D metabolites possess this structural motif. Ionization efficiency of 25(OH)D-PTAD has been reported to be 100-fold higher than the nonderivatized molecules in ESI mode [65]. A structural variation in PTAD includes a permanently charged quaternary ammonium moiety, which enhances detection sensitivity even further [66]. This reagent is commercially available. A recent chemotyping assay for multiple vitamin D metabolites has shown detection limits in the fg/mL range for vitamin D metabolites from human serum (see discussion below).

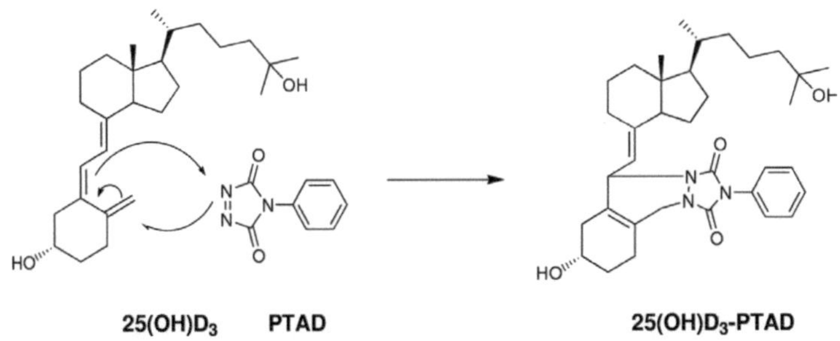

Scheme 6. Diels-Alder [4+2] derivatization of vitamin D metabolites using PTAD.

4.3 Mass analyzers for vitamin D analysis

Most clinical applications for vitamin D utilize quadrupole-based mass analyzers with low resolving powers. Because these mass spectrometers measure *m/z* signals at "unit mass resolution," they cannot be operated in full scan mode for the accurate quantification of vitamin metabolites when complex sample matrices such as serum are analyzed. Serum contains multiple components of identical nominal molecular mass to vitamin D metabolites, several of which have been shown to coelute during High Performance Liquid Chromatography (HPLC) [67]. The coeluting metabolites then generate so-called "isobaric" ions in the mass spectra after ESI, which cannot be distinguished from the vitamin D target molecules using low resolution MS. Therefore, tandem mass spectrometry (MS/MS) is implemented, which activates the ionized species further and forces them to undergo collision-induced dissociation (CID) for generating structure-specific product ions. The metabolite-specific product ions are then measured and their response used for quantification of the target metabolites.

By far the most common low resolution tandem mass spectrometer for vitamin D analysis in clinical laboratories is the triple quadrupole (QqQ) instrument. In a QqQ mass spectrometer, the $[M+H]^+$ ions of vitamin D metabolites are selected in the first quadrupole, fragmented in a quadrupole collision cell, and the products analyzed in the final quadrupole mass analyzer. The acquisition mode that is almost always applied is the so-called selected reaction monitoring (SRM) mode. In this mode, the $[M+H]^+$ precursor ion is isolated in Q1 (e.g., *m/z* 401 for 25(OH)D_3), dissociated in q2, and one specific product ion selected in Q3 (e.g., *m/z* 383). The quality of the precursor/product ion selection determines the selectivity of analysis; that is, the product ion structure should ideally directly reflect the chemical structure of the chosen precursor ion for maximum specificity. For vitamin D metabolites, the selection of an appropriate product ion is surprisingly difficult because of some inherent structural limitation of the precursor structure, leading primarily to simple dehydration reactions at low collision energies, which turn into complex mass spectra at higher collision energies (**Figure 1**) [68].

Unfortunately, because of multiple, overlapping series of fragmentation reactions, leading to "picket fence" type product ion spectra, fragment ions for vitamin D metabolites in the diagnostic *m/z* range often resemble each other.

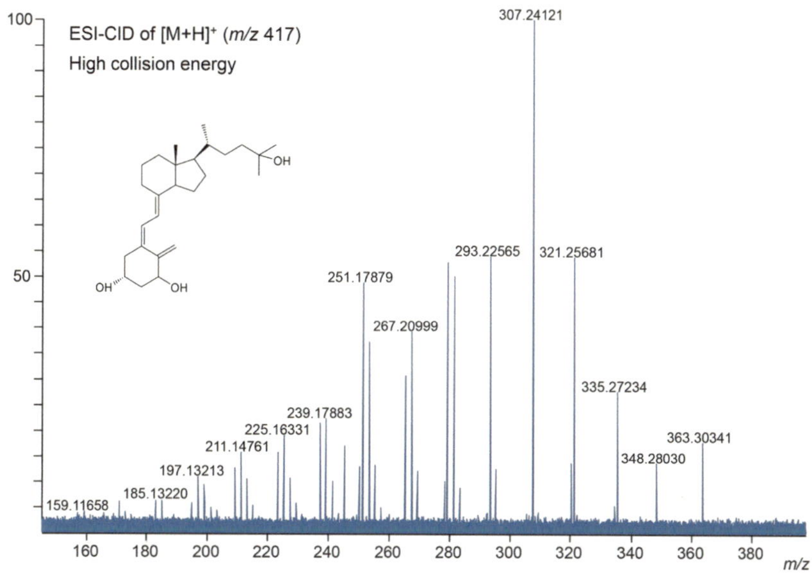

Figure 1. Collision-induced dissociation (CID) spectrum of the [M+H]+ ion of 1,25$(OH)_2D_3$ (spectrum after fully completed dehydration reactions after electrospray ionization and CID; *m/z* analysis by Fourier-transform ion cyclotron resonance, FTICR).

A few studies also report the use of a low resolution quadrupole ion trap (IT) [69, 70] in MS/MS mode for 25$(OH)D_3$ and 25$(OH)D_2$, but these instruments are rarely implemented in clinical environments. IT instruments are often not fit for purpose for quantitation of low abundant biological molecules, because of the large detrimental contributions of acquisition time overhead to the duty cycle, thus limiting detection sensitivity. There is also a general decline of this instrument type in modern mass spectrometry labs.

The most anticipated future development for vitamin D analysis is the introduction of high resolution mass spectrometry (HRMS), which is now firmly established in many other fields of modern mass spectrometry, in particular, in pharmaceutical applications and the proteomics field [71]. This trend is mostly due to the availability of robust quadrupole-quadrupole-time-of-flight (QqTOF) and orbitrap mass analyzers in recent years, which have transformed many analytical approaches to mass spectrometry. The use of mass defect as metabolite-specific property, for example, is now an integral part of many metabolite identification routines for drug metabolites [72]. While the number of applications of HRMS in the vitamin D field is still limited, the existing work has clearly demonstrated the potential of HRMS for vitamin D analysis. For example, orbitrap mass spectrometers in full scan [73, 74] and MS/MS [75, 76] modes have been applied successfully to the analysis of 25(OH)D_3 in human serum and analytical performance has been shown to be equivalent or better than triple quadrupole and immunoassays. The important role of HRMS in the separation of isobaric interferences will be shown below.

4.4. Interferences during LC-MS/MS

The LC-MS/MS analysis of vitamin D metabolites is affected by various sources of error, which can affect both precision and accuracy. As with any other LC-MS/MS analysis from biological samples, ion suppression by coeluting sample components or chemical modifiers from the sample preparation or chromatography can lead to reduced analyte signals. There are several options to assess whether or not ion suppression is present, which have been summarized in many review articles, e.g., by Matuszewski et al. [77]. Stable isotope standards of vitamin D metabolites (mostly deuterated analogs) can usually correct for accuracy errors from ion suppression effects, as long as it is guaranteed that protein binding for the isotope standard is the same as for the endogenous analyte, which requires implementation of a careful incubation routine [78]. Importantly, deuterated isotope standards are commercially available for most of the relevant vitamin D metabolites but unfortunately not for all.

A second important source of analytical error originates from isobaric noise; that is, endogenous or exogenous metabolites that coelute with the vitamin D analytes and erroneously contribute to the analytical signal, if unspecific ions are used for mass spectral analysis. This has recently been described in detail by Qi et al. [67], who clearly demonstrated the presence of multiple isobars of 25(OH)D_3 in human serum. The isobars have to be carefully removed in low resolution mass spectrometry, by application of appropriate MS/MS or ion mobility spectrometry routines [67]. A number of exogenous and endogenous molecules have been identified as relevant metabolites in serum samples [50, 67, 79].

Many of these interferences can be eliminated by application of high resolution mass spectrometry using sufficiently high resolving powers; for example, through implementation of orbitrap or Fourier-transform ion cyclotron resonance (FTICR) mass spectrometers. For example, Liebisch and Matysik demonstrated that the orbitrap MS instrument in their study was able to separate an isobaric interference of 25(OH)D_3 in the MS/MS domain; this interference was caused by fragmentation of the d6-25(OH)D_2 isotope standard [75].

Importantly, the issues relating to isobaric noise have only been studied for the 25(OH)D_3 analyte; other metabolites will likely be affected by similar interferences, the impact of which during quantification, in particular, in multimetabolite assays (see below), remains unknown.

4.5. Method accuracy and certified reference materials

The vitamin D analytical community is supported through the Vitamin D External Quality Assessment Scheme (DEQAS), a nonprofit organization that evaluates the performance of analytical assays of member laboratories for 25(OH)D_3 and 1,25$(OH)_2D_3$ [80] through round-robin analyses. DEQAS has clearly demonstrated that analytical performance of vitamin D analysis has greatly improved over the years between 1994 and 2009 [81].

The United States National Institute of Standards and Technology (NIST) provides certified standard solutions for 25(OH)D_3 and 25(OH)D_2. Furthermore, the NIST and the National Institutes of Health (NIH) Office of Dietary Supplements (ODS) have established the Vitamin D Metabolites Quality Assurance Program (VitDQAP), for interlaboratory comparison of measurement of 25(OH)D_2, 25(OH)D_3, and 3-*epi*-25(OH)D_3 in serum and plasma and provide method-appropriate control materials.

A number of commercial reference, calibration, and quality control materials are available from several companies that allow rapid implementation and validation of vitamin D analytical methodologies.

4.6. Vitamin D multimetabolite assays

One of the most important advantages of LC-MS/MS assays over clinical immunoassays is the ability to determine multiple vitamin D species independently and simultaneously. As a result, there are now several very capable LC-MS/MS assays described in the literature that provide the capability for profiling the most relevant vitamin D metabolites at the same time, within a single analytical run, and with sufficient dynamic range to allow measuring the required physiological levels, down to the picomolar range. The topic has recently been reviewed in detail and the interested reader is referred to Ref. [82]. Briefly, most of these multimetabolite assays utilize derivatization techniques that transform all vitamin D species into better responding analogs (vide supra). This procedure in turn permits analysis of both high and low abundant vitamin D species with similar analytical figures of merit [62, 83].

The simultaneous acquisition of all important metabolite levels then provides the possibility of using these vitamin D metabolite distributions ("chemotypes") as complex diagnostic or prognostic biomarkers for correlation with disease phenotype or clinical outcome of treatment. A few examples for such correlations are summarized in the last section.

5. Vitamin D fingerprints (chemotypes) in clinical applications

A number of recent studies have gone beyond the usual determination of 25(OH)D_3—as marker for vitamin D status [84]—and 1,25$(OH)_2D_3$—for diagnosis of renal diseases, hypercalcemic syndromes, and disorders of 25(OH)D_3 metabolism [85]. These broader profiling techniques are aimed at discovering dynamic effects of metabolites and catabolites, which are located further downstream the 25(OH)D_3 metabolic cascade. Current studies highlight 24,25$(OH)_2D_3$ as important diagnostic marker, which was previously only considered a clearance product of vitamin D without activity. In fact, it has been shown that 24,25$(OH)_2D_3$ has crucial roles in bone metabolism [86] and renal diseases [87].

Capturing multiple vitamin D species and their dynamic changes allows for a better understanding of interindividual variations after vitamin D supplementation. Müller et al. recently demonstrated an inverse linear correlation between baseline 25(OH)D_3 and response to supplementation for patients with chronic liver disease [83]. The study also showed that lower baseline 24,25(OH)$_2$D$_3$ levels were linked to larger changes of 25(OH)D$_3$ levels, and that those patients who exhibited greater response to vitamin D supplementation had lower levels of 3-*epi*-25(OH)D$_3$ (**Figure 2**).

Berg et al. implemented the ratio of 24,25(OH)$_2$D$_3$ and 25(OH)D$_3$ as a novel status marker for vitamin D [88]. Wagner et al. used the same ratio and demonstrated that it was predictive of 25(OH)D$_3$ response to supplementation [89]. Binkley et al. measured multiple vitamin D species and developed a model to describe interindividual variation of 25(OH)D$_3$ levels after supplementation [90]; the authors highlighted the role of absorption (as measured by the nonmetabolized vitamin D$_3$ species) and degradation (via the 24,25(OH)$_2$D$_3$ species) and presented a treat-to-target regime for tailored serum levels of 25(OH)D$_3$ [90].

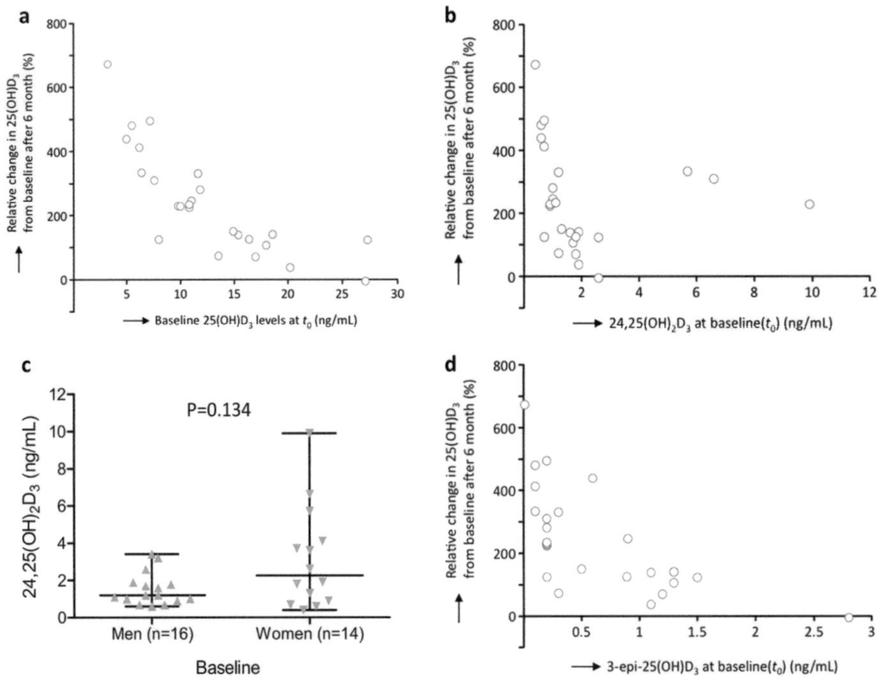

Figure 2. Nonparametric correlations between baseline vitamin D metabolites and response to vitamin D supplementation for patients with chronic liver diseases after 6-month treatment: (a) the relative change in serum 25(OH)D$_3$ (in response to vitamin D supplementation) correlated inversely with baseline 25(OH)D$_3$ concentrations; (b) similarly, an inverse correlation between relative change in serum 25(OH)D$_3$ and baseline 24,25(OH)$_2$D$_3$ was observed; (c) baseline 24,25(OH)$_2$D$_3$ concentrations were nonsignificantly higher in women as compared to men; (d) patients with lower 3-epi-25(OH)D$_3$ concentrations at baseline tended to have a larger response to vitamin D supplementation (reprinted with permission from Ref. [83]).

Other important studies include the work by Bosworth et al. [87] and Stubs et al. [91], who utilized multimetabolite LC-MS/MS methods to characterize chronic kidney disease (CKD). Similarly, Duan et al. [64] studied patients with multiple sclerosis and observed comparable levels of $25(OH)D_3$ in the healthy control subjects and the patients; however, serum levels for $1,25(OH)_2D_3$ and $24,25(OH)_2D_3$ were lower in patients than controls.

6. Conclusions

The availability of assays for simultaneous capturing multiple vitamin D metabolites combined with reliable techniques for synthesis of the required metabolite standard compounds makes accurate measurement of metabolite distributions and subsequent correlation with disease phenotype readily possible; the outcome of these strategies is expected to be useful for diagnosis and risk prediction for various diseases. It may also allow for specific supplementation strategies in the future, which consider patient-specific dosage requirements and use of selected vitamin D metabolites or analogs.

Author details

Lars Kattner[1*] and Dietrich A. Volmer[2]

*Address all correspondence to: lars.kattner@endotherm.de

1 Endotherm Life Science Molecules, Saarbrücken, Germany

2 Institute of Bioanalytical Chemistry, Saarland University, Saarbrücken, Germany

References

[1] Bouillon R, Okamura WH, Norman A: Structure and function in vitamin D endocrine system. Endocr Rev. 1995;16(2):200–257. DOI: 10.1210/edrv-16-2-200

[2] Dusso AS, Brown AJ, Slatopolsky E: Vitamin D. Am J Physiol Renal Physiol. 2005;289:F8–F28. DOI: 10.1152/ajprenal.00336.2004

[3] DeLuca H: Overview of general physiologic features and functions of vitamin D. Am J Clin Nutr. 2004;80(suppl):1689S–1696S. ISSN: 0002-9165

[4] Holick M: Vitamin D and bone health. J Nutr. 1996;126:1159S–1164S. ISSN: 0022-3166

[5] von Hurst PR, Stonehouse W, Coad J: Vitamin D supplementation reduces insulin resistance in South women living in New Zealand who are insulin resistant and vitamin D deficient – a randomized, placebo-controlled trial. Br J Nutr. 2010;103:549–555. DOI: 10.1017/S0007114509992017

[6] Hyppönen E, Läärä E, Reunanen A et al.: Intake of vitamin D and risk of type 1 diabetes: a birth-cohort study. Lancet. 2001;358:1500. ISSN: 0099-5355

[7] Lappe JM, Travers-Gustafson D, Davies KM et al: Vitamin D and calcium supplementation reduces cancer risk: results of a randomized trial. Am J Clin Nutr. 2007;85:1586–1591. Print ISSN: 0002-9165; Online ISSN: 1938-3207

[8] Souberbielle JC, Body JJ, Lappe JM et al: Vitamin D and muskoskeletal health, cardiovascular disease, autoimmunity and cancer: recommendations for clinical practice. Autoimmun Rev. 2010;9:709–715. DOI: 10.1016/j.autrev.2010.06.009

[9] Wang TJ, Pencina MJ, Booth SL et al: Vitamin D deficiency and risk of cardiovascular disease. Circulation. 2008;117:503–511. DOI: 10.1161/CIRCULATIONAHA.107.706127

[10] Farrell J, Herrmann M: Determination of vitamin D and its metabolites. Best Pract Res Clin Endocrinol Metab. 2013;27(5):675–688. DOI: 10.1016/j.beem.2013.06.001

[11] Zerwekh JE: The measurement of vitamin D: analytical aspects. Ann Clin Biochem. 2004;41:272–281. DOI: 10.1258/0004563041201464

[12] Fraser WD, Milan AM: Vitamin D assays: past and present debates, difficulties, and developments. Calcif Tissue Int. 2013;92(2):118–127. DOI: 10.1007/s00223-012-9693-3

[13] Geib T, Meier F, Schorr P, Lammert F, Stokes CS, Volmer DA: A simple micro-extraction plate assay for automated LC-MS/MS analysis of human serum 25-hydroxyvitamin D levels. J Mass Spectrom. 2015;50:275–279. DOI: 10.1002/jms.3522

[14] Granado Lorencio F, Blanco-Navarro I, Pérez-Sacrsitán B: Critical evaluation of assays for vitamin D status. Curr Opin Clin Nutr Metab Care. 2013;16(6):734–740. DOI: 10.1097/MCO.0b013e328364ca96

[15] Stokes CS, Volmer DA, Grünhage F, Lammert F: Vitamin D in chronic liver disease. Liver Int. 2013;33:338–352. DOI: 10.1111/liv.12106

[16] Higashi T, Shimadab K, Toyo'oka T: Advances in determination of vitamin D related compounds in biological samples using liquid chromatography–mass spectrometry: a review. J Chromatogr B. 2010;878:1654–1661. DOI: 10.1016/j.jchromb.2009.11.026

[17] Mena-Bravo A, Ferreiro-Vera C, Priego-Capote F, Maestro MA, Mouriño A, Quesada-Gómez JM, Luque de Castro MD: Quantitative analytical method to evaluate the metabolism of vitamin D. Clin Chim Acta. 2015;442:6–12. DOI: 10.1016/j.cca.2014.12.039

[18] Horst RL, Reinhardt TA, Reddy GS. Chapter: Vitamin D metabolism. In: Feldman D, Pike JW, Glorieux FH, editors. 20 Vitamin D. 2nd ed. Burlington: Elsevier; 2005. p. 15–36.

[19] Jones G: Metabolism and biomarkers of vitamin D. Scand J Clin Lab Invest Suppl. 2012;243:7–13. ISSN: 0036-5513

[20] Kamao M, Tatematsu S, Sawada N, Sakaki T, Hatakeyama S, Kubodera N, Okano T: Cell specificity and properties of the C-3 epimerization of vitamin D3 metabolites. J Steroid Biochem Mol Biol. 2004;89–90:39–42. DOI: 10.1016/j.jsbmb.2004.03.048

[21] Bailey D, Veljkovic K, Yazdanpanpanah M, Adeli K: Analytical measurement and clinical relevance of vitamin D3 C3-epimer. Clin Biochem. 2013;46:190–196. DOI: 10.1016/j.clinbiochem.2012.10.037

[22] Singh RJ, Taylor RL, Reddy GS, Grebe SKG: C-3 Epimers can account for a significant proportion of total circulating 25-hydroxyvitamin D in infants, complicating accurate measurement and interpretation of vitamin D status. J Clin Endocrinol Metab. 2006;91(8):3055–3061. DOI: 10.1210/jc.2006-0710

[23] Reddy GS, Rao DS, Siu-Caldera ML, Astecker N, Weiskopf A, Vouros P, Sasso GJ, Manchand PS, Uskokovic MR: 1a25-Dihydroxy-16-ene-23-yne-vitamin D3 and 1α,25-Dihydroxy-16-ene-23-yne-20-epi-vitamin D_3: analogs of 1α,25-dihydroxyvitamin D_3 that resist metabolism through the C-24 oxidation pathway are metabolized through the C-3 epimerization pathway. Arch Biochem Biophys. 2000;383(2):197–205. DOI: 10.1006/abbi.2000.2074

[24] Houghton L, Vieth R: The case against ergocalciferol (vitamin D2) as a vitamin supplement. Am J Clin Nutr. 2006;84:694–697. Print ISSN: 0002-9165; Online ISSN: 1938-3207

[25] Nadkarni S, Chodynski M, Corcoran A, Marcinkowska E, Brown G, Kutner A: Double point modified analogs of vitamin D as potent activators of vitamin D receptor. Curr Pharm Des. 2015;21:1741–1763. DOI: 10.2174/1381612821666141205125113

[26] Zhu GD, Okamura WH: Synthesis of vitamin D (calciferol). Chem Rev. 1995;95:1877–1952. DOI: 10.1021/cr00038a007

[27] Dai H, Posner GH: Synthetic approaches to vitamin D. Synthesis 1994; 1994(12): 1383–1398. DOI: 12 10.1055/s-1994-25697

[28] Milne GWA, Delander M, editors. Vitamin D Handbook: Structures, Synonyms, and Properties. Hoboken: Wiley; 2008.

[29] Kattner L, Bernardi D, Rauch E: Development of efficient chemical syntheses of vitamin D degradation products. Anticancer Res. 2015;35(2):1205–1210. ISSN: 0250-7005

[30] Nadkarni S, Chodyński M, Krajewski K, Cmoch P, Marcinkowska E, Brown G, Kutner A: Convergent synthesis of double point modified analogs of 1α,25.dihydroxyvitamin D_2 for biological evaluation. J Steroid Biochem Mol Biol. 2015 in press. DOI: 10.1016/j.jsbmb.2015.08.022

[31] Ray R, Vicchio D, Yergey A, Holick MF: Synthesis of 25-hydroxy-[6,19,19′-^{2}H3]vitamin D_3. Steroids. 1992;57:142–146. DOI: 10.1016/0039-128X(92)90072-H

[32] Mitsunobu O: The use of diethyl azodicarboxylate and triphenylphosphine in synthesis and transformation of natural products. Synthesis. Synthesis 1981; 1981(1): 1–28. DOI: 10.1055/s-1981-29317

[33] Machand PS, Yiannikouros GP, Belica P, Madan P: Nickel-mediated conjugate addition. Elaboration of calcitriol from ergocalciferol. J Org Chem. 1995;60:6574–6581. DOI: 10.1021/jo00125a051

[34] Ihara M, Suzuki S, Taniguchi T, Tokunaga Y, Fukumoto K: Modification of the Julia alkenation using SmI_2-HMPA. Synlett. 1994 (10):859–860. DOI: 10.1055/s-1994-23033

[35] Pietraszek A, Malinska M, Chodynski M et al: Synthesis and crystallographic study of 1,25-dihydroxyergocalciferol analogs. Steroids. 2013;78:1003–1014. DOI: 10.1016/j.steroids.2013.06.001

[36] Baggiolini EG, Jacobelli JA, Hennessy BM, Batcho AD, Sereno JF, Uskokovic MR: Stereocontrolled total synthesis of 1α,25-dihydroxycholecalciferol and 1α,25-dihydroxyergocalciferol. J Org Chem. 1986;51:3098–3108. DOI: 10.1021/jo00366a004

[37] Sestelo JP, Cornella I, Una O, Mouriño A, Sarandeses LA: Stereoselective convergent synthesis of 24,25-dihydroxyvitamin D_3 metabolites: a practical approach. Eur J Chem. 2002;8(12):2747–2752. ISSN: 0947-6539

[38] Achmatowicz B, Gorobets E, Marczak S, Przezdziecka A, Steinmeyer A, Wicha J, Zügel U: The first synthesis and biological testing of the enantiomer of 1α,25-dihydroxyvitamin D_3. Tetrahedron Lett. 2001;42:2891–2895. DOI: 10.1016/S0040-4039(01)00317-3

[39] Wang Z, Senn T, Kalhorn T, Zheng X, Zheng S, Davis C et al: Simultaneous measurement of plasma vitamin D_3 metabolites, including 4β,25-dihydroxyvitamin D_3, using liquid chromatography-tandem mass spectrometry. Anal Biochem. 2011;418:126–133. DOI: 10.1016/j.ab.2011.06.043

[40] Trost BM, Dumas J, Villa M: New strategies for the synthesis of vitamin D metabolites via Pd-catalyzed reactions. J Am Chem Soc. 1992;114:9836–9845. DOI: 10.1021/ja00051a016

[41] Gómez-Reino C, Vitale C, Maestro M, Mouriño A: Pd-catalized carbocyclization – Negishi cross-coupling cascade: a novel approach to 1α,25-dihydroxyvitamin D3 and analogues. Org Lett. 2005;7(26):5885–5887. DOI: 10.1021/ol052489c

[42] Gogoi P, Sigüeiro R, Eduardo S, Mouriño A: An expeditious route to 1α,25-dihydroxyvitamin D_3 and its analogues by an aqueous tandem palladium-catalyzed a-ring closure and Suzuki coupling to the C/D unit. Chem Eur J. 2010;16:1432–1435. DOI: 10.1002/chem.200902972

[43] Sibilska IK, Szybinski M, Sicinski RR, Plum LA, DeLuca HF: Highly potent 2-methylene analogs of 1,a,25-dihydroxyvitamin D_3: synthesis and biological evaluation. J Steroid Biochem Mol Biol. 2013;136:9–13. DOI: 10.1016/j.jsbmb.2013.02.001

[44] Antony P, Sigüeiro R, Huet T et al.: Structure-function relationships and crystal structures of the vitamin D receptor bound 2α-methyl(20S,23RS)- and 2α-methyl-(20S, 23R)-epoxymethano-1α,25dihydroxyvitamin D_3. J Med Chem. 2010;53:1159–1171. DOI: 10.1021/jm9014636

[45] Kato Y, Hashimoto Y, Naggasawa K: Novel heteroatom containing vitamin D_3 analogs: efficient synthesis of 1α,25-dihydroxyvitamin D_3-26,23-lactam. Molecules. 2003;8:488–499. DOI: 10.3390/80600488

[46] Honzawa S, Suhara Y, Nihei KI et al.: Concise synthesis and biological activities of 2α-Alkyl- and 2α-(ω-Hydroxyalkyl)-20-epi-1α,25-dihydroxyvitamin D3. Bioorg Med Chem Lett. 2003;13:3503–3506. DOI: 10.1016/S0960-894X(03)00739-X

[47] Moriarty RM, Kim J, Brumer III H: A general synthetic route to A-ring hydroxylated vitamin D analogs from pentoses. Tetrahedron Lett. 1995;36:51–54. DOI: 10.1016/0040-4039(94)02209-T

[48] Higashi T, Shibayama Y, Fuji M, Shimada K: Liquid chromatography tandem mass spectrometric method for the determination of salivary 25-hydroxyvitamin D3: a noninvasive tool for the assessment of vitamin D status. Anal Bioanal Chem. 2008;391:229–238. DOI: 10.1007/s00216-007-1780-3

[49] Lipkie TE, Janasch A, Cooper BR, Hohman EE, Weaver CM, Ferruzzi MG: Quantification of vitamin D and 25-hydroxyvitamin D in soft tissues by liquid chromatography-tandem mass spectrometry. J Chromatogr B Analyt Technol Biomed Life Sci. 2013;932:6–11. DOI: 10.1016/j.jchromb.2013.05.029

[50] Shah I, James R, Barker J, Petroczi A, Naughton DP: Misleading measures in vitamin D analysis: a novel LC-MS/MS assay to account for epimers and isobars. Nutr J. 2011;10:46. DOI: 10.1186/1475-2891-10-46

[51] Schleicher RL, Encisco SE, Chaudhary-Webb M, Paliakov E, McCoy LF, Pfeiffer CM: Isotope dilution ultra performance liquid chromatography tandem mass spectrometry method for simultaneous measurement of 25-hydroxyvitamin D_2, 25-hydroxyvitamin D_3 and 3-epi-25-hydroxyvitamin in human serum. Clin Chim Acta. 2011;412:1594–1599. DOI: 10.1016/j.cca.2011.05.010

[52] Baecher S, Leinenbach A, Wright JA, Pongratz S, Kobold U, Thiele R: Simultaneous quantification of four vitamin D metabolites in human serum using high performance liquid chromatography tandem mass spectrometry for vitamin D profiling. Clin Biochem. 2015;45(16–17):1491–1496. DOI: 10.1016/j.clinbiochem.2012.06.030

[53] Strathmann FG, Sadilkova K, Laha TJ, LeSourd SE, Bornhorst JA, Hoofnagle AN, Jack R: 3-epi-25 hydroxyvitamin D concentrations are not correlated with age in a cohort of infants and adults. Clin Chim Acta. 2012;413:203–206. DOI: 10.1016/j.cca.2011.09.028

[54] Tai SS-C, Bedner M, Phinney KW: Development of a candidate reference measurement procedure for the determination of 25-hydroxyvitamin D3 and 25-hydroxyvitamin D2

in human serum using isotope-dilution liquid chromatography-tandem mass spectrometry. Anal Chem. 2010;82:1942–1948. DOI: 10.1021/ac9026862

[55] Keevil B: Does the presence of 3-epi-25OHD3 affect the routine measurement of vitamin D using liquid chromatography tandem mass spectrometry? Clin Chem Lab Med. 2012;50:181–183. DOI: 10.1515/cclm.2011.755

[56] Stepman HCM, Vanderroost A, Stöckl D, Thienpont LM: Full-scan mass spectral evidence for 3-epi-25-hydroxyvitamin D3 in serum of infants and adults. Clin Chem Lab Med. 2001;49:253–256. DOI: 10.1515/CCLM.2011.050

[57] Lensmeyer G, Poquette M, Wiebe D, Binkley N: The C-3 epimer of 25-hydroxyvitamin D(3) is present in adult serum. J Clin Endocrinol Metab. 2012;97:163–168. DOI: 10.1210/jc.2011-0584

[58] Xu X, Mei H, Wang S, Zhou Q, Wang G, Broske L, Pena A, Korfmacher WA: A study of common discovery dosing formulation components and their potential for causing time-dependent matrix effects in high-performance liquid chromatography tandem mass spectrometry assays. Rapid Commun Mass Spectrom. 2005;19:2643–2650. DOI: 10.1002/rcm.2102

[59] Jessome LL, Volmer DA: Ion suppression: a major concern in mass spectrometry. LC GC N Am. 2006;24:498–510. ISSN: 1527-5949

[60] Adamec J, Jannasch A, Huang J, Hohman E, Fleet JC, Peacock M, Ferruzzi MG, Martin B, Weaver CM: Development and optimization of an LC-MS/MS-based method for simultaneous quantification of vitamin D2, vitamin D3, 25-hydroxyvitamin D2 and 25-hydroxyvitamin D3. J Sep Sci. 2001;34:11–20. DOI: 10.1002/jssc.201000410

[61] Herrmann M, Harwood T, Gaston-Parry O, Kouzios D, Wong T, Lih A, Jimenez M, Janu M, Seibel MJ: A new quantitative LC tandem mass spectrometry assay for serum 25-hydroxy vitamin D. Steroids. 2010;75:1106–1112. DOI: 10.1016/j.steroids.2010.07.006

[62] Ding S, Schoenmakers I, Jones K, Koulman A, Prentice A, Volmer DA: Quantitative determination of vitamin D metabolites in plasma using UHPLC-MS/MS. Anal Bioanal Chem. 2010;398:779–789. DOI: 10.1007/s00216-010-3993-0

[63] Aronov PA, Hall LM, Dettmer K, Stephensen CB, Hammock BD: Metabolic profiling of major vitamin D metabolites using Diels-alder derivatization and ultra-performance liquid chromatography-tandem mass spectrometry. Anal Bioanal Chem. 2008;391:1917–1930. DOI: 10.1007/s00216-008-2095-8

[64] Duan X, Weinstock-Guttman B, Wang H, Bang E, Li J, Ramanathan M, Qu J: Ultrasensitive quantification of serum vitamin D metabolites using selective solid-phase extraction coupled to microflow liquid chromatography and isotope-dilution mass spectrometry. Anal Chem. 2010;82:2488–2497. DOI: 10.1021/ac902869y

[65] Higashi T, Awada D, Shimada K: Determination of 24,25-dihydroxyvitamin D3 in human plasma using liquid chromatography-mass spectrometry after derivatization

with a Cookson-type reagent. Biomed Chromatogr. 2001;15:133–140. DOI: 10.1002/bmc.43

[66] Hedman CJ, Wiebe DA, Dey S, Plath J, Kemnitz JW, Ziegler TE: Development of a sensitive LC-MS/MS method for vitamin D metabolites: 1,25 dihydroxyvitamin D2&3 measurement using a novel derivatization agent. J Chromatogr B Analyt Technol Biomed Life Sci. 2014;953–954:62–67. DOI: 10.1016/j.jchromb.2014.01.045

[67] Qi Y, Geib T, Schorr P, Meier F, Volmer DA: On the isobaric space of 25 hydroxyvitamin D in human serum: potential for interferences in liquid chromatography/tandem mass spectrometry, systematic errors and accuracy issues. Rapid Commun Mass Spectrom. 2015;29:1–9. DOI: 10.1002/rcm.7075

[68] Volmer DA, Mendes LR, Stokes CS: Analysis of vitamin D metabolic markers by mass spectrometry: current techniques, limitations of the "gold standard" method, and anticipated future directions. Mass Spectrom Rev. 2015;34:2–23. DOI: 10.1002/mas.21408

[69] Higashi T, Awada D, Shimada K: Simultaneous determination of 25-hydroxyvitamin D2 and 25-hydroxyvitamin D3 in human plasma by liquid chromatography-tandem mass spectrometry employing derivatization with a Cookson-type reagent. Biol Pharm Bull. 2001;24:738–743. DOI: 10.1248/bpb.24.738

[70] Gören AC, Bilsel G, Bilsel M: Rapid and simultaneous determination of 25 OH vitamin D2 and D3 in human serum by LC/MS/MS: validation and uncertainty assessment. J Chem Metrl. 2007;1:1–9.

[71] Ramanathan R, Jemal M, Ramagiri S, Xia YQ, Humpreys WG, Olah T, Korfmacher WA: It is time for a paradigm shift in drug discovery bioanalysis: from SRM to HRMS. J Mass Spectrom. 2011;46:595–601. DOI: 10.1002/jms.1921

[72] Sleno L: The use of mass defect in modern mass spectrometry. J Mass Spectrom. 2012;47:226–236. DOI: 10.1002/jms.2953

[73] Bruce SJ, Rochat B, Béguin A, Pesse B, Guessous I, Boulat O, Henry H: Analysis and quantification of vitamin D metabolites in serum by ultra-performance liquid chromatography coupled to tandem mass spectrometry and high-resolution mass spectrometry – a method comparison and validation. Rapid Commun Mass Spectrom. 2013;27:200–206. DOI: 10.1002/rcm.6439

[74] Abdel-Khalik J, Crick PJ, Carter GD, Makin HL, Wang Y, Griffiths WJ: Studies on the analysis of 25-hydroxyvitamin D_3 by isotope-dilution liquid chromatography-tandem mass spectrometry using enzyme-assisted derivatisation. Biochem Biophys Res Commun. 2014;446:745–750. DOI: 10.1016/j.bbrc.2014.01.088

[75] Liebisch G, Matysik S: Accurate and reliable quantification of 25-hydroxy-vitamin D species by liquid chromatography high-resolution tandem mass spectrometry. J Lipid Res. 2015;56:1234–1239. DOI: 10.1194/jlr.D058511

[76] Raml R, Ratzer M, Obermayer-Pietsch B, Mautner A, Pieber TR, Sinner FM, Magnes C: Quantifying vitamin D and its metabolites by LC/Orbitrap MS. Anal Methods. 2015;7:8961–8966. DOI: 10.1039/C5AY01583A

[77] Matuszewski BK, Constanzer ML, Chavez-Eng CM: Strategies for the assessment of matrix effect in quantitative bioanalytical methods based on HPLC-MS/MS. Anal Chem. 2003;75:3019–3030. DOI: 10.1021/ac020361s

[78] Vogeser M, Kyriatsoulis A, Huber E, Kobold U: Candidate reference method for the quantification of circulating 25-hydroxyvitamin D3 by liquid chromatography-tandem mass spectrometry. Clin Chem. 2004;50:1415–1417. DOI: 10.1373/clinchem.2004.031831

[79] Maunsell Z, Wright DJ, Rainbow SJ: Routine isotope-dilution liquid chromatography-tandem mass spectrometry assay for simultaneous measurement of the 25-hydroxy metabolites of vitamins D_2 and D_3. Clin Chem. 2005;51:1683–1690. DOI: 10.1373/clinchem.2005.052936

[80] Carter GD, Carter CR, Gunter E, Jones J, Jones G, Makin HLJ, Sufi S: Measurement of vitamin D metabolites: an international perspective on methodology and clinical interpretation. J Steroid Biochem Mol Biol. 2004;89–90:467–471. DOI: 10.1016/j.jsbmb.2004.03.055

[81] Carter GD, Berry JL, Gunter E, Jones G, Jones JC, Makin HLJ, Sufi S, Wheeler MJ: Proficiency testing of 25-hydroxyvitamin D (25-OHD) assays. J Steroid Biochem Mol Biol. 2010;121:176–179. DOI: 10.1016/j.jsbmb.2010.03.033

[82] Müller MJ, Volmer DA: Mass spectrometric profiling of vitamin D metabolites beyond 25-hydroxyvitamin D. Clin Chem. 2015;61:1033–1048. DOI: 10.1373/clinchem.2015.241430

[83] Müller MJ, Stokes CS, Lammert F, Volmer DA: Chemotyping the distribution of vitamin D metabolites in human serum. Sci Rep. 2016;6:21080. DOI: 10.1038/srep21080

[84] Hollick MF: Vitamin D status: measurement, interpretation, and clinical application. Ann Epidemiol. 2009;19:73–78. E-ISSN: 1873-2585

[85] Hollick MF: The D-lemma: to screen or not to screen for 25-hydroxyvitamin D concentrations. Clin Chem. 2010;56:729–731. DOI: 10.1373/clinchem.2009.139253

[86] van Leeuwen JP, van den Bemd GJ, van Driel M, Buurman CJ, Pols HA: 24,25-dihydroxyvitamin D3 and bone metabolism. Steroids. 2011;66:375–80. DOI: 10.1016/S0039-128X(00)00155-0

[87] Bosworth CR, Levin G, Robinson-Cohen C, Hoofnagle AN, Ruzinski J, Young B, Schwartz SM, Himmelfarb J, Kestenbaum B, de Boer IH: The serum 24,25-dihydroxyvitamin D concentration, a marker of vitamin D catabolism, is reduced in chronic kidney disease. Kidney Int. 2012;82:693–700. DOI: 10.1038/ki.2012.193

[88] Berg AH, Powe CE, Evans MK, Wenger J, Ortiz G, Zonderman AB, Suntharalingam P, Lucchesi K, Powe NR, Karumanchi SA, Thadhani RI: 24,25-dihydroxyvitamin D3 and

vitamin D status of community-dwelling black and white americans. Clin Chem. 2015;61:877–884. DOI: 10.1373/clinchem.2015.240051

[89] Wagner D, Hanwell HE, Schnabl K, Yazdanpanah M, Kimball S, Fu L, Sidhom G, Rousseau D, Cole DE, Vieth R: The ratio of serum 24,25-dihydroxyvitamin D3 to 25-hydroxyvitamin D3 is predictive of 25-hydroxyvitamin D3 response to vitamin D3 supplementation. J Steroid Biochem Mol Biol. 2011;126:72–77. DOI: 10.1016/j.jsbmb.2011.05.003

[90] Binkley N, Lappe J, Singh RJ, Khosla S, Krueger D, Drezner MK, Blank RD: Can vitamin D metabolite measurements facilitate a "treat-to-target" paradigm to guide vitamin D supplementation? Osteoporos Int. 2015;26:1655–1660. DOI: 10.1007/s00198-014-3010-0

[91] Stubbs JR, Zhang S, Friedman PA, Nolin TD: Decreased conversion of 25-hydroxyvitamin D3 to 24,25-dihydroxyvitamin D3 following cholecalciferol therapy in patients with CKD. Clin J Am Soc Nephrol. 2014;9:1965–1973. DOI: 10.2215/CJN.03130314

Chapter 6

Vitamin D and Sphingolipids: Role in Bone and Neural System

Alessia Frati, Mercedes Garcia Gil,
Federica Pierucci and Elisabetta Meacci

Additional information is available at the end of the chapter

http://dx.doi.org/10.5772/66648

Abstract

1-Alpha,25-dihydroxyvitamin D3 (1,25(OH)$_2$D$_3$) is known to play an important physiological role on growth and differentiation in a variety of nonmalignant and malignant cell types through classical actions, mediated by its specific receptor (VDR), and nongenomic actions resulting in the activation of specific signalling pathways. Due to the broad distribution of Vitamin D Receptor (VDR) in many tissues and the ability of 1,25(OH)$_2$D$_3$ to regulate fundamental processes, such as cell proliferation and differentiation, this steroid hormone has been suggested in the treatment of different diseases, from cancer to neurodegenerative diseases. In fact, structural 1,25(OH)$_2$D$_3$ analogues, with weaker collateral effects, have recently entered in clinical trials. Other interesting molecules due to their pleiotropic actions are the bioactive sphingolipids (SLs), in particular ceramide (Cer) and sphingosine 1-phosphate (S1P). Cells maintain a dynamic balance of these metabolites since Cer and sphingoid bases mediate cell death, while S1P exerts mitogenic effects and promotes differentiation of several cell types including osteogenic and neural cells. The biological actions of 1,25(OH)$_2$D$_3$ and SLs, in particular S1P, share many common effectors, including calcium regulation, growth factor expression, inflammatory cytokines, etc., but whether they could act synergistically is still unknown and deserves further investigation.

Keywords: vitamin D, sphingosine 1-phosphate, ceramide, neurodegeneration, bone

1. Introduction

1-Alpha,25-dihydroxyvitamin D3 (1,25(OH)$_2$D$_3$), a known regulator of calcium and phosphorus homeostasis, has also important physiological effects on growth and differentiation in a variety of nonmalignant and malignant cell types [1–4] and a central role in host defense against infections [5]. The classical actions of 1,25$(OH)_2D_3$ start with the hormone binding to the 1,25$(OH)_2D_3$ receptor/retinoic X receptor (VDR/RXR) heterodimeric complex to specific DNA sequences, whereas the rapid nongenomic actions result in the activation of specific signal transduction pathways [6, 7]. The broad distribution of VDR in the human body and the ability of 1,25$(OH)_2D_3$ to control cell growth and differentiation make this hormone a potentially useful agent in the treatment of diseases, including cancer and neurodegenerative diseases. However, the systemic application of 1,25$(OH)_2D_3$ is limited because of its hypercalcemic side effects [8]. Therefore, 1,25$(OH)_2D_3$ analogues, with potent cell regulatory effects, but with weaker effects on calcium metabolism, have recently been obtained and some of them have entered in clinical trials [9, 10]. Sphingolipids (SLs) constitute a biologically active lipid class that is significantly important from both structural and regulatory aspects [11–14]. Indeed, they regulate fundamental cellular processes that are important in determining cellular fate, such as proliferation, apoptosis senescence, and inflammation [15–19]. Cells maintain a dynamic balance of distinct SL metabolites [20, 21], with ceramide (Cer) and sphingoid bases acting in opposite manner with respect to sphingosine 1-phosphate (S1P) that exerts mitogenic effects and promote differentiation of several cell types including skeletal muscle cells [22, 23] and neural cells [24–27]. The manipulation of SL metabolism is currently being studied as a novel strategy to regulate cell proliferation/inflammation [18, 19, 28, 29]. Interestingly, S1P can be released from many cell types including neuronal cells and osteoblasts, thus, acting as ligand of specific S1P receptors, triggers paracrine and autocrine signalling [13, 19, 30]. In the present chapter, we review the potential contribution of the biological effects of sphingolipids, 1,25$(OH)_2D_3$ and its structural analogues in bone and neural disorders.

2. Vitamin D in bone physiology, osteogenesis, and osteoporosis

2.1. Bone remodelling

Bone remodelling consists in the balance between the elimination of bone due to osteoclast death and the formation of new bone by osteoblast proliferation [31]. The migration of mesenchimal stem cells (MSCs) to areas of new bone formation is also a fundamental process for skeleton maintenance. In fact, in these functional locations, MSCs, under the influence of bone morphogenetic protein, differentiate into osteoblasts. The majority of osteoblasts become osteocytes, the fully differentiated cells, within the bone matrix, where they also help in tissue repair [32]. Osteoclasts are multinucleated cells, abundant in mitochondria, vacuoles, and lysosomes [33, 34]. These cells derive from the fusion of preosteoclasts through a mechanism that is regulated by the dendritic cell-specific transmembrane protein (DC-STAMP) and the osteoclast-stimulatory transmembrane protein (OC-STAMP) [35, 36]. Osteoclasts are characterized by a ruffled border in contact with the bone surface, where the vacuolar H^+-ATPase, responsible of the maintenance of acid pH that favors the dissolution of the bone minerals, is localized. The zone beneath the ruffled borders is called resorption lacunae and is isolated from the surrounding by the sealing zone of osteoclasts that attaches the cells to the bone surface. Cathepsin K, MMP9, and tartrate resistant-acid phosphatase (TRAP), the main enzymes responsible for the degradation of bone matrix, are released in the resorption lacunae. Matrix degradation products are endocytosed from the central portion of the ruffled border, packaged into transcytotic vesicles and secreted from the functional secretory domain [37]. The bone matrix represents an important storage of factors secreted by the osteoblasts during bone formation. Among them are the transforming growth factor β (TGFβ), the bone morphogenetic proteins (BMPs), fibroblast growth factors (FGFs), etc. They are stored in the bone matrix and serve as osteoblast-promoting components once liberated and activated by the osteoclasts.

In 1981, Rodan and Martin [38] observed that osteoblasts, but not osteoclasts, express the receptors of bone-resorbing factors, such as Parathyroid hormone (PTH) and prostaglandin E2, and proposed that osteoblasts may interfere in the process of osteoclastic resorption. Experimental evidence of the role of the microenvironment provided by osteoblastics for cell differentiation of splenic precursors, confirms this hypothesis [39]. In response to bone-resorption stimulating factors, osteoblasts produce Receptor Activator of Nuclear Factor kappa-B Ligand (RANKL), a membrane associate factor that, by binding to its receptor constitutively expressed on the surface of osteoclast precursors, stimulates cell differentiation and the activation of bone resorption. Another potential physiological regulator of bone mass is the prolin/arginine-rich end leucine-rich repeat protein (PRELP), a heparin/heparan sulfate-binding protein expressed in developing bone, cartilage, and basement membranes. PRELP inhibits osteoclast formation with a mechanism that affects the RANKL-dependent late stage of osteoclastogenesis, and its administration is reported to reduce bone loss in ovariectomized and tumor-bearing mice [40].

2.2. Effect of vitamin D on calcium homeostasis and bone remodeling

$1,25(OH)_2D_3$ was historically discovered as an anti-rachitic agent due to its effects on the demineralization process carried out by osteoclasts and intestinal calcium absorption [41]. The hormonally active form is the dihydroxylated metabolite 1,25-dihydroxyvitamin D3, or $1,25(OH)_2D_3$. It is generated by two enzymatic hydroxylation reactions, which occur first in the liver to produce 25-hydroxyvitamin D3 and second in the kidney, where both $1,25(OH)_2D_3$ and its sister metabolite 24,25-dihydroxyvitamin D3 ($24,25[OH]_2D_3$) are produced. Thus, most tissues have the ability to convert $1,25(OH)_2D_3$ into its active form, which, in turn, will bind to hormone nuclear receptor (VDR) [42]. The presence of the VDR in many tissues that are not involved in mineral metabolism indicates a wider physiological role for $1,25(OH)_2D_3$ able to positively or negatively influence target gene expression via binding of the hormone/VDR complex to specific receptor response elements, i.e., RANKL gene [43]. $1,25(OH)_2D_3$ ability to maintain serum calcium homeostasis is due to VDR-mediated signaling affecting bone physiology and intestinal calcium absorption.

In the intestine, $1,25(OH)_2D_3$-mediated processes can occur by two different mechanisms: (a) the *paracellular pathway* predominates when dietary levels of calcium are high, and it is a passive, nonsaturable diffusion process [44], calcium is transported through tight junctions, and involves several claudins proteins upregulated by $1,25(OH)_2D$ [45]; (b) the *transcellular pathway* occurs in the jejunum and duodenum in circumstances of low calcium dietary intake. It requires an active transport of calcium through the thickness of the enterocytes, and it is regulated by the subsequent involvement of transient receptor potential vanilloid

6 (TRPV6), calcium-binding protein calbindin D_{9k} (CaBP-9K), and a calcium ATPase. The transcription of these proteins is stimulated by the $1,25(OH)_2D$ signal [46]. When intestinal calcium absorption decreases, the transcellular transport in the kidney, a process similar to the active transport through the enterocytes, increases calcium reabsorption. This pathway is also stimulated by $1,25(OH)_2D_3$ [44].

In addition, in hypocalcemia conditions, parathyroid glands release PTH which, in the kidney, stimulate the production of the active form of $1,25(OH)_2D_3$. Whenever intestinal and renal calcium fluxes are insufficient to maintain the correct calcium levels in the serum, the bone represents an additional pool of calcium and the bone-resorbing process permits to adjust serum calcium concentration.

$1,25(OH)_2D_3$ activates multiple signaling pathways in bone precursor cells [42]—*nongenomic pathways* that involve: (a) the activation of voltage-sensitive calcium channels located at plasma membrane; (b) the release of calcium from intracellular stores; (c) a shift in the charge state of the matrix protein osteopontin (OPN)—*a classical nuclear receptor-mediated event* that leads to the upregulation of the OPN gene at 48 h after hormone addition. Nanomolar concentration of $1,25(OH)_2D_3$ is necessary for the nonclassical effects, while physiological $1,25(OH)_2D_3$ serum concentrations are necessary in the picomolar range [42].

Calcium balance and cell differentiation stage affect VDR action in osteogenic cells. In fact during a positive calcium balance, $1,25(OH)_2D_3$ signaling acts in a different way in osteoblasts at diverse differentiation stage: in immature osteoblasts, the hormone causes an increase in RANKL expression leading to a catabolic function on bone mass [47], whereas in mature osteoblasts, VDR stimulation produces a decrease in RANKL expression and an increase in the production of the osteoclastogenic inhibitor osteoprotegerin (OPG), which inhibits RANKL-RANK interaction [48].

During a negative calcium balance, the high RANKL/OPG ratio and the increased levels of mineralization inhibitors, such as OPN, are the two events that permit the mobilization of calcium from the bone to the serum in response to $1,25(OH)_2D_3$ and PTH signaling. By increasing calcium absorption, hormone signaling indirectly preserves bone mass mineralization. In osteocytes, VDR signaling upregulates the transcription of the bone-derived fibroblast growth factor-23 (FGF23) that stimulates TRPV5 expression, thus, increasing renal phosphate excretion [49]. Notably, FGF23 also decreases renal CYP27B1 activity, the enzyme that catalyzes the hydroxylation of $25(OH)_2D_3$ to $1,25(OH)_2D_3$ [42], avoiding overstimulation of the hormone pathway.

Osteoporosis is a systemic skeletal disease characterized by a reduction of bone mass and microarchitectural deterioration that leads to increase in bone fragility and susceptibility to fracture [50]. Osteoporosis results from the imbalance between bone resorption and bone formation, and $1,25(OH)_2D_3$ can regulate both aspects of bone turnover. Nearly in all studies, the treatment with $1,25(OH)_2D_3$ or its precursor, $1\alpha(OH)D_3$, is found to increase bone mineral density [51, 52]. More recently, the combination of $1\alpha(OH)D_3$ with an anti-resorptive *bis*-phosphonate (alendronate) enhanced bone mass with fewer falls and fractures [53].

2.3. Vitamin D deficiency and supraphysiological dose

The lack of 1,25(OH)$_2$D$_3$ has obvious implications for human diseases. Severe 1,25(OH)$_2$D$_3$ deficiency causes nutritional rickets in children [54] that can easily be prevented and cured by 1,25(OH)$_2$D$_3$ supplementation [55]. Moreover, deficiency of 1,25(OH)$_2$D$_3$ or reduced level of VDR can lead to osteomalacia disease and the recovery of a normal hormone level can resolve the disease [56]. Less severe 1,25(OH)$_2$D$_3$ deficiency produces an increased bone turnover and an accelerated bone loss and it is associated with osteoporosis. A correlation between the sun exposure and bone physiology in regions insufficiently exposed to sunlight may involve 1,25(OH)$_2$D$_3$ deficiency [57]. However, since 1,25(OH)$_2$D$_3$ insufficiency is widespread, it is difficult to provide evidence regarding the direct role of sunlight on osteogenesis and osteoporosis without taking into account many other variables, such as personal bone physiology, age, weight, clothing habits, medication, and others.

Supraphysiological doses of the 1,25(OH)$_2$D$_3$ induce calcemic side effects. In order to preserve or augment the beneficial effects of the hormone and to minimize its collateral consequences, structural analogues of 1,25(OH)$_2$D$_3$ have been synthesized by introducing chemical modifications in the A-ring, central CD-ring region, or side chain of the hormone [58].

Some of these analogues have tissue-specific actions, exert prodifferentiating and antiproliferative effects on keratinocytes, and also possess important anti-inflammatory properties. The recently approved eldecalcitol (1a,25[OH]$_2$-2b-(3-hydroxypropyloxy)vitamin D3; ED-71; Edirol®) is an orally administered analogue of calcitriol that binds to VDR [59]. Edirol® is available for the treatment of osteoporosis [60]. The effects of this compound on bone metabolism have been reported in a randomized, open-label study in postmenopausal women, in which reductions in the markers of bone reabsorption were observed [61]. Similarly, in a randomized, noncomparative study of patients with osteoporosis, the structural analogue suppresses the biochemical markers of bone turnover in a dose-dependent manner.

Another 1,25(OH)$_2$D$_3$ analogue with some effects on bone is ZK191784, a compound characterized by 22,23-double bond, 24R-hydroxy group, 25-cyclopropyl ring, and 5-butyloxazole group [10]. It has been shown to exert therapeutic potential in T cell-mediated immune disorders and to significantly counteract acute and chronic intestinal inflammation [62]. While the 1,25(OH)$_2$D$_3$ agonist effect of ZK191784 in kidney and antagonistic effect in intestine were clear, its effect on bone is still Ongoing: preliminay analyses appeared to suggest a tendency toward restoration of the reduced bone thickness in mice lacking the renal epithelial calcium channel TRPV (Trpv5$^{-/-}$) [37].

Another 1,25(OH)$_2$D$_3$ analogue, Seocalcitol, is able to reduce the number and growth of metastasis originating from various types of cancer cells, such as bone metastasis originating from intracardially injected breast cancer cells [63]. Since 1,25(OH)$_2$D$_3$ and its analogues possess cytostatic properties, many in vivo studies have focused on hormone analog cancer treatment combined with radiotherapy and/or chemotherapy. However, while the combination of Seocalcitol, with radiotherapy in a xenograft model for breast cancer, lead to more effective anti-cancer effects [64, 65], the combination of the analogues with chemotherapy does not always result in additive or synergistic effects. In view of the promising results that certain hormone analogs show against cancer in vitro and in vivo animal models, some of them have been tested in cancer patients, such as Seocalcitol, which however has given rather insufficient results in clinical trials.

2.4. Animal model for bone remodeling study

Several animal models have been developed and used to understand the pathogenesis of osteoporosis and osteogenesis for the preclinical testing of new treatment options [66]. Only few of them were used in combination with $1{,}25(OH)_2D_3$. treatment.

The senescence accelerated mouse (SAM/P6) is a mouse model for severe osteoporosis that has low level of bone mass and develops fractures in old age [67]. It is a unique model for the study of age-related osteopenia and severe osteoporosis mimicking many aspects of the age-related changes seen in bones of humans and offering the opportunity to study relevant genes that contribute to this process.

Other studies have been performed in ovariectomy (OVX) on rats for either osteoporotic induction or fracture healing. Notably, an additive effect on bone loss was reported by combined OVX-deficient calcium or OVX-deficient $1{,}25(OH)_2D_3$. In addition, the rat model subjected to ovariectomy and multideficiency diet (depletion of $1{,}25(OH)_2D_3$, calcium, vitamin K, and phosphorus), and, thus, characterized by increased bone turnover could contribute to the study of bone- and energy metabolism in early and late stages of osteoporosis. Several ovariectomized large animals might be also used as models of osteoporosis, such as the dog, the pig, the sheep, and the nonhuman primates [68]. In particular, the sheep is also well established as a model for human bone loss/osteoporosis in orthopedic research [69]. Some limits can be the largeness in size and the difficulty to manage relatively expensive experiments.

Mouse can be a reliable animal model of glucocorticoid-induced osteopenia/osteoporosis and mimic the changes seen in humans [70]. Mice receiving glucocorticoid for a week showed an early increase in bone resorption, decreased bone mineral density, and bone mass.

Transgenic mice showing bone alterations have been also developed. Klotho mouse is a transgenic mouse model obtained by an insertion mutation that disrupts the Klotho gene locus. Klotho is a gene encoding a transmembrane protein that forms a complex with multiple fibroblast growth factor receptors and functions as coreceptor for FGF23, an osteocyte-derived hormone that induces negative phosphate balance. Defects in either Klotho or FGF23 gene expression result in osteopenia [71]. In these animals, similarly to what occurs in human senile osteoporosis, the reduction in bone formation occurs faster than bone resorption. TghuRANKL (Tg5519) is a transgenic mouse overexpressing human RANKL resulting in the spontaneous development of osteoporosis similar to human pathology [71, 72]. The overexpression of huRANKL results in the spontaneous development of early onset osteoporosis characterized by lack of trabecular bone, increased osteoclastogenesis, increased bone remodeling, and decreased bone strength. Another model, already mentioned in this paragraph is $TRPV5^{-/-}$ mice that display hypercalciuria due to a primary renal failure to reabsorb calcium and hypervitaminosis D, leading to calcium hyperabsorption in the intestine and reduced bone mass [37, 73]. Other models are useful to evaluate osteocyte functions. They have been obtained, for example, by osteocyte-specific disruption of gap junction protein, *Gja1*. Osteocyte-specific *Gja1* conditional knockout mice show an increase in apoptosis of osteocytes [74].

Animal models for disused osteoporosis were also developed. Methods to reduce skeletal biomechanical loading include nerve, spinal cord, or tendon resections, casting, bandaging of one limb or suspension of both hindlimbs in rats. Immobilization-induced osteopenia/osteoporosis in rat skeletal model with the highly predictable pattern of bone loss and the hormone in plasma is significantly decreased [75].

Regarding osteogenesis, most of the studies have focused on osteogenesis imperfecta (OI), an autosomal dominant disorder caused by mutations in type I collagen, the most abundant protein of bone, skin, and tendon extracellular matrices. OI is characterized by increased bone fragility and low bone mass. Tansgenic mice expressing a premature stop codon or glycine substitution in the COL1A1 gene encoding the chains of type I collagen are good model for OI [76]. Very recently, the effect of high dose of hormones on bone density in OI patients has been reported [77]. Notably, new genes implicated in autosomal recessive forms of OI have been described and one of each is sphingomyelin phosphodiesterase (SMPD3), the gene encoding for neutral sphingomyelinase [78].

Some conflicting results are reported in literature, but the animal models developed so far have given valuable information on the pathogenesis of osteoporosis as well as on other pathological conditions of the skeleton and bone. The development of new animal models will help to better understand what have been poorly investigated in the past.

3. Vitamin D in nervous system physiology, neuroprotection, and neurogenesis

VDR is expressed in both neurons and glial cells (i.e., microglia, astrocytes, and oligodendrocytes) in different regions of the nervous system [3]. Vitamin D Response Element (VRE) response elements modulate gene expression. For example, it increases the expression genes codifying growth factors such as Nerve growth factor (NGF), glial cell line–derived neurotrophic factor (GDNF), NT3, and enzymes involved in the synthesis of neurotransmitters (tyrosine hydroxylase, tryptophan hydroxylase 2, and glutamate decarboxylase), whereas it decreases expression of voltage-dependent calcium channel [3, 79]. VDR is also expressed in the caveolae and induces nongenomic effects that include activation of cAMP-dependent protein kinase (PKA), Ca^{2+}/calmodulin-dependent protein kinase, phosphatidylinositol 3-kinase, and mitogen-activated protein kinase p38 leading to phosphorylation of neurofilaments, modulation of chloride, potassium, and voltage-dependent calcium channel in rat cortical neurons [80]. In addition, 25-hydroxylase and 1-a-hydroxylase activity are also found in the nervous tissue.

The combination of in vitro, ex vivo, and animal model data provides compelling evidence that $1,25(OH)_2D_3$ has a crucial role in neuronal proliferation, differentiation, neurotransmission, neuroplasticity, and neuroprotection. Increasing evidence derived from studies of $1,25(OH)_2D_3$ deficiency and from VDR polymorphisms implicates $1,25(OH)_2D_3$ as a candidate in influencing susceptibility to a number of psychiatric and neurological diseases, such as schizophrenia, autism, Parkinson disease (PD), amyotrophic lateral sclerosis, epilepsy, Alzheimer disease (AD), and is especially strong for multiple sclerosis (MS) [8, 81, 82]. In epigenetic studies, maternal dietary deprivation of $1,25(OH)_2D_3$ has induced vitamin D deficiency (VDD) in rats prior to mating and maintained it during pregnancy. The $1,25(OH)_2D_3$-deficient rats showed modifications in brain morphology including increased overall brain size and larger lateral ventricles. Interestingly, some changes persist despite the addition of $1,25(OH)_2D_3$ to the diet of the pups. In adult life, these rats tend to demonstrate subtle alterations in learning and memory and impaired attentional processing [81]. Prenatal VDD induces similar alterations in fetal mouse brain morphology and mouse behavior [83, 84]. Notably, it has been reported that maternal $1,25(OH)_2D_3$ insufficiency during pregnancy in humans is also significantly associated with offspring's language impairment [85]. Interestingly, prenatal $1,25(OH)_2D_3$-depleted rats showed a significant impairment of latent inhibition, a feature often associated with schizophrenia [81].

The neuroprotective effect of $1,25(OH)_2D_3$ has been recently reported in cognitive decline of aging rats [86], and it has been extensively studied in the animal model of MS and the experimental allergic encephalomyelitis (EAE). The hormone prevents onset and reversibly blocks progression of clinical signs, but such a protective effect is absent in VDR knockout mice [81]. The effect of $1,25(OH)_2D_3$ might not be due exclusively to its neuroimmunomodulatory properties [81] since recently it has been reported that the hormone enhances neural stem cell proliferation and differentiation into neurons and oligodendrocytes, the myelinating cells of central nervous system [4, 87]. Neural stem cells constitutively express VDR, which can be upregulated by $1,25(OH)_2D_3$ [4].

$1,25(OH)_2D_3$ regulates the expression of many AD-related genes. It attenuates Aβ peptide accumulation by stimulating phagocytosis of Aβ peptide probably by modulating transcription of Toll-like receptors and cytokines together with enhancing brain-to-blood efflux transport by increasing P-glycoprotein expression [88].

Adult neurogenesis is limited to specific brain regions in the mammalian brain, such as the hippocampal dentate gyrus and the subventricular zone. Alterations in adult neurogenesis appear to be a common hallmark in different neurodegenerative diseases including PD and AD [89]. Therefore, factors that stimulate neurogenesis have been indicated as possible treatments of neurodegenerative disorders. The antiproliferative and prodifferentiating effects of $1,25(OH)_2D_3$ in neural cells were described more than 10 years ago [90, 91]. $1,25(OH)_2D_3$ decreases the expression of G1/S and G2/M cellular gatekeeper components, such as cyclins D1 and B1 [91, 92] and decreases the percentage of cultured hippocampal cells undergoing mitosis in conjunction with increases in both neurite outgrowth and NGF production [90]. Recently, ceramide kinase signalling pathway has been involved in the antiproliferative action of $1,25(OH)_2D_3$ human neuroblastoma cells [18]. Accumulated evidence indicates that $1,25(OH)_2D_3$ has complex effects on neurogenesis of neural stem cells. Cui et al. [93] have investigated the effect of Developmental vitamin D (DVD) deficiency on neuroprogenitor formation in the neonatal brain, and they have shown an increase in the number of neurospheres formed in cultures from the neonatal subventricular zone. Exogenous $1,25(OH)_2D_3$ added to the culture medium reduced neurosphere number in control (in agreement with the putative antiproliferative effect of 1,25$(OH)_2D_3$), but not in cultures from the deprived pups [93]. In contrast, neurogenesis in adult subgranular zone of the hippocampus is decreased [94]. In another model of 1,25$(OH)_2D_3$ deficiency, Zhu et al. [95] have reported increased proliferation, but decreased survival of newborn neurons in the dentate gyrus of adult mice lacking 1,25$(OH)_2D_3$, the 1α-hydroxylase knockout mice [95]. The different effects probably depend on the time window of exposition and/or the different sensibility to the hormone of distinct neurogenic niches.

The neuroprotective effect of high intake of 1,25$(OH)_2D_3$ has been confirmed in some AD trials, but not in others [88]. It is not clear whether hypovitaminosis D triggers AD or it removes protection in the Central nervous system (CNS) against AD. However, the combination of antineurodegenerative drugs with 1,25$(OH)_2D_3$ supplementation might be useful. Indeed, the supplementation of the combination nemantidine plus 1,25$(OH)_2D_3$ has been shown to prevent cognitive decline more efficiently than that of the single compounds [96].

4. Sphingolipids

Sphingolipids (SLs) have long been regarded as inactive and stable structural components of the membrane, but they are biologically active molecules [11–14, 17, 19]. They are formed by the attachment of different polar headgroups at the primary alcohol group of a ceramide molecule. As reported in **Figure 1A** and **B**, there are two main pathways of Cer production: (1) *de novo* biosynthesis and (2) hydrolysis of sphingolipids, such as Sphingomielin (SM). Which of these pathways dominates depends on the cell type, stimulus, and developmental stage of the cell [97].

SLs are the main components of specific membrane platforms that function in membrane signaling and trafficking, named lipid rafts [98, 99]. The most studied SLs are the bioactive lipid S1P and Cer [14], even if in the last few years relevant role has been demonstrated also for ceramide 1-phosphate (C1P) [18, 99, 100] and Cer species [101].

S1P plays a crucial physiological function. In fact, it is present at different concentrations in plasma as well in tissues, constituting a gradient that drives the trafficking of various immune cells [102]. The S1P gradient is due to the constitutive activity of Sphingosine-1-phosphate lyase (SPL) in the cells, leading to a low concentration of S1P in most tissues (0.5–75 pmol/mg) and low-micromolar range in plasma (0.2–0.9 μM) where S1P, produced and released by platelets and erythrocytes, is complexed with albumin and lipoproteins, particularly High-density lipoprotein (HDL) [103]. Of note, it has been proposed that cell fate is regulated by the ratio between S1P and Cer/Sphingosine (Sph) described the first time by Cuvillier et al. [20]. (**Figure 1C**) S1P enhances cell growth and survival, whereas its precursors, Cer and Sph, are generally associated with cell growth arrest and death. However, recent results indicate that Cer species (short or long chain Cer) have different functions [104].

S1P has a peculiar mechanism of action. In fact, S1P, produced inside the cell, can act as intracellular mediator or can be exported outside by the putative transporter Spinster 2, spns2 [105, 106]. Outside the cells, S1P can act as ligand of specific G-protein-coupled receptors. Five

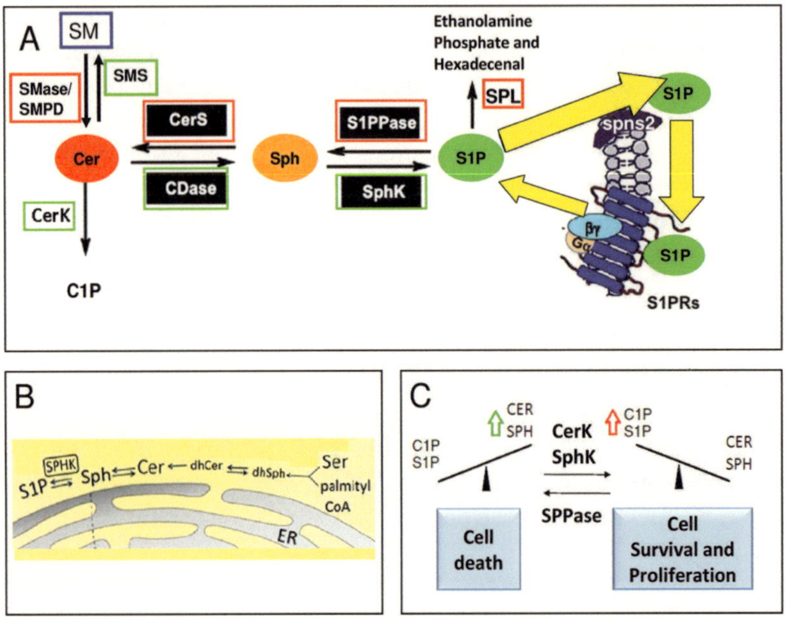

Figure 1. Schematic representation of sphingolipid metabolism pathways and rheostat model. (A) Cers are produced from Sphyngomyelin (SM) by activation of sphingomyelinase (SMase/SMPD) that produces Cer and phosphocholine. The reverse reaction is catalyzed by SM synthase (SMS). Ceramides can also be phosphorylated by ceramide kinase (CerK) to form ceramide-1-phosphate (C1P). S1P is derived by Cer deacylation by ceramidases (CDase) to Sph followed by its phosphorylation by Sph kinases (SphK). The degradation of S1P requires S1P phosphatases activity that produces Sph, and by S1P lyase (Spl) that produces hexadecenal and phosphoethanolamine. (B) *De novo* synthesis of SLs. The *de novo* synthesis is initiated at the cytosolic membranes of the endoplasmatic reticulum and in mitochondria, via the condensation of l-serine with palmitoyl-CoA to form 3-keto-dihydrospingosine, and it is catalyzed by serine palmitoyltransferase (SPT). The product of SPT is reduced by a hydroxyl by 3-keto-dihydrosphingosine reductase in a NADPH-dependent manner producing dihydrosphingosine that is subsequently N-acylated to dihydroceramide (dhCer) by ceramide synthases (CerS). In mammals, CerS is encoded by six distinct genes, and each enzyme has a distinct, but overlapping acyl CoA preference. Dihydroceramide desaturase converts the DHSph backbone within Cer into Sph. (C) Sphingolipids rheostat model.

S1P specific named EDG/S1PR$_{1-5}$ have been described to mediate S1P signaling [11, 13, 14, 19]. S1P1, S1P2, and S1P3 show broad tissue gene expression, while S1P4 shows gene expression primarily in immune system cells, and S1P5 is primarily expressed in the spleen (natural killer cells and other lymphocytes) and central nervous system [107]. Regarding the intracellular S1P effects, the bioactive lipid induces calcium release from the ER, alters the function of intracellular proteins, such as E3 ligase activity of TNF receptor associated factor 2 (TRAF2), binds the mitochondrial protein, and regulates mitochondrial assembly and function. When produced in the nuclei, S1P modulates gene expression inhibiting histone deacetylases [107].

4.1. Sphingolipids in bone physiology

Various SLs play a crucial role in the development of normal skeletogenesis acting in three different skeletal cell types: chondrocytes in cartilage and osteoblasts and osteoclasts in bone. Abnormal tissue development of genetically modified animal models, such as mouse lacking

http://dx.doi.org/10.5772/66648

sphingomyelin phosphatase 3 (SMP3) also named sphingomyelinase 2 (SMase2), have been reported [108]. As described in the first part of this chapter, SMase are a family of different isoenzymes. Among them, the nSMSase2/SMPD3 is largely expressed in bone and cartilage, and it is involved in Cer-mediated signaling events [109]. In other studies, the role of SLs in bone have been performed by using synthetic analogues of ceramide (i.e., C2-Cer), specific inhibitors of the rate limiting enzyme (i.e., SphK), and agonists and antagonists that mimic or inhibit, respectively, the function of S1P, as ligand, of specific receptor subtypes. In this paragraph, we examine the involvement of Sph/S1P and Cer/C1P axis in the remodelling and physiology of bone.

4.1.1. S1P—sphingosine

Several studies demonstrate that osteoblasts at different stages of differentiation may respond differently to SLs [110]. S1P has been shown to promote the proliferation of rat primary chondrocytes, whereas in preosteoblast MC3T3-E1, it negatively regulates the synthesis of the osteoblast marker osteocalcin [111, 112]. Notably, S1P can be released by osteoclasts [113] and secreted S1P can promote bone formation by enhancing the differentiation of osteoblast precursors to functional osteoblasts and by the recruitment of MSCs leading to high bone mass phenotype [114]. Recently, Keller et al. [115] report that the release of S1P by osteoclasts is regulated by calcitonin. S1P induces the upregulation of osteoptotegerin and osteoblast differentiation markers in two osteoblast-like cell lines promoting the translocation of β-catenin [116].

In the same cell type, Sph and S1P induce intracellular calcium release [117, 118] and in rat osteoblasts and in human osteosarcome cells; both sphingoids prevent the apoptotic process elicited by serum deprivation [119]. A functional cross talk exists between S1P and platelet-derived growth factor (PDGF) signaling. In fact, S1P limits, whereas PDGF promotes the migration of preosteoblasts, and it is the balance between these two bioactive molecules that allow only the differentiated osteoblasts to reach the site of bone formation. Very recently, it has been confirmed that in humans the detrimental effects of S1P on bone metabolism depend on the S1P gradient between blood and bone marrow cavity and on S1P receptor subtypes, mainly S1P1 and S1P2, play a crucial role in this control [112, 120]. In particular, S1PR1 exerts positive chemotaxis action on an S1P gradient, whereas S1PR2 counteracts this positive effect [121]. The deletion of S1PR2 led to moderate osteopetrosis by affecting the homing of osteoclast precursors into bone [112, 120]. Higher circulating S1P levels are associated with lower bone mineral density, higher levels of bone resorption markers, and higher prevalence and severity of osteoporotic vertebral fracture in Koreans [122]. Therefore, S1P appears to be an important osteoclast-derived anabolic factor that couples bone resorption to bone formation, but it preferentially influences bone resorption rather than bone formation in humans.

Notably, increase in the secretion of BMP-2, OPN, and osteocalcin as well as highest extracellular matrix mineralization and osteonodules formation were observed when MSCs are cultured on thin titania dioxide coatings (TiO_2) on stainless steel substrate doped with S1P [TiO_2/S1P(CII_m)] [123]. Therefore, S1P, also in an appropriate combination with other sphingoid, such as C1P, may find wide application in regenerative medicine, particularly in bone regeneration with the use of MSCs [123].

4.1.2. C1P—ceramide

Abnormal cartilage development and bone mineralization defects are observed in animal models in which the enzyme inactivation is generated by gene targeting (Smpd3−/− model) or chemically induced deletion in the Smpd3 locus (fro/fro model) [124, 125]. Several evidence underlines the importance of phosphocholine in bone mineralization [126]; however, at present no experimental data directly links Cer or choline to Extracellular matrix (ECM) mineralization. Studies in animal models are in some way in conflict with studies performed in cell culture: the inhibition of nSMase2/SMPD3 by GW4869 accelerates the mineralization of chondrogenic ATDC5 cultures [127], whereas either the fro/fro mice or mice lacking choline kinase presents an expansion of hypertrophic zone likely due to a delay in apoptosis of hypertrophic chondrocytes [128]. The role of Cer in bone tissue has been examined by other approaches leading also to conflicting results, such as short-chain synthetic cell-permeable forms of Cer, preferentially C-2 Cer [129]. In fact, it is not clear whether exogenous Cer has the same cellular targets as the endogenous species and the dose used in cell culture may not relate to the endogenous levels found under physiological or pathological conditions. Therefore, it is reported that Cer may be able to promote cell apoptosis or cell survival, and it may depend on the dose and cell type used. The treatment of cells with high-dose of C-2 Cer (1–10 μM) leads to apoptosis of mouse primary osteoblasts. Contrary, the low-dose treatment (10 time less or more) of C-2 Cer promotes an antiapoptotic effect in other cells [130]. Regarding cell types, C-2 Cer treatment leads to increased apoptosis in both osteoblasts and chondrocytes, while there was no effect on the apoptosis of rabbit osteoclasts [131]. Recently, it has been demonstrated that short-chain C-6 Cer induces anti-osteosarcoma activity in vitro and in vivo [132]. In other studies, Cer has been shown to be mitogenic in preosteoblast MC3T3-E1 cells [133]. Recently, Cer species, in particular long-chain Cer (C-22 and C-24 Cer), have been demonstrated to mediate the proapoptotic effect of sodium nitroprusside, a nitric oxide donor, in MC3T3-E1 cells [134], suggesting that the controversial results may also be due to different levels of Cer species. Detectable increase of endogenous Cer levels is observed when cells are induced to apoptosis by TNF-alpha [135]. This upregulation leads to the modulation of NF-kappa B localization and function [136]. Different intracellular signaling pathways are involved in osteoblast death and survival: Cer induces osteoblast apoptosis through protein phosphatase 1 and protein kinase C (PKC) δ [137, 138], whereas the ability of Cer to promote osteoblast survival is prevented by PKC ζ inhibitor. At present, it is not clear whether Cer acts up- or downstream to or independently of caspases [139]. Interestingly, in mice lacking the enzyme that converts dihydroceramide to ceramide (dihydroceramide desaturase 1) Cer reduction does not lead to bone defects, indicating that the maintenance of Cer levels is not essential for normal bone mineralization. Recently, it has been found a role also for C1P in the osteogenesis of multipotent stromal cells derived from MSCs. C1P can affect the growth and expanded intercellular connections, thereby, increasing the proliferative activity, acting in opposite manner of S1P. Therefore, it has been suggested that an appropriate combination of C1P and S1P may be a useful strategy in bone regeneration with the use of MSCs [123].

4.2. Sphingolipids in nervous system physiology

SLs are particularly abundant in the central nervous system. Mutations in genes coding for enzymes involved in their metabolism cause sphingolipidosis many of which show alterations in the nervous system (reviewed by Sabourdy et al. [140]). In addition, modifications in SL metabolism are found in neurodegenerative diseases, such as AD, PD, Huntington disease (HD), MS, and major depression [141–146]. Various SLs are crucial in regulating neural physiological functions, including cell survival, apoptosis, differentiation, inflammation, excitability, and neurotransmitter release [15, 137–149].

4.2.1. S1P—sphingosine

Like Cer, Sph acts as a proapoptotic signal as well as an inhibitor of several enzymes such as protein kinase C (PKC), phospholipase D, and of the transcription factor SF-1 [15, 150]. Sph also directly modulates voltage-activated calcium channels in pituitary cells and several components of the melastatin-like transient receptor potential channel subfamily, such as Transient receptor potential cation channel subfamily M member 3 (TRPM3) [151, 152]. Sph alters the integrity of membranes and induces the release of lysosomal cysteine proteases, such as cathepsins and of cytochrome *c*, which in turn activates the intrinsic pathway of apoptosis. Increasing evidence indicates that Sph regulates vesicle fusion and trafficking and, therefore, the strength and reliability of synaptic transmission [153, 154]. In addition, Sph is a competitive antagonist of the type 1 cannabinoid receptors (CB1Rs) [155] and possibly it contributes to SL regulation of nociception [156]. S1P modulates survival, proliferation, differentiation, cell migration, calcium homeostasis, neurite retraction, angiogenic vascular maturation, and cytoskeleton dynamics [13, 107, 149]. The bioactive lipid can also induce neuroprotection through many mechanisms that include production of growth factors, decrease of oxidative stress, increase of Mitogen-activated protein kinase (MAPK) activation, activation of PI3K/AKT pathway, modulation of antiapoptotic proteins, pigenomic effects by direct inhibition of deacetylases, and affecting mitochondrial functionality (for a recent review see Ref. [149].

S1P receptors are expressed in CNS cells (neurons, oligodendrocytes, astrocytes, and microglia) and their expression levels change during development. The role of S1PR depends on cell types and on their different expression and localization during development or following stimulation. The studies of knockout mice with deficits in both Sphk1 and Sphk2 highlight the importance of S1P in the development of the nervous system. These mice show severely disturbed neurogenesis, including neural tube closure and angiogenesis, and they die at an early embryonic stage [157]. Migration of neural stem progenitor cells toward injury sites is promoted by S1P via S1P1 [158] and inhibited on the oligodendrocyte progenitor cells via S1P5 [159]. S1P receptors are expressed in CNS cells (neurons, oligodendrocytes, astrocytes, and microglia) and their expression levels change during development. Analogously to the double SphK knockout, the S1P1 knockout mouse shows altered neurogenesis and angiogenesis, while the knockout of the other receptors have less severe consequences. Maclennan et al. [160] reported that mice lacking S1P2 show significant increase in excitatory postsynaptic potentiation, resulting in spontaneous seizures. The role of S1PR depends on their different expression and localization during development or following stimulation. For example, activation of S1P5 causes activation of Rho and retraction of processes in immature oligodendrocytes and survival in mature cells [161]. NGF induces differentiation in PC12 cells through a relocalization of S1P receptors; S1P1 that induces neurite growth is expressed at the plasma membrane, while S1P2 is internalized [162] becoming unable to cause cell rounding and loss of neurites. S1P and neurotrophic factors have mutual effects on expression of each other:

NGF and GDNF are able to stimulate S1P generation, and S1P increases GDNF production by astrocytes [144]. S1P plays different roles in synaptic transmission. It increases glutamate release in hippocampus via S1P3 [163], it enhances excitability in rat sensory neurons trough S1P1 and S1P3 [155, 158, 164], and it is involved in the recruitment of vesicles in the presynaptic membrane [165] and also in endocytic membrane trafficking [166].

4.2.2. C1P—ceramide

Cer regulates cell growth, differentiation, apoptosis, inflammation, exosome release, and neural excitability [15, 157, 158, 167, 168]. Accumulating evidence indicates that different Cer species might have different functions [169]. Cer 18:0 is synthesized by CerS1, an enzyme abundant in the brain, and appears to have a protective role [167, 168]. Serum deprivation that increases apoptosis in embryonic hippocampal cells increases Cer 16:0 and decreases Cer 24:0 content [27]. Cer has been involved in synaptic regulation and plasticity [170, 171]. For example, Cer is able to increase dopamine release and uptake [172] and to modulate excitatory postsynaptic currents by controlling the insertion and clustering of NMDA receptors [173]. Moreover, SMase2 inhibition delays formation of spatial memory in mice [174]. Regarding inflammation, astrocytes display increased Cer following ischemia/reperfusion leading to generation of pro-inflammatory cytokines [175]. Recently, it has been reported that Cer induces ciliogenesis, a critical step in differentiation in embryonic stem cells and neural progenitors [176]. CerK, the enzyme generating C1P, was first observed in brain synaptic vesicles [177] and found to be highly expressed in brain [178] suggesting a role of C1P in neurotransmitter release. C1P induces proliferation or survival in several types of cells including macrophages and fibroblasts [179] while inhibition or downregulation of CerK decreases proliferation in human neuroblastoma cells [18]. C1P plays a regulatory role in inflammation since it directly binds and activates α-type cytosolic phospholipase A2 stimulating arachidonic acid release [180]. Recently, it has been demonstrated that C1P plays an important role in recruitment of stem/progenitor cells to damaged organs [181]. Whether C1P is also released from the injured nervous system or whether it induces migration in nervous stem cells is unknown.

5. Cross talk between vitamin D and sphingolipid metabolism: a potential role in Alzheimer's disease

$1,25(OH)_2D_3$ and SL metabolism cross talk at different levels. For example, $1,25(OH)_2D_3$ modulates the expression of genes involved in S1P degradation, such as Sphingosine-1-phosphate phosphatase 2 (SGPP2) [182] and of growth factors involved in differentiation and neuroprotection, such as NT-3, Brain-derived neurotrophic factor (BDNF) and ciliary neurotrophic factor (CNTF) [3, 4]. On the other hand, many neuroprotective actions of $1,25(OH)_2D_3$ have been reported to be due to stimulation of SphK and increased levels of S1P [183, 184]. In addition, $1,25(OH)_2D_3$ also is able to decrease the expression of CerK, the enzyme that generates C1P [18]. C1P activates directly Secretory PLA2 enzymes (sPLA2) [180], producing arachidonic acid that can be further metabolized to proinflammatory mediators. C1P and S1P have a crucial role in migration. C1P is released from damaged cells and chemoattracts bone marrow-derived multipotent stromal cells, endothelial progenitor cells, and very small embryonic-like stem cell. The migration of osteoclast precursors is controlled by S1P and, recently, it has been found that 1,25(OH)$_2$D$_3$ reduces the expression of the chemorepulsive receptor S1P2 on circulating precursors [185]. AD is a neurodegenerative disorder of the central nervous system and the most common form of dementia [186]. The pathogenic hallmarks of AD include extracellular amyloid-containing plaques, intracellular neurofibrillary tangles consisting of hyperphosphorylated tau protein and death of cholinergic neurons of the basal forebrain. Amyloid plaques are mainly formed by aggregated amyloid beta peptide (Aβ). Alterations in the enzymes involved in SL metabolism and content have been observed in brains and cerebrospinal fluid of AD patients [187–192], leading to an increase of Cer and loss of S1P. Many studies in culture cells and animal models have demonstrated that Aβ affects SL metabolism. For example, Aβ42 directly binds and activates nSMase in vitro decreasing SM content [193]. In addition, Aβ can activate both neutral and acidic SMases through increased ROS accumulation via Nicotinamide adenine dinucleotide phosphate (NADPH) oxidase activation and glutatione (GSH) depletion [194]. Increased Cer could then induce apoptosis [186, 195, 196]. In addition, Cer increases stability of β-amyloid precursor protein cleaving enzyme (BACE1) activity [197], while S1P binds to and increases proteolytic activity of BACE1 [189] and SM decreases Aβ production by inhibition of the γ-secretase [193]. Summarizing, some SLs might be protective by enhancing Aβ clearance or decreasing Aβ production, others increase Aβ toxicity or Aβ oligomerization and at the same time, amyloid precursor protein processing also affects lipid metabolism, resulting in complex regulatory feed-back cycles, which appear to be dysregulated in AD. It is worth noting that recently, a novel mechanism of Cer-enriched exosomes released by Aβ-treated astrocytes has been proposed to be responsible for Aβ-induced apoptosis [167].

The neuroprotective action of SL analogues, such as FTY720, has been tested in some neurodegenerative diseases. FTY720 is a prodrug that is converted to an analogue of S1P when phosphorylated by SphK. It has been approved for the treatment of multiple sclerosis (MS) and acts both as an immunomodulatory drug and on different cells of the CNS (neurons, astrocytes, oligodendrocytes, and microglia), all of which express S1P receptors [198]. Administration of FTY720 in a rat model of AD obtained by injection of Aβ decreases death in hippocampus and cortex and increases memory compared with control rats [199, 200]. FTY720 decreases production of Aβ in cultured neuronal cells [189]. Increasing evidence derived from epidemiological studies indicate that 1,25(OH)$_2$D$_3$ deficiency and VDR polymorphisms influence susceptibility to AD [201], whereas Aβ may disrupt the hormone-VDR pathway and cause defective utilization of 1,25(OH)$_2$D$_3$ by suppressing the level of the VDR and elevating the level of 24OHase [202]. In addition to neuroprotective effects involving calcium, Reactive Oxygen Species (ROS), and inflammation, 1,25(OH)$_2$D$_3$ is able to exert other specific effects important for AD, by regulating the expression of many AD-related genes. It attenuates Aβ peptide accumulation by stimulating phagocytosis of Aβ peptide probably by modulating transcription of Toll-like receptors and cytokines together with enhancing brain-to-blood efflux transport by increasing P-glycoprotein expression and likely by altering Amyloid Precursor Protein (APP) processing [88] and prevents the acetylcholine defect by increasing the activity of choline acetyltransferase (thus acetylcholine synthesis) in the brain [203]. Part of VDR is also located in lipid microdomains in the nuclear membrane [2], and this localization is modified by altering SL metabolism and has been associated with embryonic hippocampal cell differentiation. Neuroprotective actions of SLs, in particular S1P, and $1,25(OH)_2D_3$ include many common effectors such as calcium regulation, synaptic modulation, growth factor expression, regulation of inflammation, etc., but whether $1,25(OH)_2D_3$ and SLs, in particular S1P, could act synergistically on neuroprotection and/or neurogenesis in AD is still unknown and deserves further investigation. A study in our lab indicates that the cross talk between SLs and $1,25(OH)_2D_3$ leads to a specific balance between neurodegeneration/neuroprotection in neuronal cells [204]. The neuroprotective effect of high intake of $1,25(OH)_2D_3$ has been found in some AD trials, but not in others [88]. It is not clear whether hypovitaminosis D triggers AD or it removes protection in the CNS against AD. However, the combination of antineurodegenerative drugs with $1,25(OH)_2D_3$ supplementation might be useful. In fact, the supplementation with nemantidine plus $1,25(OH)_2D_3$ has been shown to prevent cognitive decline more efficiently than that with the single compounds [96]. Regarding MS, very limited evidence suggests a potential benefit of $1,25(OH)_2D_3$ supplementation for the prevention of MS and this needs to be further verified by future studies [3].

Acknowledgements

This work was supported by grants from Fondazione Cassa di Risparmio di Pistoia e Pescia, Regione Toscana to E.M, MIUR (ex 60%) to EM and MM and from A.I.L. for contributing to AF fellowship. The authors are grateful to Dr. Maria Martinesi and Dr. Elena Bruzzone from the Department of Experimental and Clinical Biomedical Sciences for their contribution.

Author details

Alessia Frati[1, 2], Mercedes Garcia Gil[3, 4], Federica Pierucci[1, 2] and Elisabetta Meacci[1, 2*]

*Address all correspondence to: elisabetta.meacci@unifi.it

1 Department of Experimental and Clinical Biomedical Sciences, Molecular and Applied Biology Research Unit, University of Firenze, Firenze, Italy

2 Interuniversity Institute of Myology, Italy

3 Department of Biology, University of Pisa, Pisa, Italy

4 Interdepartmental Research Center Nutrafood "Nutraceuticals and Food for Health," University of Pisa, Pisa, Italy

References

[1] Marini F, Bartoccini E, Cascianelli G, Voccoli V, Baviglia MG, Magni MV, Garcia-Gil M, Albi E. Effect of 1alpha,25-dihydroxyvitamin D3 in embryonic hippocampal cells. Hippocampus. 2010 Jun;20(6):696–705. doi: 10.1002/hipo.20670.

[2] Bartoccini E, Marini F, Damaskopoulou E, Lazzarini R, Cataldi S, Cascianelli G, Gil Garcia M, Albi E. Nuclear lipid microdomains regulate nuclear vitamin D3 uptake and influence embryonic hippocampal cell differentiation. Mol Biol Cell. 2011 Sep;22(17):3022–31. doi: 10.1091/mbc.E11-03-0196.

[3] DeLuca GC, Kimball SM, Kolasinski J, Ramagopalan SV, Ebers GC. Review: the role of vitamin D in nervous system health and disease. Neuropathol Appl Neurobiol. 2013 Aug;39(5):458–84. doi: 10.1111/nan.12020.

[4] Shirazi HA, Rasouli J, Ciric B, Rostami A, Zhang GX. 1,25-Dihydroxyvitamin D3 enhances neural stem cell proliferation and oligodendrocyte differentiation. Exp Mol Pathol. 2015 Apr;98(2):240–5. doi: 10.1016/j.yexmp.2015.02.004.

[5] Chun RF et al. Vitamin D supplementation and antibacterial immune responses in adolescents and young adults with HIV/AIDS. J Steroid Bioch Mol Biol. 2015 Apr;148:290–7. doi: 10.1016/j.jsbmb.2014.07.013. Epub 2014 Aug 1.

[6] DeLuca HF. Overview of general physiologic features and functions of vitamin D. Am J Clin Nutr. 2004 Dec;80(6 Suppl):1689S–96S.

[7] Christakos S, Dhawan P, Verstuyf A, Verlinden L, Carmeliet G. Vitamin D: metabolism, molecular mechanism of action, and pleiotropic effects. Physiol Rev. 2016 Jan;96(1):365–408. doi: 10.1152/physrev.00014.2015.

[8] Trump DL, Deeb KK, Johnson CS. Vitamin D: considerations in the continued development as an agent for cancer prevention and therapy. Cancer J. 2010 Jan-Feb;16(1):1–9. doi: 10.1097/PPO.0b013e3181c51ee6.

[9] Masuda S, Jones G. Promise of vitamin D analogues in the treatment of hyperproliferative conditions. Mol Cancer Ther. 2006 Apr;5(4):797–808.

[10] Zügel U, Steinmeyer A, Giesen C, Asadullah K. A novel immunosuppressive 1alpha,25-dihydroxyvitamin D3 analog with reduced hypercalcemic activity. J Invest Dermatol. 2002 Dec;119(6):1434–42.

[11] Spiegel S, Milstien S. Sphingosine 1-phosphate, a key cell signaling molecule. J Biol Chem. 2002 Jul 19;277(29):25851–4. doi: 10.1038/nrm1103.

[12] Gault CR, Obeid LM, Hannun YA. An overview of sphingolipid metabolism: from synthesis to breakdown. Adv Exp Med Biol. 2010;688:1–23.

[13] Maceyka M, Harikumar KB, Milstien S, Spiegel S. Sphingosine-1-phosphate signaling and its role in disease. Trends Cell Biol. 2012 Jan;22(1):50–60. doi: 10.1016/j.tcb.2011.09.003.

[14] Newton J, Lima S, Maceyka M, Spiegel S. Revisiting the sphingolipid rheostat: evolving concepts in cancer therapy. Exp Cell Res. 2015 May 1;333(2):195–200. doi: 10.1016/j.yexcr.2015.02.025.

[15] Colombaioni L, Garcia-Gil M. Sphingolipid metabolites in neural signalling and function. Brain Res Brain Res Rev. 2004 Nov;46(3):328–55.

[16] Bartke N, Hannun YA. Bioactive sphingolipids: metabolism and function. J Lipid Res. 2009 Apr;50 Suppl:S91–6. doi: 10.1194/jlr.R800080-JLR200.

[17] Fyrst H, Saba JD. An update on sphingosine-1-phosphate and other sphingolipid mediators. Nat Chem Biol. 2010 Jul;6(7):489–97. doi: 10.1038/nchembio.392.

[18] Bini F, Frati A, Garcia-Gil M, Caleo M, Peretto P, Gomez-Muñoz A, Meacci E. New signalling pathway involved in the anti-proliferative action of vitamin D3 and its analogues in human neuroblastoma cells. A role for ceramide kinase. Neuropharmacology. 2012 Sep;63(4):524–37. doi: 10.1016/j.neuropharm.2012.04.026.

[19] Pyne NJ et al. Role of sphingosine 1-phosphate receptors, sphingosine kinases and sphingosine in cancer and inflammation. Adv Biol Regul. 2016 Jan;60:151–9. doi: 10.1016/j.jbior.2015.09.001.

[20] Cuvillier O, Pirianov G, Kleuser B, Vanek PG, Coso OA, Gutkind S, Spiegel S. Suppression of ceramide-mediated programmed cell death by sphingosine-1-phosphate. Nature. 1996 Jun 27;381(6585):800–3. doi: 10.1038/381800a0.

[21] Catarzi S, Giannoni E, Meacci E, Iantomasi T, Vincenzini MT. Sphingosine 1-phosphate stimulation of NADPH oxidase activity: relationship with platelet-derived growth factor receptor and c-Src kinase. Biochim Biophys Acta. 2007 Jun;1770(6):872–83.

[22] Meacci E, Bini F, Sassoli C, Martinesi M, Squecco R, Chellini F, Zecchi S, Francini F, Formigli L. Functional interaction between TRPC1 channel and connexin-43 protein: a novel pathway underlying S1P action on skeletal myogenesis. Cell Mol Life Sci. 2010 Dec;67(24):4269–85. doi: 10.1007/s00018-010-0442-3.

[23] Sassoli C, Formigli L, Bini F, Squecco R, Battistini C, Zecchi S, Francini F, Meacci E. Effects of S1P on skeletal muscle repair/regeneration during eccentric contraction. J Cell Mol Med. 2011 Nov;15(11):2498–511. doi: 10.1111/j.1582-4934.2010.01250.x.

[24] Tavarini S, Colombaioni L, Garcia-Gil M. Sphingomyelinase metabolites control survival and apoptotic death in SH-SY5Y neuroblastoma cells. Neurosci Lett. 2000 May 19;285:185–8.

[25] Colombaioni L, Frago LM, Varela-Nieto I, Pesi R, Garcia-Gil M. Serum deprivation increases ceramide levels and induces apoptosis in undifferentiated HN9.10e cells. Neurochem Int. 2002 Apr;40(4):327–36.

[26] Albi E et al. Nuclear sphingomyelin pathway in serum deprivation-induced apoptosis of embryonic hippocampal cells. J Cell Physiol. 2006 Jan;206(1):189–95.

[27] Garcia-Gil M, Lazzarini A, Lazzarini R, Floridi E, Cataldi S, Floridi A, Albi E. Serum deprivation alters lipid profile in HN9.10e embryonic hippocampal cells. Neurosci Lett. 2015 Mar 4;589:83–7. doi: 10.1016/j.neulet.2014.12.059.

[28] Bornancin F. Ceramide kinase: the first decade. Cell Signal. 2011 Jun;23(6):999–1008. doi: 10.1016/j.cellsig.2010.11.012. Epub 2010 Nov 25.

[29] Frati A, Ricci B, Pierucci F, Nistri S, Bani D, Meacci E. Role of sphingosine kinase/S1P axis in ECM remodeling of cardiac cells elicited by relaxin. Mol Endocrinol. 2015 Jan;29(1):53–67. doi: 10.1210/me.2014-1201.

[30] Sassoli C, Frati A, Tani A, Anderloni G, Pierucci F, Zecchi Orlandini S, Formigli L, Meacci E. Mesenchymal stromal cell secreted sphingosine 1-phosphate (S1P) exerts a stimulatory effect on skeletal myoblast proliferation. PLoS One. 2014 Sep 29;9(9):e108662. doi: 10.1371/journal.pone.0108662.

[31] Jilka RL. Cytokines, bone remodeling, and estrogen deficiency: a 1998 update. Bone. 1998 Aug;23(2):75–81.

[32] Weinstein RS, Manolagas SC. Apoptosis and osteoporosis. Am J Med. 2000;108:153–64.

[33] Nakamura I, Takahashi N, Jimi E, Udagawa N, Suda T. Regulation of osteoclast function. Mod Rheumatol. 2012 Apr;22(2):167–77. doi: 10.1007/s10165-011-0530-8.

[34] Väänänen HK, Laitala-Leinonen T. Osteoclast lineage and function. Arch Biochem Biophys. 2008 May 15;473(2):132–8. doi: 10.1016/j.abb.2008.03.037. Epub 2008 Apr 6.

[35] Yagi M et al. DC-STAMP is essential for cell-cell fusion in osteoclasts and foreign body giant cells. J Exp Med. 2005 Aug 1;202(3):345–51. PubMed PMID: 16061724; PubMed Central PMCID: PMC2213087.

[36] Miyamoto H et al. Osteoclast stimulatory transmembrane protein and dendritic cell–specific transmembrane protein cooperatively modulate cell–cell fusion to form osteoclasts and foreign body giant cells. J Bone Miner Res. 2012 Jun;27(6):1289–97. doi: 10.1002/jbmr.1575. PubMed PMID: 22337159.

[37] Nijenhuis T, van der Eerden BC, Zügel U, Steinmeyer A, Weinans H, Hoenderop JG, van Leeuwen JP, Bindels RJ. The novel vitamin D analog ZK191784 as an intestine-specific vitamin D antagonist. FASEB J. 2006 Oct;20(12):2171–3. PubMed PMID: 17012263.

[38] Rodan GA, Martin TJ. Role of osteoblasts in hormonal control of bone resorption—a hypothesis. Calcif Tissue Int. 1982 May;34(3):311. PubMed PMID: 6809295.

[39] Takahashi N, Akatsu T, Udagawa N, Sasaki T, Yamaguchi A, Moseley JM, Martin TJ, Suda T. Osteoblastic cells are involved in osteoclast formation. Endocrinology. 1988 Nov;123(5):2600–2. PubMed PMID: 2844518.

[40] Rucci N, Teti A. The "love–hate" relationship between osteoclasts and bone matrix. Matrix Biol. 2016 May–Jul;52–54:176–90. doi: 10.1016/j.matbio.2016.02.009. Review.

[41] Roodman GD, Ibbotson KJ, MacDonald BR, Kuehl TJ, Mundy GR. 1,25-Dihydroxyvitamin D3 causes formation of multinucleated cells with several osteoclast characteristics in cultures of primate marrow. Proc Natl Acad Sci U S A. 1985 Dec;82(23):8213–7. PubMed PMID: 3865222.

[42] Miller WL. Genetic disorders of vitamin D biosynthesis and degradation. J Steroid Biochem Mol Biol. 2016 April 6;S0960-0760(16):30097–8.

[43] Pike JW, Lee SM, Meyer MB. Regulation of gene expression by 1,25-dihydroxyvitamin D3 in bone cells: exploiting new approaches and defining new mechanisms. Bonekey Rep. 2014 Jan 8;3:482. doi: 10.1038/bonekey.2013.216. eCollection 2014 Jan 8. Review. PubMed PMID: 24466413; PubMed Central PMCID: PMC3899561.

[44] Bouillon R, Carmeliet G, Verlinden L, van Etten E, Verstuyf A, Luderer HF, Lieben L, Mathieu C, Demay M. Vitamin D and human health: lessons from vitamin D receptor null mice. Endocr Rev. 2008 Oct;29(6):726–76. doi: 10.1210/er.2008-0004. Epub 2008 Aug 11. Review. PubMed PMID: 18694980; PubMed Central PMCID: PMC2583388.

[45] Fujita H et al. Tight junction proteins claudin-2 and -12 are critical for vitamin D-dependent Ca^{2+} absorption between enterocytes. Mol Biol Cell. 2008 May;19(5):1912–21. doi: 10.1091/mbc.E07-09-0973.

[46] Van Cromphaut SJ et al. Duodenal calcium absorption in vitamin D receptor-knockout mice: functional and molecular aspects. Proc Natl Acad Sci U S A. 2001 Nov 6;98(23):13324–9. Epub 2001 Oct 30. PubMed PMID: 11687634; PubMed Central PMCID: PMC60869.

[47] Wong BR et al. TRANCE is a novel ligand of the tumor necrosis factor receptor family that activates c-Jun N-terminal kinase in T cells. J Biol Chem. 1997;272(40):25190–4.

[48] Anderson DM et al. A homologue of the TNF receptor and its ligand enhance T-cell growth and dendritic-cell function. Nature. 1997 Nov 13;390(6656):175–9.

[49] Lacey DL et al. Osteoprotegerin ligand is a cytokine that regulates osteoclast differentiation and activation. Cell. 1998 Apr 17;93(2):165–76. PubMed PMID: 9568710.

[50] O'Connor KM. Evaluation and treatment of osteoporosis. Med Clin North Am. 2016 Jul;100(4):807–26. doi: 10.1016/j.mcna.2016.03.016. Epub 2016 Apr 20. Review.

[51] Gallagher JC, Jerpbak CM, Jee WS, Johnson KA, DeLuca HF, Riggs BL. 1,25-Dihydroxyvitamin D3: short- and long-term effects on bone and calcium metabolism in patients with postmenopausal osteoporosis. Proc Natl Acad Sci U S A. 1982 May;79(10):3325–9.

[52] Dechant KL, Goa KL. Calcitriol. A review of its use in the treatment of postmenopausal osteoporosis and its potential in corticosteroid-induced osteoporosis. Drugs Aging. 1994 Oct;5(4):300–17.

[53] Gillespie LD, Robertson MC, Gillespie WJ, Sherrington C, Gates S, Clemson LM, Lamb SE. Interventions for preventing falls in older people living in the community. Cochrane Database Syst Rev. 2012 Sep 12;9:1–18.

[54] Glorieux FH, Pettifor JM. Vitamin D/dietary calcium deficiency rickets and pseudo-vitamin D deficiency rickets. Bonekey Rep. 2014 Mar 19;3:524. doi: 10.1038/bonekey.2014.19. Review.

[55] Lips P, Gielen E, van Schoor NM. Vitamin D supplements with or without calcium to prevent fractures. Bonekey Rep. 2014 Mar 5;3:512. doi: 10.1038/bonekey.2014.7.

[56] Schafer AL et al. Intestinal calcium absorption decreases dramatically after gastric bypass surgery despite optimization of vitamin D status. J Bone Miner Res. 2015 Aug;30(8):1377–85. doi: 10.1002/jbmr.2467.

[57] Iuliano-Burns S, Ayton J, Hillam S, Jones G, King K, Macleod S, Seeman E. Skeletal and hormonal responses to vitamin D supplementation during sunlight deprivation in Antarctic expeditioners. Osteoporos Int. 2012 Oct;23(10):2461–7. doi: 10.1007/s00198-011-1858-9.

[58] Leyssens C, Verlinden L, Verstuyf A. The future of vitamin D analogs. Front Physiol. 2014 Apr 3;5:122. doi: 10.3389/fphys.2014.00122. eCollection 2014. Review. PubMed PMID: 24772087; PubMed Central PMCID: PMC3982071.

[59] Kubodera N, Tsuji N, Uchiyama Y, Endo K. A new active vitamin D analog, ED-71, causes increase in bone mass with preferential effects on bone in osteoporotic patients. J Cell Biochem. 2003 Feb 1;88(2):286–9. PubMed PMID: 12520528.

[60] Uchiyama Y, Higuchi Y, Takeda S, Masaki T, Shira-Ishi A, Sato K, Kubodera N, Ikeda K, Ogata E. ED-71, a vitamin D analog, is a more potent inhibitor of bone reabsorption than alfacalcidol in an estrogen-deficient rat model of osteoporosis. Bone. 2002 Apr;30(4):582–8. PubMed PMID: 11934649.

[61] Matsumoto T, Takano T, Yamakido S, Takahashi F, Tsuji N. Comparison of the effects of eldecalcitol and alfacalcidol on bone and calcium metabolism. J Steroid Biochem Mol Biol. 2010 Jul;121(1–2):261–4. doi: 10.1016/j.jsbmb.2010.03.035.

[62] Strauch UG, Obermeier F, Grunwald N, Dunger N, Rath HC, Schölmerich J, Steinmeyer A, Zügel U, Herfarth HH. Calcitriol analog ZK191784 ameliorates acute and chronic dextran sodium sulfate-induced colitis by modulation of intestinal dendritic cell numbers and phenotype. World J Gastroenterol. 2007 Dec 28;13(48):6529–37.

[63] Park MR, Lee JH, Park MS, Hwang JE, Shim HJ, Cho SH, Chung IJ, Bae WK. Suppressive effect of 19-nor-1α-25-dihydroxyvitamin D2 on gastric cancer cells and peritoneal metastasis model. J Korean Med Sci. 2012 Sep;27(9):1037–43. doi: 10.3346/jkms.2012.27.9.1037. Epub 2012 Aug 22. PubMed PMID: 22969249.

[64] Ma Y, Yu WD, Hidalgo AA, Luo W, Delansorne R, Johnson CS, Trump DL. Inecalcitol, an analog of 1,25D3, displays enhanced antitumor activity through the induction of apoptosis in a squamous cell carcinoma model system. Cell Cycle. 2013 Mar 1;12(5):743–52. doi: 10.4161/cc.23846. Epub 2013 Feb 6. PubMed PMID: 23388458.

[65] Sundaram S, Sea A, Feldman S, Strawbridge R, Hoopes PJ, Demidenko E, Binderup L, Gewirtz DA. The combination of a potent vitamin D3 analog, EB 1089, with ionizing radiation reduces tumor growth and induces apoptosis of MCF-7 breast tumor xenografts in nude mice. Clin Cancer Res. 2003 Jun;9(6):2350–6. PubMed PMID: 12796405.

[66] Komori T. Animal models for osteoporosis. Eur J Pharmacol. 2015 Jul 15;759:287–94. doi: 10.1016/j.ejphar.2015.03.028. Review.

[67] Matsushita M et al. Age-related changes in bone mass in the senescence-accelerated mouse (SAM). Am J Pathol. 1986;125:276–83.

[68] Aref M, Gallant MA, Organ JM, Wallace JM, Newman CL, Burr DB, Brown DM, Allen MR. In vivo reference point indentation reveals positive effects of raloxifene on mechanical properties following 6 months of treatment in skeletally mature beagle dogs. Bone. 2013 Oct;56(2):449–53. doi: 10.1016/j.bone.2013.07.009.

[69] Shapiro IM, Hickok NJ, Parvizi J, Stewart S, Schaer TP. Molecular engineering of an orthopaedic implant: from bench to bedside.Eur Cell Mater. 2012 May 24;23:362–70. Review.

[70] Weinstein RS, Jilka RL, Parfitt AM, Manolagas SC. Inhibition of osteoblastogenesis and promotion of apoptosis of osteoblasts and osteoclasts by glucocorticoids: potential mechanisms of their deleterious effects on bone. J Clin Invest. 1998;102:274–82.

[71] Klotho and β Klotho." Adv Exp Med Biol. 2012;728:25-40. doi: 10.1007/978-1-4614-0887-1_2

[72] Rinotas V et al. Novel genetic models of osteoporosis by overexpression of human RANKL in transgenic mice. J Bone Miner Res. 2014;29(5):1158–69.

[73] Renkema KY et al. Hypervitaminosis D mediates compensatory Ca^{2+} hyperabsorption in TRPV5 knockout mice. J Am Soc Nephrol. 2005 Nov;16(11):3188–95.

[74] Bivi N, Condon KW, Allen MR, Farlow N, Passeri G, Brun LR, Rhee Y, Bellido T, Plotkin LI. Cell autonomous requirement of connexin 43 for osteocyte survival: consequences for endocortical resorption and periosteal bone formation. J Bone Miner Res. 2012 Feb;27(2):374–89. doi: 10.1002/jbmr.548.

[75] Thierry-Palmer M, Cephas S. Plasma 25-hydroxycholecalciferol and 1,25-dihydroxycholecalciferol concentrations are decreased in hind limb unloaded Dahl salt-sensitive female rats. J Steroid Biochem Mol Biol. 2010 Feb 15;118(3):188–93. doi: 10.1016/j.jsbmb.2009.12.012. Epub 2010 Jan 4.

[76] Kamoun-Goldrat AS, Le Merrer MF. Animal models of osteogenesis imperfecta and related syndromes. J Bone Miner Metab. 2007;25(4):211–8. Epub 2007 Jun 25.

[77] Plante L, Veilleux LN, Glorieux FH, Weiler H, Rauch F. Effect of high-dose vitamin D supplementation on bone density in youth with osteogenesis imperfecta: a randomized controlled trial. Bone. 2016 May;86:36–42. doi: 10.1016/j.bone.2016. 02.013.

[78] Aubin I, Adams CP, Opsahl S, Septier D, Bishop CE, Auge N, Salvayre R, Negre-Salvayre A, Goldberg M, Guénet JL, Poirier C. A deletion in the gene encoding sphingomyelin phosphodiesterase 3 (Smpd3) results in osteogenesis and dentinogenesis imperfecta in the mouse. Nat Genet. 2005 Aug;37(8):803–5.

[79] Patrick RP, Ames BN. Vitamin D hormone regulates serotonin synthesis. Part 1: relevance for autism. FASEB J. 2014 Jun;28(6):2398–413. doi: 10.1096/fj.13-246546.

[80] Zanatta L et al. 1α,25-Dihydroxyvitamin D(3) mechanism of action: modulation of L-type calcium channels leading to calcium uptake and intermediate filament phosphorylation in cerebral cortex of young rats. Biochim Biophys Acta. 2012 Oct;1823(10):1708–19. doi: 10.1016/j.bbamcr.2012.06.023.

[81] Eyles DW, Burne TH, McGrath JJ. Vitamin D, effects on brain development, adult brain function and the links between low levels of vitamin D and neuropsychiatric disease. Front Neuroendocrinol. 2013 Jan;34(1):47–64.

[82] Jiang P et al. Association between vitamin D receptor gene polymorphisms with childhood temporal lobe epilepsy. Int J Environ Res Public Health. 2015;12(11):13913–22.

[83] Fernandes de Abreu DA, Eyles D, Féron F. Vitamin D, a neuroimmunomodulator: implications for neurodegenerative and autoimmune diseases. Psychoneuroendocrinology. 2009 Dec;34 Suppl 1:S265–77. doi: 10.1016/j.psyneuen.2009.05.023.

[84] Hawes JE, Tesic D, Whitehouse AJ, Zosky GR, Smith JT, Wyrwoll CS. Maternal vitamin D deficiency alters fetal brain development in the BALB/c mouse. Behav Brain Res. 2015 Jun 1;286:192–200.

[85] Whitehouse AJ, Holt BJ, Serralha M, Holt PG, Kusel MM, Hart PH. Maternal serum vitamin D levels during pregnancy and offspring neurocognitive development. Pediatrics. 2012 Mar;129(3):485–93.

[86] Latimer CS et al. Vitamin D prevents cognitive decline and enhances hippocampal synaptic function in aging rats. Proc Natl Acad Sci U S A. 2014 Oct 14;111(41):E4359–66. doi: 10.1073/pnas.1404477111.

[87] De la Fuente AG et al. Vitamin D receptor-retinoid X receptor heterodimer signaling regulates oligodendrocyte progenitor cell differentiation. J Cell Biol. 2015 Dec 7;211(5):975–85. doi: 10.1083/jcb.201505119.

[88] Landel V, Annweiler C, Millet P, Morello M, Féron F. Vitamin D, cognition, and Alzheimer's disease: the therapeutic benefit is in the D-tails. J Alzheimer's Dis. 2016 May 11. doi: 10.3233/JAD-150943. [Epub ahead of print].

[89] Winner B, Winkler J. Adult neurogenesis in neurodegenerative diseases. Cold Spring Harb Perspect Biol. 2015 Apr 1;7(4):a021287. doi: 10.1101/cshperspect.a021287.

[90] Brown J, Bianco JI, McGrath JJ, Eyles DW. 1,25-dihydroxyvitamin D3 induces nerve growth factor, promotes neurite outgrowth and inhibits mitosis in embryonic rat hippocampal neurons. Neurosci Lett. 2003 Jun 5;343(2):139–43.

[91] Ko P, Burkert R, McGrath J, Eyles D. Maternal vitamin D3 deprivation and the regulation of apoptosis and cell cycle during rat brain development. Brain Res Dev Brain Res. 2004 Oct 15;153(1):61–8.

[92] Osborne JE, Hutchinson PE. Vitamin D and systemic cancer: is this relevant to malignant melanoma?. Br J Dermatol. 2002 Aug;147(2):197–213.

[93] Cui X, McGrath JJ, Burne TH, Mackay-Sim A, Eyles DW. Maternal vitamin D depletion alters neurogenesis in the developing rat brain. Int J Dev Neurosci. 2007 Jun;25(4):227–32.

[94] Keilhoff G, Grecksch G, Becker A. Haloperidol normalized prenatal vitamin D depletion-induced reduction of hippocampal cell proliferation in adult rats. Neurosci Lett. 2010 May 31;476(2):94–8. doi: 10.1016/j.neulet.2010.04.010.

[95] Zhu Y et al. Abnormal neurogenesis in the dentate gyrus of adult mice lacking 1,25-dihydroxy vitamin D_3 (1,25-$(OH)_2D_3$). Hippocampus. 2012 Mar;22(3):421–33. doi: 10.1002/hipo.20908.

[96] Annweiler C, Karras SN, Anagnostis P, Beauchet O. Vitamin D supplements: a novel therapeutic approach for Alzheimer patients. Front Pharmacol. 2014 Jan 28;5:6.

[97] Andhoff BKS, Sandhoff K. My journey into the world of sphingolipids and sphingolipidoses. Proc Jpn Acad Ser B Phys Biol Sci. 2012;88(10):554–82. doi: 10.2183/pjab.88.554.

[98] Lingwood D, Simons K. Lipid rafts as a membrane-organizing principle. Science. 2010 Jan 1;327(5961):46–50. doi: 10.1126/science.1174621.

[99] Zhou K, Blom T. Trafficking and functions of bioactive sphingolipids: lessons from cells and model membranes. Lipid Insights. 2015;8(S1):11–20. doi: 10.4137/LPI.S31615.

[100] Gomez-Muñoz A, Frago LM, Alvarez L, Varela-Nieto I. Stimulation of DNA synthesis by natural ceramide 1-phosphate. Biochem J. 1997 Jul 15;325(Pt 2):435–40. doi: 10.1042/bj3250435.

[101] Saddoughi SA, Ogretmen B. Diverse functions of ceramide in cancer cell death and proliferation. Adv Cancer Res. 2013;117:37–58. doi: 10.1016/B978-0-12-394274-6.00002-9.

[102] Schwab SR, Pereira JP, Matloubian M, Xu Y, Huang Y, Cyster JG. Lymphocyte sequestration through S1P lyase inhibition and disruption of S1P gradients. Science. 2005 Sep 9;309(5741):1735–9. doi: 10.1126/science.1113640.

[103] Podbielska M, Krotkiewski H, Hogan EL. Signaling and regulatory functions of bioactive sphingolipids as therapeutic targets in multiple sclerosis. Neurochem Res. 2012 Jun;37(6):1154–69. doi: 10.1007/s11064-012-0728-y.

[104] Tidhar R, Futerman AH. The complexity of sphingolipid biosynthesis in the endoplasmic reticulum. Biochim Biophys Acta. 2013 Nov;1833(11):2511–8. doi: 10.1016/j.bbamcr.

[105] Sato K et al. Critical role of ABCA1 transporter in sphingosine 1-phosphate release from astrocytes. J Neurochem. 2007 Dec;103(6):2610–9. doi: 10.1111/j.1471-4159.2007.04958.x.

[106] Hisano Y, Kobayashi N, Kawahara A, Yamaguchi A, Nishi T. The sphingosine 1-phosphate transporter, SPNS2, functions as a transporter of the phosphorylated form of the immunomodulating agent FTY720. J Biol Chem. 2011;286(3):1758–66. doi: 10.1074/jbc.M110.171116.

[107] Proia RL, Hla T. Emerging biology of sphingosine-1-phosphate: its role in pathogenesis and therapy. J Clin Invest. 2015 Apr;125(4):1379–87. doi: 10.1172/JCI76369.

[108] Khavandgar Z, Poirier C, Clarke CJ, Li J, Wang N, McKee MD, Hannun YA, Murshed. A cell-autonomous requirement for neutral sphingomyelinase 2 in bone mineralization. J Cell Biol. 2011;194(2):277–89. doi: 10.1083/jcb.201102051.

[109] Stoffel W. Functional analysis of acid and neutral sphingomyelinases in vitro and in vivo. Chem Phys Lipids. 1999 Nov;102(1–2):107–21. doi: 10.1016/S0009-3084(99)00079-1.

[110] Dziak R. The role of sphingosine-1-phosphate (S1P) and lysophosphatidic acid (LPA) in regulation of osteoclastic and osteoblastic cells. Immunol Invest. 2013;42(7):510–8. doi: 10.3109/08820139.2013.823804.

[111] Kozawa O, Niwa M, Matsuno H, Tokuda H, Miwa M, Ito H, Kato K, Uematsu T. Sphingosine 1-phosphate induces heat shock protein 27 via p38 mitogen-activated protein kinase activation in osteoblasts. J Bone Miner Res. 1999 Oct;14(10):1761–7. doi: 10.1359/jbmr.1999.14.10.1761.

[112] Ishii M, Egen JG, Klauschen F, Meier-Schellersheim M, Saeki Y, Vacher J, Proia RL, Germain RN. Sphingosine-1-phosphate mobilizes osteoclast precursors and regulates bone homeostasis. Nature. 2009 Mar 26;458(7237):524–8. doi: 10.1038/nature07713.

[113] Lotinun S et al. Osteoclast-specific cathepsin K deletion stimulates S1P-dependent bone formation. J Clin Invest. 2013 Feb;123(2):666–81. doi: 10.1172/JCI64840.

[114] Quint P, Ruan M, Pederson L, Kassem M, Westendorf JJ, Khosla S, Oursler MJ. Sphingosine 1-phosphate (S1P) receptors 1 and 2 coordinately induce mesenchymal cell migration through S1P activation of complementary kinase pathways. J Biol Chem. 2013 Feb 22;288(8):5398–406. doi: 10.1074/jbc.M112.413583.

[115] Keller J et al. Calcitonin controls bone formation by inhibiting the release of sphingosine 1-phosphate from osteoclasts. Nat Commun. 2014 Oct 21;5:5215. doi: 10.1038/ncomms6215.

[116] Matsuzaki E et al. Sphingosine-1-phosphate promotes the nuclear translocation of β-catenin and thereby induces osteoprotegerin gene expression in osteoblast-like cell lines. Bone. 2013 Aug;55(2):315–24. doi: 10.1016/j.bonc.2013.04.008.

[117] Liu R, Farach-Carson MC, Karin NJ. Effects of sphingosine derivatives on MC3T3-E1 pre-osteoblasts: psychosine elicits release of calcium from intracellular stores. Biochem Biophys Res Commun. 1995;214(2):676–84. doi: 10.1006/bbrc.1995.2339.

[118] Lyons JM, Karin NJ. A role for G protein-coupled lysophospholipid receptors in sphingolipid-induced Ca^{2+} signaling in MC3T3-E1 osteoblastic cells. J Bone Miner Res. 2001;16(11):2035–42. doi: 10.1359/jbmr.2001.16.11.2035.

[119] Grey A, Chen Q, Callon K, Xu X, Reid IR, Cornish J. The phospholipids sphingosine-1-phosphate and lysophosphatidic acid prevent apoptosis in osteoblastic cells via a signaling pathway involving G(i) proteins and phosphatidylinositol-3 kinase. Endocrinology. 2002;143(12):4755–63. doi: 10.1210/en.2002-220347.

[120] Ishii M, Kikuta J. Sphingosine-1-phosphate signaling controlling osteoclasts and bone homeostasis. Biochim Biophys Acta. 2013 Jan;1831(1):223–7. doi: 10.1016/j.bbalip.2012.06.002.

[121] Okamoto H, Takuwa N, Yokomizo T, Sugimoto N, Sakurada S, Shigematsu H, Takuwa Y. Inhibitory regulation of Rac activation, membrane ruffling, and cell migration by the G protein-coupled sphingosine-1-phosphate receptor EDG5 but not EDG1 or EDG3. Mol Cell Biol. 2000 Dec;20(24):9247–61. doi: 10.1128/MCB.20.24.9247-9261.2000.

[122] Kim BJ et al. Plasma sphingosine 1-phosphate levels and the risk of vertebral fracture in postmenopausal women. J Clin Endocrinol Metab. 2012 Oct;97(10):3807–14. doi: 10.1210/jc.2012-2346.

[123] Marycz K, Krzak J, Marędziak M, Tomaszewski KA, Szczurek A, Moszak K. The influence of metal-based biomaterials functionalized with sphingosine-1-phosphate on the cellular response and osteogenic differentiation potential of human adipose derived mesenchymal stem cells in vitro. J Biomater Appl. 2016 Jan 22. doi: 10.1177/0885328216628711.

[124] Aubin I et al. A deletion in the gene encoding sphingomyelin phosphodiesterase3 results in osteogenesis and dentinogenesis imperfecta in the mouse. Nat Genet. 2005;37(8):803–5.

[125] Stoffel W, Jenke B, Block B, Zumbansen M, Koebke J. Neutral sphingomyelinase 2 (smpd3) in the control of postnatal growth and development. Proc Natl Acad Sci USA. 2005;102(12):4554–9. doi: 10.1073/pnas.0406380102.

[126] Yadav MC, Simao AM, Narisawa S, Huesa C, McKee MD, Farquharson C, Millan JL. Loss of skeletal mineralization by the simultaneous ablation of PHOSPHO1 and alkaline phosphatase function: a unified model of the mechanisms of initiation of skeletal calcification. J Bone Miner Res. 2011;26(2):286–97. doi: 10.1002/jbmr.195.

[127] Kakoi H et al. Bone morphogenic protein (BMP) signaling upregulates neutral sphingomyelinase 2 to suppress chondrocyte maturation via the Akt protein signaling pathway as a negative feedback mechanism. J Biol Chem. 2014;289(12):8135–50. doi: 10.1074/jbc.M113.509331.

[128] Li Z et al. Choline kinase beta is required for normal endochondral bone formation. Biochim Biophys Acta. 2014;1840(7):2112–22. doi: 10.1016/j.bbagen.2014.03.008.

[129] Haimovitz-Friedman A, Kan CC, Ehleiter D, Persaud RS, McLoughlin M, Fuks Z, Kolesnick RN. Ionizing radiation acts on cellular membranes to generate ceramide and initiate apoptosis. J Exp Med. 1994 Aug 1;180(2):525–35. non lo trovo.

[130] Hill PA, Tumber A. Ceramide-induced cell death/survival in murine osteoblasts. J Endocrinol. 2010 Aug;206(2):225–33. doi: 10.1677/JOE-10-0068.

[131] Takeda H, Ozaki K, Yasuda H, Ishida M, Kitano S, Hanazawa S. Sphingomyelinase and ceramide inhibit formation of F-actin ring in and bone resorption by rabbit mature osteoclasts. FEBS Lett. 1998 Jan 30;422(2):255–8. doi: 10.1016/S0014-5793(98)00005-2.

[132] Zhai L, Sun N, Han Z, Jin HC, Zhang B. Liposomal short-chain C6 ceramide induces potent anti-osteosarcoma activity in vitro and in vivo. Biochem Biophys Res Commun. 2015 Dec 4;468(1–2):274–80. doi: 10.1016/j.bbrc.2015.10.113.

[133] Takeshita A, Watanabe A, Takada Y, Hanazawa S. Selective stimulation by ceramide of the expression of the alpha isoform of retinoic acid and retinoid X receptors in osteoblastic cells. A role of sphingosine 1-phosphate-mediated AP-1 in the ligand-dependent

transcriptional activity of these receptors. J Biol Chem. 2000 Oct 13;275(41):32220–6. doi: 10.1074/jbc.M002569200.

[134] Olivier S, Fillet M, Malaise M, Piette J, Bours V, Merville MP, Franchimont N. Sodium nitroprusside-induced osteoblastapoptosis is mediated by long chain ceramide and is decreased by raloxifene. Biochem Pharmacol. 2005 Mar 15;69(6):891–901. doi: 10.1016/j.bcp.2004.11.030.

[135] Kitajima I, Soejima Y, Takasaki I, Beppu H, Tokioka T, Maruyama I. Ceramide-induced nuclear translocation of NF-kappa B is a potential mediator of the apoptotic response to TNF-alpha in murine clonal osteoblasts. Bone. 1996;19(3):263–70. doi: 10.1016/8756-3282(96)00181-0.

[136] Chae HJ et al. Dexamethasone suppresses tumor necrosis factor-alpha-induced apoptosis in osteoblasts: possible role for ceramide. Endocrinology. 2000 Aug;141(8):2904–13. doi: 10.1210/endo.141.8.7604.

[137] Kolesnick RN, Krönke M. Regulation of ceramide production and apoptosis. Annu Rev Physiol. 1998;60:643–65. doi: 10.1146/annurev.physiol.60.1.643.

[138] Ghosh N, Patel N, Jiang K, Watson JE, Cheng J, Chalfant CE, Cooper DR. Ceramide-activated protein phosphatase involvement in insulin resistance via Akt, serine/arginine-rich protein 40, and ribonucleic acid splicing in L6 skeletal muscle cells. Endocrinology. 2007 Mar;148(3):1359–66. doi: 10.1210/en.2006-0750.

[139] Hannun YA. Functions of ceramide in coordinating cellular responses to stress. Science. 1996 Dec;274(5294):1855–9. doi: 10.1126/science.274.5294.1855.

[140] Sabourdy F et al. Monogenic neurological disorders of sphingolipid metabolism. Biochim Biophys Acta. 2015 Aug;1851(8):1040–51. doi: 10.1016/j.bbalip.2015.01.010.

[141] Haughey NJ, Bandaru VV, Bae M, Mattson MP. Roles for dysfunctional sphingolipid metabolism in Alzheimer's disease neuropathogenesis. Biochim Biophys Acta. 2010 Aug;1801(8):878–86. doi: 10.1016/j.bbalip.2010.05.003.

[142] Grimm MO, Zimmer VC, Lehmann J, Grimm HS, Hartmann T. The impact of cholesterol, DHA, and sphingolipids on Alzheimer's disease. Biomed Res Int. 2013;2013:814390. doi: 10.1155/2013/814390.

[143] Murphy KE, Gysbers AM, Abbott SK, Tayebi N, Kim WS, Sidransky E, Cooper A, Garner B, Halliday GM. Reduced glucocerebrosidase is associated with increased α-synuclein in sporadic Parkinson's disease. Brain. 2014 Mar;137(Pt 3):834–48. doi: 10.1093/brain/awt367.

[144] Abbott SK, Li H, Muñoz SS, Knoch B, Batterham M, Murphy KE, Halliday GM, Garner B. Altered ceramide acyl chain length and ceramide synthase gene expression in Parkinson's disease. Mov Disord. 2014 Apr;29(4):518–26. doi: 10.1002/mds.25729.

[145] Halmer R, Walter S, Faßbender K. Sphingolipids: important players in multiple sclerosis. Cell Physiol Biochem. 2014;34(1):111–8. doi: 10.1159/000362988.

[146] Gulbins E et al. A central role for the acid sphingomyelinase/ceramide system in neurogenesis and major depression. J Neurochem. 2015 Jul;134(2):183–92. doi: 10.1111/jnc.13145.

[147] Mencarelli C, Martinez-Martinez P. Ceramide function in the brain: when a slight tilt is enough. Cell Mol Life Sci. 2013 Jan;70(2):181–203.

[148] Shamseddine AA, Airola MV, Hannun YA. Roles and regulation of neutral sphingomyelinase-2 in cellular and pathological processes. Adv Biol Regul. 2015 Jan;57:24–41. doi: 10.1016/j.jbior.2014.10.002.

[149] Ghasemi R, Dargahi L, Ahmadiani A. Integrated sphingosine-1 phosphate signaling in the central nervous system: from physiological equilibrium to pathological damage. Pharmacol Res. 2016 Jan 7;104:156–64. doi: 10.1016/j.phrs.2015.11.006.

[150] Lucki NC, Sewer MB. Multiple roles for sphingolipids in steroid hormone biosynthesis. Subcell Biochem. 2008;49:387–412. doi: 10.1007/978-1-4020-8831-5_15.

[151] Titievsky A, Titievskaya I, Pasternack M, Kaila K, Törnquist K. Sphingosine inhibits voltage-operated calcium channels in GH4C1 cells. J Biol Chem. 1998 Jan 2;273(1):242–7.

[152] Qin X et al. Sphingosine and FTY720 are potent inhibitors of the transient receptor potential melastatin 7 (TRPM7) channels. Br J Pharmacol. 2013 Mar;168(6):1294–312. doi: 10.1111/bph.12012.

[153] Rohrbough J, Rushton E, Palanker L, Woodruff E, Matthies HJ, Acharya U. Ceramidase regulates synaptic vesicle exocytosis and trafficking. J Neurosci. 2004 Sep 8;24(36):7789–803.

[154] Darios F et al. Sphingosine facilitates SNARE complex assembly and activates synaptic vesicle exocytosis. Neuron. 2009 Jun 11;62(5):683–94. doi: 10.1016/j.neuron.2009.04.024.

[155] Paugh SW, Cassidy MP, He H, Milstien S, Sim-Selley LJ, Spiegel S, Selley DE. Sphingosine and its analog, the immunosuppressant 2-amino-2-(2-[4-octylphenyl]ethyl)-1,3-propanediol, interact with the CB1 cannabinoid receptor. Mol Pharmacol. 2006 Jul;70(1):41–50.

[156] Selley DE, Welch SP, Sim-Selley LJ. Sphingosine lysolipids in the CNS: endogenous cannabinoid antagonists or a parallel pain modulatory system? Life Sci. 2013 Aug 14;93(5–6):187–93. doi: 10.1016/j.lfs.2013.06.004.

[157] Mizugishi K, Yamashita T, Olivera A, Miller GF, Spiegel S, Proia RL. Essential role for sphingosine kinases in neural and vascular development. Mol Cell Biol. 2005 Dec;25(24):11113–21.

[158] Kimura A et al. Essential roles of sphingosine1-phosphate/S1P1 receptor axis in the migration of neural stem cells toward a site of spinal cord injury. Stem Cells. 2007 Jan;25(1):115–24. doi: 10.1634/stemcells.2006-0223.

[159] Novgorodov AS, El-Alwani M, Bielawski J, Obeid LM, Gudz TI. Activation of sphingosine-1-phosphate receptor S1P5 inhibits oligodendrocyte progenitor migration. FASEB J. 2007 May;21(7):1503–14.

[160] MacLennan AJ et al. An essential role for the H218/AGR16/Edg-5/LP(B2) sphingosine 1-phosphate receptor in neuronal excitability. Eur J Neurosci. 2001 Jul;14(2):203–9.

[161] Jaillard C et al. Edg8/S1P5: an oligodendroglial receptor with dual function on process retraction and cell survival. J Neurosci. 2005 Feb 9;25(6):1459–69.

[162] Toman RE, Payne SG, Watterson KR, Maceyka M, Lee NH, Milstien S, Bigbee JW, Spiegel S. Differential transactivation of sphingosine-1-phosphate receptors modulates NGF-induced neurite extension. J Cell Biol. 2004 Aug 2;166(3):381–92.

[163] Kanno T, Nishizaki T, Proia RL, Kajimoto T, Jahangeer S, Okada T, Nakamura S. Regulation of synaptic strength by sphingosine 1-phosphate in the hippocampus. Neuroscience. 2010 Dec 29;171(4):973–80. doi: 10.1016/j.neuroscience.2010.10.021.

[164] Li C, Li JN, Kays J, Guerrero M, Nicol GD. Sphingosine 1-phosphate enhances the excitability of rat sensory neurons through activation of sphingosine 1-phosphate receptors 1 and/or 3. J Neuroinflamm. 2015 Apr 12;12:70. doi: 10.1186/s12974-015-0286-8.

[165] Chan JP, Sieburth D. Localized sphingolipid signaling at presynaptic terminals is regulated by calcium influx and promotes recruitment of priming factors. J Neurosci. 2012 Dec 5;32(49):17909–20. doi: 10.1523/JNEUROSCI.2808-12.2012.

[166] Shen H et al. Coupling between endocytosis and sphingosine kinase 1 recruitment. Nat Cell Biol. 2014 Jul;16(7):652–62. doi: 10.1038/ncb2987.

[167] Wang G, Dinkins M, He Q, Zhu G, Poirier C, Campbell A, Mayer-Proschel M, Bieberich E. Astrocytes secrete exosomes enriched with proapoptotic ceramide and prostate apoptosis response 4 (PAR-4): potential mechanism of apoptosis induction in Alzheimer disease (AD). J Biol Chem. 2012 Jun 15;287(25):21384–95. doi: 10.1074/jbc.M112.340513.

[168] Scarlatti F, Bauvy C, Ventruti A, Sala G, Cluzeaud F, Vandewalle A, Ghidoni R, Codogno P. Ceramide-mediated macroautophagy involves inhibition of protein kinase B and up-regulation of beclin 1. J Biol Chem. 2004 Apr 30;279(18):18384–91.

[169] Ben-David O, Futerman AH. The role of the ceramide acyl chain length in neurodegeneration: involvement of ceramide synthases. Neuromol Med. 2010 Dec;124:341–50. doi: 10.1007/s12017-010-8114-x.

[170] Zhao L et al. A deficiency of ceramide biosynthesis causes cerebellar purkinje cell neurodegeneration and lipofuscin accumulation. PLoS Genet. 2011 May;7(5):e1002063. doi: 10.1371/journal.pgen.1002063.

[171] Ginkel C et al. Ablation of neuronal ceramide synthase 1 in mice decreases ganglioside levels and expression of myelin-associated glycoprotein in oligodendrocytes. J Biol Chem. 2012 Dec 7;287(50):41888–902. doi: 10.1074/jbc.M112.413500.

[172] Kim SK et al. Neutral sphingomyelinase 2 induces dopamine uptake through regulation of intracellular calcium. Cell Signal. 2010 May;22(5):865–70. doi: 10.1016/j.cellsig.2010.01.012.

[173] Wheeler D et al. Tumor necrosis factor-alpha-induced neutral sphingomyelinase-2 modulates synaptic plasticity by controlling the membrane insertion of NMDA receptors. J Neurochem. 2009 Jun;109(5):1237–49. doi: 10.1111/j.1471-4159.2009.06038.x.

[174] Tabatadze N, Savonenko A, Song H, Bandaru VV, Chu M, Haughey NJ. Inhibition of neutral sphingomyelinase-2 perturbs brain sphingolipid balance and spatial memory in mice. J Neurosci Res. 2010 Oct;88(13):2940–51. doi: 10.1002/jnr.22438.

[175] Gu L, Huang B, Shen W, Gao L, Ding Z, Wu H, Guo J. Early activation of nSMase2/ceramide pathway in astrocytes is involved in ischemia-associated neuronal damage via inflammation in rat hippocampi. J Neuroinflamm. 2013 Sep 3;10:109. doi: 10.1186/1742-2094-10-109.

[176] He Q, Wang G, Wakade S, Dasgupta S, Dinkins M, Kong JN, Spassieva SD, Bieberich E. Primary cilia in stem cells and neural progenitors are regulated by neutral sphingomyelinase 2 and ceramide. Mol Biol Cell. 2014 Jun;25(11):1715–29. doi: 10.1091/mbc.E13-12-0730.

[177] Bajjalieh SM, Martin TF, Floor E. Synaptic vesicle ceramide kinase. A calcium-stimulated lipid kinase that co-purifies with brain synaptic vesicles. J Biol Chem. 1989 Aug 25;264(24):14354–60.

[178] Sugiura M, Kono K, Liu H, Shimizugawa T, Minekura H, Spiegel S, Kohama T. Ceramide kinase, a novel lipid kinase. Molecular cloning and functional characterization. J Biol Chem. 2002 Jun 28;27726:23294–300.

[179] Presa N, Gomez-Larrauri A, Rivera IG, Ordoñez M, Trueba M, Gomez-Muñoz A. Regulation of cell migration and inflammation by ceramide 1-phosphate. Biochim Biophys Acta. 2016 Feb 11. pii: S1388-1981(16)30028-2. doi: 10.1016/j.bbalip.2016.02.007.

[180] Pettus BJ et al. Ceramide 1-phosphate is a direct activator of cytosolic phospholipase A2. J Biol Chem. 2004 Mar 19;279(12):11320–6.

[181] Kim C et al. Ceramide-1-phosphate regulates migration of multipotent stromal cells and endothelial progenitor cells—implications for tissue regeneration. Stem Cells. 2013 Mar;313:500–10. doi: 10.1002/stem.1291.

[182] Reardon BJ et al. Vitamin D-responsive SGPP2 variants associated with lung cell expression and lung function. BMC Med Genet. 2013 Nov 25;14:122.

[183] Kleuser B, Cuvillier O, Spiegel S. 1Alpha,25-dihydroxyvitamin D3 inhibits programmed cell death in HL-60 cells by activation of sphingosine kinase. Cancer Res. 1998 May 1;58(9):1817–24.

[184] Manggau M, Kim DS, Ruwisch L, Vogler R, Korting HC, Schäfer-Korting M, Kleuser B. 1Alpha,25-dihydroxyvitamin D3 protects human keratinocytes from apoptosis by the formation of sphingosine-1-phosphate. J Invest Dermatol. 2001 Nov;117(5):1241–9.

[185] Kikuta J, Kawamura S, Okiji F, Shirazaki M, Sakai S, Saito H, Ishii M. Sphingosine-1-phosphate-mediated osteoclast precursor monocyte migration is a critical point of control in antibone-resorptive action of active vitamin D. Proc Natl Acad Sci U S A. 2013 Apr 23;110(17):7009–13.

[186] Selkoe DJ. The therapeutics of Alzheimer's disease: where we stand and where we are heading. Ann Neurol. 2013 Sep;74(3):328–36. doi: 10.1002/ana.24001.

[187] Satoi H, Tomimoto H, Ohtani R, Kitano T, Kondo T, Watanabe M, Oka N, Akiguchi I, Furuya S, Hirabayashi Y, Okazaki T. Astroglial expression of ceramide in Alzheimer's disease brains: a role during neuronal apoptosis. Neuroscience. 2005;130(3):657–66.

[188] Katsel P, Li C, Haroutunian V. Gene expression alterations in the sphingolipid metabolism pathways during progression of dementia and Alzheimer's disease: a shift toward ceramide accumulation at the earliest recognizable stages of Alzheimer's disease?. Neurochem Res. 2007 Apr–May;32(4–5):845–56.

[189] Takasugi N et al. BACE1 activity is modulated by cell-associated sphingosine-1-phosphate. J Neurosci. 2011 May 4;31(18):6850–7. doi: 10.1523/JNEUROSCI.6467 10.2011.

[190] Grimm MO, Rothhaar TL, Hartmann T. The role of APP proteolytic processing in lipid metabolism. Exp Brain Res. 2012 Apr;217(3–4):365 75. doi: 10.1007/s00221-011-2975-6.

[191] Couttas TA, Lim XY, Don AS. A three-step assay for ceramide synthase activity using a fluorescent substrate and HPLC. Lipids. 2015 Jan;50(1):101–9.

[192] Ceccom J et al. Reduced sphingosine kinase 1 and enhanced sphingosine 1-phosphate lyase expression demonstrate deregulated sphingosine 1-phosphate signaling in Alzheimer's disease. Acta Neuropathol Commun. 2014;2(1):12. doi: 10.1186/2051-5960-2-12.

[193] Grimm C, Kraft R, Schultz G, Harteneck C. Activation of the melastatin-related cation channel TRPM3 by D-erythro-sphingosine. Mol Pharmacol. 2005 Mar;67(3):798–805. Epub 2004 Nov 18.

[194] Jazvinšćak Jembrek M, Hof PR, Šimić G. Ceramides in Alzheimer's disease: key mediators of neuronal apoptosis induced by oxidative stress and Aβ accumulation. Oxid Med Cell Longev. 2015;2015:346783. doi: 10.1155/2015/346783.

[195] Jana A, Pahan K. Fibrillar amyloid-beta peptides kill human primary neurons via NADPH oxidase-mediated activation of neutral sphingomyelinase. Implications for Alzheimer's disease. J Biol Chem. 2004 Dec 3;279(49):51451–9.

[196] Malaplate-Armand C et al. Soluble oligomers of amyloid-beta peptide induce neuronal apoptosis by activating a cPLA2-dependent sphingomyelinase-ceramide pathway. Neurobiol Dis. 2006 Jul;23(1):178–89.

[197] Puglielli L, Ellis BC, Saunders AJ, Kovacs DM. Ceramide stabilizes beta-site amyloid precursor protein-cleaving enzyme 1 and promotes amyloid beta-peptide biogenesis. J Biol Chem. 2003 May 30;278(22):19777–83.

[198] Brunkhorst R, Vutukuri R, Pfeilschifter W. Fingolimod for the treatment of neurological diseases-state of play and future perspectives. Front Cell Neurosci. 2014 Sep 12;8:283. doi: 10.3389/fncel.2014.00283.

[199] Asle-Rousta M, Kolahdooz Z, Oryan S, Ahmadiani A, Dargahi L. FTY720 (fingolimod) attenuates beta-amyloid peptide (Aβ42)-induced impairment of spatial learning and memory in rats. J Mol Neurosci. 2013 Jul;50(3):524–32. doi: 10.1007/s12031-013-9979-6.

[200] Hemmati F, Dargahi L, Nasoohi S, Omidbakhsh R, Mohamed Z, Chik Z, Naidu M, Ahmadiani A. Neurorestorative effect of FTY720 in a rat model of Alzheimer's disease: comparison with memantine. Behav Brain Res. 2013 Sep 1;252:415–21. doi: 10.1016/j.bbr.2013.06.016.

[201] Gezen-Ak D et al. Vitamin D receptor gene haplotype is associated with late-onset Alzheimer's disease. Tohoku J Exp Med. 2012;228(3):189–96.

[202] Dursun E, Gezen-Ak D, Yilmazer S. Beta amyloid suppresses the expression of the vitamin d receptor gene and induces the expression of the vitamin d catabolic enzyme gene in hippocampal neurons. Dement Geriatr Cogn Disord. 2013;36:76–86.

[203] Durk MR, Han K, Chow EC, Ahrens R, Henderson JT, Fraser PE, Pang KS. 1α,25-Dihydroxyvitamin D3 reduces cerebral amyloid-β accumulation and improves cognition in mouse models of Alzheimer's disease. J Neurosci. 2014 May 21;34(21):7091–101.

[204] Meacci E, Bini F, Cecchi M, Garcia-Gil M. Role of sphingolipids on the citotoxicity induced by 1–42 beta amyloid peptide in differentiated SH-SY5Y cells. Acta Physiol. 2008;194:665.

Chapter 7

Immunomodulatory Effect of Vitamin D in Children with Allergic Diseases

Iwona Stelmach, Joanna Jerzyńska and
Daniela Podlecka

Additional information is available at the end of the chapter

http://dx.doi.org/10.5772/65072

Abstract

The discovery that many cells express vitamin D receptors and the recognition of widespread vitamin D insufficiency has stimulated interest in the potential role of vitamin D in nonskeleton conditions. There is an increasing evidence to support the role of vitamin D pathway in the regulation of the function of both innate and adoptive immune systems. Vitamin D regulates immune function by inhibiting the differentiation and maturation of human dendritic cells, enhancing interleukin (IL)-10 and tumor growth factor-β (TGF-β) secretion and inhibiting T-cell functions. Vitamin D has the ability to suppress inflammatory cytokines, such as tumor necrosis factor (TNF), interleukin-1 (IL-1), interferon gamma (IFN-γ), and interleukin-2 (IL-2), while it increases the generation of anti-inflammatory cytokines IL-4 and IL-10. In B cells, vitamin D3 has also been shown to suppress immunoglobulin E (IgE) antibody class switch partly through the inhibition of nuclear factor kappa-light-chain-enhancer of activated B cells (NF-κB).

Keywords: allergic diseases, vitamin D, vitamin's D role, methabolism, implication on immune system

1. Introduction

Vitamin D deficiency has been proposed as a potential contributing factor in patients with allergic diseases. It has been shown that serum levels of vitamin D correlate with pulmonary function, asthma onset, and the development of allergic diseases. Based on clinical and observational data, the plasma level of 25-hydroxyvitamin D may serve as a "marker" to detect or define a subclinical deficiency of vitamin D.

Its possible role in immunotherapy has also been studied. Natural activity of vitamin D seems to be highly attractive in the context of the mechanism and clinical effect of subcutaneous immunotherapy (SIT); however, its role in sublingual immunotherapy (SLIT) is still undefined. Vitamin D enhances interleukin (IL)-10 production by CD4+ T cells and enhances sublingual immunotherapy efficacy in a murine asthma model. Recently, Heine et al. showed that 25-hydroxvitamin D3 promotes the long-term effect of specific immunotherapy in a murine allergy model; their findings were paralleled by reduced Th2 cytokine expression in the lungs. They observed that vitamin D deficiency promotes the development of type I sensitization and correction of its serum concentrations enhances the benefit of specific immunotherapy.

2. Vitamin D metabolism

Vitamin D refers to a group of fat-soluble corticosteroids responsible for enhancing intestinal absorption of calcium, iron, magnesium, phosphate, and zinc. In humans, the most important compounds in this group are vitamin D3 (also known as cholecalciferol) and vitamin D2 (ergocalciferol).

There are two ways in which vitamin D3 (cholecalciferol) is normally provided to human body:

1. It may be produced in the skin as a result of ultraviolet irradiation of 7-dehydrocholesterol (previtamin D), which is slowly isomerized to vitamin D3 [1, 2].
2. It is also derived from the diet in fish (vitamin D3) and in plants (mainly vitamin D2); vitamin D2 is derived from irradiation of ergosterol, which occurs to some degree in plankton or plants under natural conditions [1].

Both vitamin D3 and vitamin D2 are inactive in many biological systems and must undergo a series of metabolic transformations before exerting effects in target tissues [1–3].

The prohormone vitamin D3 is produced in the skin through ultraviolet irradiation of 7-dehydrocholesterol, than to make it biologically active the prohormone vitamin D is transported in the blood by the vitamin D binding protein (DBP) to the liver, where it is metabolized to 25-hydroxyvitamin [4]. In the liver, vitamin D is hydroxylated at C-25 by cytochrome P 450 vitamin D 25- hydroxylases, resulting in the formation of 25-hydroxyvitamin D (25(OH)D3). This is the major circulating metabolite of vitamin D in plasma, and its measurements are used to provide an index of vitamin D nutritional status [2, 3]. Based on the evidence in the clinical trials and meta-analyses, the workgroup concluded that a serum 25 hydroxyvitamin D concentration of 30 ng/mL (75 nmol/L) should be a minimum goal to achieve in adults and children [5–7]. The best outcomes were seen with 25(OH) levels from 36 to 40 ng/mL (90–100 nmol/L) [6–8].

However, 25(OH)D3 itself is metabolically inactive and must be modified before function [3]. On the next metabolic step, 25(OH)D3 is transported by DBP to the kidney, where in the proximal renal tube it is hydroxylated at the position C1 (position of carbon 1), resulting in the hormonally active form of vitamin D 1,25-dihydroxyvitamin D3 (1,25$(OH)_2D_3$), which is responsible for most of the biological actions of vitamin D [4]. As a result of 1 and 25 hydroxylation, the prohormone vitamin D is being structurally transformed into an active hormone.

The best known target organ effects of 1,25$(OH)_2D_3$ are on the intestine to stimulate absorption of calcium and phosphate, next on bone to cause release of calcium and phosphate. It has been proved that there exists a negative feedback mechanism with the hormonal systems (parathyroid hormone, prolactin, and growth hormone) and the effect of 1,25 $(OH)_2D_3$ [1, 2].

Beside 1,25$(OH)_2D_3$, the kidney can also produce 24,25 dihydroxyvitamin D3 (24,25$(OH)_2D_3$), a relatively inactive metabolite comparing to 1,25 $(OH)_2D_3$. The metabolite 24,25 $(OH)_2D_3$ is produced by 24 hydroxyvitamin D3 24 hydroxylase from both substrates: 25(OH)D3 and 1,25 $(OH)_2D_3$ [4, 9]. This enzyme 24(OH)D3 24 hydroxylase limits the amount of 1,25$(OH)_2D_3$ in target tissues by accelerating the catabolism of 1,25 $(OH)_2D_3$ and by decreasing the amount of 25$(OH)D_3$ available to 1 hydroxylation [4]. Summarizing the main role of 24(OH)lase is vitamin D inactivation.

3. The mechanism of action of vitamin D

The vitamin D functions through a single vitamin D receptor (VDR). It is classified as a steroid hormone, related to the retinoic acid receptors and as most of the receptors has a DNA-binding domain (called C-Domain), a ligand-binding domain (called E-domain), and finally activating domain (called F-domain) [1]. It has been proved that a single receptor mediates all of the functions of vitamin D. This receptor is a 427 amino acid peptide and acts through vitamin D-responsive elements (VDREs), which are placed on the start site of the target gene [1].

VDR was proved to be found in almost all tissues and cells. It has been established that the diverse biological actions of 1,25-dihydroyxyvitamin D3 are initiated through precise changes in gene expression, which are mediated by an intracellular VDR. Activation of the VDR through direct interaction with 1,25$(OH)_2D_3$ prompts the receptor's rapid binding to regulatory regions of target genes, where it acts to nucleate the formation of large protein complexes whose functional activities are essential for directed changes in transcription. When VDR interacts with the ligand, the repressor is no longer able to bind the receptor and the receptor changes conformation, forming heterodimer at the VDREs [1, 10]. At the same time, it binds several other proteins required in the transcription complex and acquires an activator [1, 10]. Once the complex is formed, DNA bends, biochemical processes take place, and transcription is either initiated or suppressed depending on the gene [1, 10, 11]. These responses are tissue-specific and range from highly complex actions essential for homeostatic control of mineral metabolism to focal actions that control the growth, differentiation, and functional activity of numerous cell types including those of the immune system, skin, the pancreas, and bone.

4. Physiological effects of vitamin D

Vitamin D is well known as a hormone involved in mineral metabolism and bone growth. Its main effect is to facilitate intestinal absorption of calcium; however, it also stimulates absorption of phosphate and magnesium ions. In cases when there is a lack of vitamin D, dietary calcium is not absorbed efficiently. Vitamin D stimulates the expression of a number of proteins involved in transporting calcium from the lumen of the intestine, across the epithelial cells and into blood. The best studied of these calcium transporters is calbindin, an intracellular protein that carries calcium across the intestinal epithelial cell [12].

Vitamin D-dependent calcium-binding proteins (calbindin-D) are the major proteins involved in intracellular calcium transport in the intestine and the kidneys. In mammals there are two classes of calbindin-D: the 9000-dalton calbindin (calbindin-D9k) and the 28,000-dalton calbindin (calbindin-D28k). First identified in the intestine, calbindin-D9k is also expressed in the kidneys, the placenta, the yolk sac, and the lungs. The expression of calbindin-D9k and calbindin-D28k in the intestine and kidneys has been shown to be regulated by 1,25-$(OH)_2D_3$ in animal models as well as in cell and organ cultures [13].

In several studies it has been shown that the regulation of calbindin-D9k and calbindin-28k by 1,25-$(OH)_2D_3$ involve both transcriptional and posttranscriptional mechanisms [12, 13]. It can also be found in other tissues (brain, placenta, and lungs); however, calbindin-D genes have not been shown to be regulated by 1,25-$(OH)_2D_3$ [14], suggesting that the regulation of calbindin-D gene expression by 1,25-$(OH)_2D_3$ involves tissue-specific factors.

There is evidence that calcium may also modulate the actions of 1,25-$(OH)_2D_3$ on calbindin gene expression. *In vitro* experiments demonstrate that the induction of renal calbindin-D28k by 1,25-$(OH)_2D_3$ is enhanced by high levels of extracellular calcium [12–14]. *In vivo* studies of the receptor-dependent actions of 1,25-(OH)2D3 on calbindin-D gene expression have been complicated by the inability to differentiate the effects of hormone deficiency from those of hypocalcemia.

Vitamin D response elements have been found in both the murine calbindin-D9k and calbindin-28k genes [12, 14]; however, Chun Li et al. demonstrated that the VDR dependence of calbindin-D28k gene expression varies among the tissues examined. Consistent with the observation that the kidney and lung express the VDR, calbindin-D28k mRNA levels in these tissues were induced in the control mice treated with 1,25-$(OH)_2D_3$.

There is evidence in the literature on the effects of vitamin D on bone tissue. As a transcriptional regulator of bone matrix proteins, it induces the expression of osteocalcin and suppresses synthesis of type I collagen. In cell cultures, vitamin D stimulates differentiation of osteoclasts. The crucial effect of vitamin D on bone is to provide the proper balance of calcium and phosphorus to support mineralization [12].

It turns out that vitamin D receptors are present in most if not all cells in the body. Additionally, experiments using cultured cells have demonstrated that vitamin D has potent effects on the growth and differentiation of many types of cells. These findings suggest that vitamin D has

http://dx.doi.org/10.5772/65072

physiologic effects much broader than that of a role in mineral homeostasis and bone function. This is an active area of research and a much better understanding of this area will likely be available in the near future [12].

5. Polymorphisms of the vitamin D receptor

It is known that there exist interindividual differences in the vitamin D endocrine system. This is due to the influence of variations in the DNA sequence of important proteins of this system [14]. For example, different mutations in VDR gene may cause obesity, 1,25-dihydroxyresistant rickets, or cancer [14–16]. The interpretation of VDR polymorphic variations is quite difficult, because the VDR gene is a large one, and the most of these are anonymous restriction fragment length polymorphisms (RLFP) that have unknown functional effect. It is believed that in the population there exist a lot of different mutations, some of them linked with the so-called complex diseases (as osteoporosis), other clinically insignificant or just unexplored [14].

To understand the mechanisms underlying in the associations, it is necessary to study the genomic organization of VDR locus, to determine haplotypes across the gene, and to analyze their relationship with RFLP and finally with a disease.

The VDR gene contains several polymorphisms, including three single nucleotide polymorphisms (SNPs) located near the 39 untranslated region are identified by their restriction endonuclease sites: Apa1 [15, 17], Bsm [15, 18], and Taq1 [15, 19]. The Apa1 and the Bsm1 polymorphisms of the VDR gene are considered to be silent single nucleotide polymorphisms and do not change the amino acid sequence of the end-coded protein but can affect gene expression through regulation of mRNA stability [15]. Analyzing the polymorphism of Taq1, it has been shown that depending on the presence or absence of Taq1 restriction site in each allele, products are described as follows: T allele—absence of the restriction site, and t—presence of the restriction site. According to this individuals are classified as TT, Tt, and tt. The TT genotype has been shown to be associated with lower circulating levels of active vitamin D3 [15].

6. The role of vitamin D in the immune system- lymphocytes T and B

There is evidence in the literature that there is an important role for vitamin D (more specifically, calcitriol) in the body's immune system. Most of these researches have been done in cultured cells and in animals with either severe vitamin D deficiency or whose genes have been altered to "knock out" proteins that control vitamin D metabolism or active vitamin D action. Vitamin D receptor is found in significant concentrations in the lymphocyte T and macrophages populations, but the highest concentration is observed in the immature immune cells of the thymus and the mature CD-8 T lymphocytes. This suggests that phagocytes may communicate with T and B cells through calcitriol [20, 21]. When vitamin D is present, it blocks the features of the adaptive immune system that would lead to autoimmunity. Animal studies show that when calcitriol is absent, the cells of the immune system are more likely to attack the healthy cells of the body (autoimmunity) [20]. There are two ways that vitamin D influences the immunity. First is by avoiding to trigger and arm the T cells during autoimmunity. Vitamin D can stimulate production of transforming growth factor (TGF)-β1 and interleukin (IL)-4, which may suppress inflammatory T cells activity. Blocking the production of these cells results in diminished ability of T cells to recognize the native protein as foreign. This leads to decrease in killer T cells production [20–22]. In other words, the presence of adequate levels of vitamin D and calcitriol keeps the T cells from attacking the body's own tissues. Second, as the number of T cells decreases, calcitriol also diminishes the role of B cells in producing chemicals to destroy native tissue. What is more, vitamin D inhibits B cell proliferation and blocks its differentiation and immunoglobulin secretion [21–23]. These mechanisms are considered to be involved in triggering/suppressing the autoimmune process in rheumatoid disease, diabetes type I, and systemic lupus erythematosus [21, 22].

The microenvironment in which naive Th cells develop determines which of the two subtypes (Th1 or Th2) will predominate. In normal immune response, both subtypes balance. Th1 produce INF-α, IL-2, and tumor necrosis factor (TNF)-α, and this response is strongly addressed to tumors or intracellular pathogens as viruses [24]. Th2 lymphocytes produce mainly TGF-β1and IL-4 and IL-5, and these mediators are linked with extracellular pathogens as bacteria and parasites [24]. It has been proved that Th1 and Th2 are targets of 1,25 $(OH)_2D_3$. 1,25 $(OH)_2D_3$ decreases the production of (IFN-γ), IL-2, and IL-5, and in Th2 cells it increases the production of IL-4 [24, 25]. In conclusion, vitamin D additionally suppresses T cell proliferation and results in shift from a Th1 to a Th2 phenotype [23].

7. The effect of vitamin D on monocytes/macrophages and dendritic cells

In the beginning of the century, several investigators concluded that VDR can be found also on the promyelocytes and that vitamin D can suppress proliferation of promyelocytes and cause their differentiation to monocytes [24]. What is more, it inhibits monocyte production of inflammatory cytokines such as IL-1, IL-6, IL-8, IL-12, and TNF-α [23, 26]. It additionally inhibits dendritic cell (DC) differentiation and maturation, directing to immature phenotype. The inhibition of DC differentiation and maturation plays a crucial role in the developing of the autoimmune processes and self-tolerance to own tissues. When mature DC presents an antigen to a T cell, it directs the immune response against the antigen. In reverse, when the antigen is presented by an immature DC it facilitates tolerance [23]. A decreased concentrations of vitamin D have been reported from different researchers concerning autoimmune diseases as bowel inflammatory disease and diabetes mellitus type I [22–24].

8. Association of allergic sensitization and vitamin D

Epidemiological and laboratory investigations have convincingly shown that vitamin D deficiency is associated with several common diseases, including rickets and other bone diseases, diabetes, cardiovascular diseases, autoimmune diseases, tuberculosis, and cancer. Sensitization to food allergen is a precursor and a risk factor for the development of allergic diseases later in life, which is why earliest possible detection and modification of risk indicators for food allergen sensitization might prevent the development of allergic diseases [27–29].

Vitamin D deficiency has been proposed as a potential contributing factor in patients with allergic diseases. It has been shown that serum levels of vitamin D correlate with pulmonary function, asthma onset, and the development of allergic diseases. Based on clinical and observational data, the plasma level of 25(OH)D may serve as a "marker" to detect or define a subclinical deficiency of vitamin D [30]. It has recently been demonstrated that lower levels of vitamin D are associated with reduced asthma control. Several studies confirmed that vitamin D interfere favorably with bone turnover as shown by a significant reduction in serum phosphorus of our patients. It is a well-known fact that asthmatic people with lower serum vitamin D concentrations do not respond as well to inhaled glucocorticosteroids therapy as those with optimal vitamin D concentrations. In several studies [31–33], it was proved that supplementation of vitamin D has a beneficial effect on glucocorticosteroids therapy response in patients with severe asthma. A work of Chen Wu et al. in children has supported a role for vitamin D in preventing asthma exacerbations in children in the CAMP trial, especially those children treated with inhaled corticosteroids [34]. Another study found that children with asthma who are deficient in vitamin D levels have less improvement in prebronchodilator FEV1 over the course of 1 year when treated with inhaled corticosteroids as compared with children who are sufficient in vitamin D [35]. These findings support the hypothesis that vitamin D supplementation may enhance the anti-inflammatory function of corticosteroids in patients with asthma, and what is more, we can use vitamin D concentrations as a "marker" concerning the potential effect of treatment in allergic diseases.

Sutherland et al. [36] also found that reduced vitamin D levels are associated with impaired lung function, especially in adults who are not being treated with inhaled corticosteroids. The authors suggested that vitamin D supplementation could be especially beneficial in patients who are not treated with inhaled corticosteroids [36].

9. Vitamin D and atopic dermatitis

The symptoms of atopic dermatitis do not always coincide with the levels of sensitization. To this date, the mechanism of how the vitamin D status is related to the development of atopic dermatitis stays unclear. However, there are two pathophysiological mechanisms in the development of atopic dermatitis: first, immunological mechanisms that influence the Th1/Th2 adaptive immune responses; vitamin D receptor agonists have been shown to impact on Th1 and Th2 cell function, suppress allergen-specific IgE synthesis, inhibit dendritic cell maturation and induce tolerogenic dendritic cells, and finally to induce regulatory CD4+CD25+25Foxp3+T cells. Second, injury of the skin barrier function [26, 29, 31]. Several recent studies have shown that infants with high level of sensitization had a decreased concentration of 25(OH)D levels [27, 29, 37]. What is more, the severity of atopic dermatitis was independently influenced by the serum 25(OH)D levels and by food allergen sensitization. Promising results of improved eczema symptoms have been found using oral vitamin D supplementation (1000 IU vitamin D for 30 days) in children between 2 and 17 years in two double-blinded placebo-controlled studies performed during wintertime [38, 39].

The prevalence of food allergy has increased dramatically over the past decade and has now reached epidemic levels in countries like Australia or United States, with up to 10% of 12-month-old infants having a clinically confirmed food allergy. As food allergies have increased, vitamin D levels in the population appear to have concurrently decreased. There is no strict evidence in the literature that food allergens could react with vitamin D; however, vitamin D3 supplements have the potential to trigger an allergic reaction in some people, usually when the body mistakenly recognizes the vitamin D3 as a potentially harmful chemical and mounts an immune response against it.

10. Vitamin D and asthma

Maternal vitamin D intake during pregnancy has previously been associated with asthma symptoms in several childhood epidemiological studies [40, 41], while variants in the vitamin D receptor have been associated with asthma in genetic studies [42, 43].

The results of some, but not all, epidemiological studies suggest that vitamin D deficiency is associated with increased risk of wheezing, respiratory tract infections, and asthma symptoms [31–33].

In the past decade of twentieth century there was evidence that there is higher incidence of recurrent respiratory tract infection and nutritional rickets [48, 49]. The evidence that the peak of viral infection is in the winter time when synthesis of vitamin D across the skin is naturally impaired supported that association. In addition, other studies have proved that vitamin D deficiency in pregnant women may result in increased risk of respiratory tract infections in their infants [44–49]. The majority of immune cells express VDRs, mainly after they themselves have been stimulated [40, 41, 48]. The mechanism by which vitamin D regulates inflammation and immunity is complicated. It controls macrophage and dendritic cell activities and various Toll-like receptor mediated events in neutrophils [40, 41], and it diminishes the function of human dendritic cells by decreasing maturation, antigen presentation, and the production of cytokines such as interleukin (IL)-12 and IL-23 [42, 48].

Furthermore, treating macrophages with vitamin D result in the expression of various cytokines and chemokines, including CXCL8, IL-6, and IL-12, and tumor necrosis factor-α [48]. Additionally, vitamin D induces the expression of two antimicrobial peptides—cathelicidin and β-defensin—that are widely expressed in the body and play a key role in innate immunity owing to their chemotactic action and toxin neutralization [48].

The allergic phenotype of asthma is determined by an increased activity of Th2 cells resulting in the production of IgE and inflammatory cytokines causing airway hyperresponsiveness with a predominantly eosinophilic inflammation. Vitamin D would also modulate various

http://dx.doi.org/10.5772/65072

cytokine-induced effects through different cells of the immune system with dose-dependent action. Moderate doses of vitamin D can inhibit the production of both Th1 and Th2 cytokine response, while high concentrations may intensify the Th2 response [49]. Airway inflammation and hyperreactivity are contributed mainly by the regulatory T cells through the production of cytokines such as IL-10 and TGF-β.

What is more, there is evidence in the literature that vitamin D can affect remodeling of the airways in asthma through a direct effect on the proliferation of smooth muscle cells, influence their growth, and contractility [50].

In a study from 2010, it was shown that there exists an association between low levels of vitamin D and impaired lung function, increase of airway hyperresponsiveness, and reduction of glucocorticosteroid response in patients with moderate or severe asthma [51].

Acute lower respiratory tract infections are well-known factor leading to wheezing and asthma-like symptoms. It concerns mainly to bronchiolitis caused by respiratory syncytial virus (RSV). An *in vitro* study has shown that vitamin D can increase the inflammatory response of airway epithelial cells to RSV infection [52], what is more, there is evidence that genetic polymorphisms in VDR are associated with hospitalization due to acute bronchiolitis in infancy [53].

Meta-analysis of randomized controlled trials confirmed the fact that prophylactic supplementation of vitamin D in children significantly reduces the odds of contracting respiratory tract infections [54]. These findings were supported by another study to demonstrate that a higher maternal intake of vitamin D during pregnancy may decrease the recurrent wheeze in early childhood [55]. Additionally, cord blood concentrations of 25(OH)D in neonates correlated with increased risk of lower respiratory tract infections in the first 2 years of life [56].

Vitamin D is also considered to play a big role in allergy and asthma treating as influencing the immunotherapy.

Studies have documented the efficacy and safety of sublingual immunotherapy in pediatric patients with allergies to grass pollen [57 60]. However, an important issue in mucosal immunotherapy is how to improve efficacy. There are a few studies on assessing modification of the effectiveness of sublingual immunotherapy in children. In two studies from Poland [32, 33] reduction in combined symptom-medication score, nasal symptoms in children receiving vitamin D supplementation to SLIT was observed compared to those receiving SLIT and placebo, as well as lower asthma symptoms.

Up to this date, there is evidence in the literature that vitamin D supplementation has a beneficial effect on allergy treatment. It has been proved that vitamin D supplementation in patients with allergic rhinitis treated with SLIT results in overall clinical improvement; however, vitamin D supplementation is more effective in the reduction of nasal and asthma symptoms and improvement in symptoms observed during SLIT were fairly clearly correlated with the serum level of vitamin D [31]. In a different study of Stelmach et al., it was shown that clinical and immunological efficacy of allergen-specific immunotherapy in children with asthma allergic to house dust mites was correlated with 25(OH)D serum concentration [32].

Another study of Stelmach et al. demonstrated that combined administration of a systemic corticosteroid and allergen extract suppressed early clinical and immunological effects of SIT and that vitamin D3 prevented this "adverse" effect of corticosteroids. Therefore, a favorable effect of vitamin D in immunotherapy is very encouraging [31]. Since there is an evidence that the efficacy of allergen-specific immunotherapy correlates with 25(OH)D serum concentration, it seems to be reasonable to monitor the serum level of 25(OH)D in children undergoing allergen immunotherapy, especially in those at risk of vitamin D insufficiency. It seems that the serum level of 25(OH)D above 30 ng/mL facilitates the optimal effect of allergen immunotherapy. In a study of Stelmach et al. [31–33], administration of 1000 IU vitamin D once daily with SLIT significantly reduced phosphorus in serum of children, which suggests its beneficial effect on calcium-phosphorus metabolism and on collagen turnover in children. Therefore, the clinical implication of that study suggests that the supplementation of SLIT with vitamin D should be recommended for treatment of allergic rhinitis in children.

11. Recommendations

In different countries there exist different recommendations concerning vitamin D supplementation. Currently, European Food Safety Authority (EFSA) proposed a daily intake of 100 mcg (equal to 400 IU) vitamin D for adults including pregnant and breastfeeding women, 50 mcg for children 11–17 years, and 25 mcg for infants [61, 62]. The institute of Medicines Committee in its report from 2011 recommended at least 600 IU for infants with maximum upper limit of 2500 IU for children 1–3 years and 3000 IU for children 4–8 years and 4000 IU for 9 years and older [61]. By this date, it is common practice for all physitians to supplement vitamin D under the verification of 25(OH)D concentration. The last recommendations for vitamin D supplementation in Central Europe state that 30–50 ng/mL are considered as optimal [62].

Based on its review of data of vitamin D needs, a committee of the Institute of Medicine concluded that persons are at risk of vitamin D deficiency at serum 25(OH)D concentrations <30 nmol/L (<12 ng/mL). Some are potentially at risk for inadequacy at levels ranging from 30 to 50 nmol/L (12–20 ng/mL). Practically, all people are sufficient at levels ≥50 nmol/L (≥20 ng/mL); the committee stated that 50 nmol/L is the serum 25(OH)D level that covers the needs of 97.5% of the population. Serum concentrations >125 nmol/L (>50 ng/mL) are associated with potential adverse effects [63].

Intake reference values for vitamin D and other nutrients are provided in the Dietary Reference Intakes (DRIs) developed by the Food and Nutrition Board (FNB) at the Institute of Medicine of The National Academies (formerly National Academy of Sciences) [63].

The FNB established an RDA for vitamin D representing a daily intake that is sufficient to maintain bone health and normal calcium metabolism in healthy people. RDAs for vitamin D are listed in both International Units (IUs) and micrograms (mcg); the biological activity of 40 IU is equal to 1 mcg (**Table 1**). Even though sunlight may be a major source of vitamin D for some, the vitamin D RDAs are set on the basis of minimal sun exposure [63].

http://dx.doi.org/10.5772/65072

Age	Male	Female	Pregnancy	Lactation
0–12 months*	400 IU (10 mcg)	400 IU (10 mcg)		
1–13 years	600 IU (15 mcg)	600 IU (15 mcg)		
14–18 years	600 IU (15 mcg)	600 IU (15 mcg)	600 IU (15 mcg)	600 IU (15 mcg)
19–50 years	600 IU (15 mcg)	600 IU (15 mcg)	600 IU (15 mcg)	600 IU (15 mcg)
51–70 years	600 IU (15 mcg)	600 IU (15 mcg)		
>70 years	800 IU (20 mcg)	800 IU (20 mcg)		

*Adequate intake (AI).

Table 1. Recommended dietary allowances (RDAs) for vitamin D [63].

Author details

Iwona Stelmach*, Joanna Jerzyńska and Daniela Podlecka

*Address all correspondence to: iwona.stelmach@kopernik.lodz.pl

Department of Pediatrics and Allergy, Copernicus Memorial Hospital in Lodz, Medical University of Lodz, Lodz, Poland

References

[1] DeLuca HF. Overview of general physiologic features and functions of vitamin D. Am J Clin Nutr. 2004 Dec;80(6 Suppl):1689S–1696S.

[2] Kanis JA. Vitamin D metabolism and its clinical application. J Bone Joint Surg Br. 1982;64(5):542–560.

[3] Bikle DD. Vitamin D metabolism, mechanism of action, and clinical applications. Chem Biol. 2014 Mar 20;21(3):319–329. doi: 10.1016/j.chembiol.2013.12.016. Epub 2014 Feb 13.

[4] How JA, Hazewinkle HAW, Mol JA. Dietary vitamin D dependence of cat and dog due to inadequate cutaneous synthesis of vitamin D. Gen Comp Endocrinol 1994;96:12–18.

[5] American Geriatrics Society Workgroup on Vitamin D Supplementation for Older Adults. Recommendations abstracted from the American Geriatrics Society Consensus Statement on vitamin D for Prevention of Falls and Their Consequences. J Am Geriatr Soc. 2014 Jan;62(1):147–152. doi: 10.1111/jgs.12631. Epub 2013 Dec 18.

[6] Nguyen VT, Li X, Castellanos KJ, Fantuzzi G, Mazzone T, Braunschweig CA. The accuracy of vitamin D assays of circulating 25-hydroxyvitamin D values: influence of

25-hydroxylated ergocalciferol concentration. AOAC Int. 2014 Jul-Aug;97(4):1048–1055.

[7] Glendenning P. Measuring vitamin D. Aust Prescr. 2015 Feb;38(1):12–15. Epub 2014 Nov 24. Review.

[8] Bischoff-Ferrari HA, Giovannucci E, Willett WC, Dietrich T, Dawson-Hughes B. Estimation of optimal serum concentrations of 25-hydroxyvitamin D for multiple health outcomes. Am J Clin Nutr. 2006 Jul;84(1):18–28.

[9] Omdahl JL, Morris HA, May BK. Hydroxylase enzymes of the vitamin D pathway: expression, function, and regulation. Annu Rev Nutr. 2002;22:139–166.

[10] Jones G, Strugnell SA, DeLuca HF. Current understanding of the molecular actions of vitamin D. Physiol Rev. 1998 Oct;78(4):1193–1231.

[11] Kimmel-Jehan C, Darwish HM, Strugnell SA, Jehan F, Wiefling B, DeLuca HF. DNA bending is induced by binding of vitamin D receptor-retinoid X receptor heterodimers to vitamin D response elements. J Cell Biochem. 1999 Aug 1;74(2):220–228.

[12] Lanjouw E, Branković I, Pleijster J, Spaargaren J, Hoebe CJ, van Kranen HJ, Ouburg S, Morré SA.Specific polymorphisms in the vitamin D metabolism pathway are not associated with susceptibility to Chlamydia trachomatis infection in humans. Pathog Dis. 2016 Apr;74(3). pii: ftw010. doi: 10.1093/femspd/ftw010. Epub 2016 Feb 10.

[13] Chun Li Y, Pirro A, Demay M. Analysis of vitamin D-dependent calcium-binding protein messenger ribonucleic acid expression in mice lacking the vitamin D receptor. Endocrinology. 1998;139:847–851.

[14] Uitterlinden AG, Fang Y, Van Meurs JB, Pols HA, Van Leeuwen JP. Genetics and biology of vitamin D receptor polymorphisms. Gene. 2004 Sep 1;338(2):143–156

[15] Köstner K, Denzer N, Müller CS, Klein R, Tilgen W, Reichrath J. The relevance of vitamin D receptor (VDR) gene polymorphisms for cancer: a review of the literature. Anticancer Res. 2009 Sep;29(9):3511–3536.

[16] Al-Daghri NM, Guerini FR, Al-Attas OS, Alokail MS, Alkharfy KM, Draz HM, Agliardi C, Costa AS, Saulle I, Mohammed AK, Biasin M, Clerici M. Vitamin D receptor gene polymorphisms are associated with obesity and inflammosome activity. PLoS One. 2014 Jul 14;9(7):e102141.

[17] Faraco JH, Morrison NA, Baker A, Shine J, Frossard PM. ApaI dimorphism at the human vitamin D receptor gene locus. Nucleic Acids Res. 1989 Mar 11;17(5):2150.

[18] Morrison NA, Yeoman R, Kelly PJ, Eisman JA. Contribution of trans-acting factor alleles to normal physiological variability: vitamin D receptor gene polymorphism and circulating osteocalcin. Proc Natl Acad Sci U S A. 1992 Aug 1;89(15):6665–6669.

[19] Morrison NA, Qi JC, Tokita A, Kelly PJ, Crofts L, Nguyen TV, Sambrook PN, Eisman JA. Prediction of bone density from vitamin D receptor alleles. Nature. 1994 Jan 20;367(6460):284–287. Erratum in: Nature 1997 May 1;387(6628):106.

[20] von Essen MR, Kongsbak M, Schjerling P, Olgaard K, Odum N, Geisler C. Vitamin D controls T cell antigen receptor signaling and activation of human T cells. Nat Immunol. 2010 Apr;11(4):344–349.

[21] Deluca HF, Cantorna MT. Vitamin D: its role and uses in immunology. FASEB J. 2001 Dec;15(14):2579–2585.

[22] Hewison M. An update on vitamin D and human immunity. Clin Endocrinol (Oxf). 2012 Mar;76(3):315–325.

[23] Aranow C. Vitamin D and the immune system. J Investig Med. 2011 Aug;59(6):881–886.

[24] Cantorna MT, Zhu Y, Froicu M, Wittke A. Vitamin D status, 1,25-dihydroxyvitamin D3, and the immune system. Am J Clin Nutr. 2004 Dec;80(6 Suppl):1717S–1720S.

[25] Mahon BD, Wittke A, Weaver V, Cantorna MT. The targets of vitamin D depend on the differentiation and activation status of CD4 positive T cells. J Cell Biochem. 2003 Aug 1;89(5):922–932.

[26] Almerighi C, Sinistro A, Cavazza A, Ciaprini C, Rocchi G, Bergamini A. 1Alpha,25-dihydroxyvitamin D3 inhibits CD40L-induced pro-inflammatory and immunomodulatory activity in human monocytes. Cytokine. 2009 Mar;45(3):190–197. doi: 10.1016/j.cyto.2008.12.009. Epub 2009 Jan 30.

[27] Baek JH, Shin YH, Chung IH, Kim HJ, Yoo EG, Yoon JW, Jee HM, Chang YE, Han MY. The link between serum vitamin D level, sensitization to food allergens, and the severity of atopic dermatitis in infancy. J Pediatr. 2014 Oct;165(4):849–854.e1.

[28] Dean T, Venter C, Pereira B, Arshad SH, Grundy J, Clayton CB, Higgins B. Patterns of sensitization to food and aeroallergens in the first 3 years of life. J Allergy Clin Immunol. 2007 Nov;120(5):1166–1171. Epub 2007 Sep 7.

[29] Nwaru BI, Ahonen S, Kaila M, Erkkola M, Haapala AM, Kronberg-Kippilä C, Veijola R, Ilonen J, Simell O, Knip M, Virtanen SM. Maternal diet during pregnancy and allergic sensitization in the offspring by 5 yrs of age: a prospective cohort study. Pediatr Allergy Immunol. 2010 Feb;21(1 Pt 1):29–37.

[30] Jones AP, Palmer D, Zhang G, Prescott SL. Cord blood 25-hydroxyvitamin D3 and allergic disease during infancy. Pediatrics. 2012 Nov;130(5):e1128-e1135.

[31] Jerzynska J, Stelmach W, Rychlik B, Lechańska J, Podlecka D, Stelmach I. The clinical effect of vitamin D supplementation combined with grass-specific sublingual immunotherapy in children with allergic rhinitis. Allergy Asthma Proc. 2016 Mar;37(2):105–114.

[32] Stelmach I, Majak P, Jerzynska J, Podlecka D, Stelmach W, Polańska K, Gromadzińska J, Wąsowicz W, Hanke W. Cord serum 25-hydroxyvitamin D correlates with early childhood viral-induced wheezing. Respir Med. 2015 Jan;109(1):38–43.

[33] Majak P, Jerzyńska J, Smejda K, Stelmach I, Timler D, Stelmach W. Correlation of vitamin D with Foxp3 induction and steroid-sparing effect of immunotherapy in asthmatic children. Ann Allergy Asthma Immunol. 2012 Nov;109(5):329–335.

[34] Brehm JM, Schuemann B, Fuhlbrigge AL, Hollis BW, Strunk RC, Zeiger RS, Weiss ST, Litonjua AA. Childhood Asthma Management Program Research Group Serum vitamin D levels and severe asthma exacerbations in the Childhood Asthma Management Program study. J Allergy Clin Immunol 2010;126:52–58e5.

[35] Chen Wu A, Tantisira K, Lingling Li, Fuhlbrigge A, Weiss S, Litonjua A. Effect of vitamin D and inhaled corticosteroid treatment on lung function in children. Am J Respir Crit Care Med. 2012 Sep 15;186(6):508–513.

[36] Sutherland ER, Goleva E, Jackson LP, Stevens AD, Leung DY. Vitamin D levels, lung function, and steroid response in adult asthma. Am J Respir Crit Care Med 2010;181:699–704.

[37] Peroni DG, Boner AL. Food allergy: the perspectives of prevention using vitamin D. Curr Opin Allergy Clin Immunol. 2013 Jun;13(3):287–292.

[38] Camargo CA Jr, Ganmaa D, Sidbury R, Erdenedelger KH, Radnaakhand N, Khandsuren B. Randomized trial of vitamin D supplementation for winter-related atopic dermatitis in children. J Allergy Clin Immunol. 2014 Oct;134(4):831–835.e1.

[39] Sidbury R, Sullivan AF, Thadhani RI, Camargo CA Jr. Randomized controlled trial of vitamin D supplementation for winter-related atopic dermatitis in Boston: a pilot study. Br J Dermatol. 2008 Jul;159(1):245–247.

[40] Camargo CA Jr, Rifas-Shiman SL, Litonjua AA, Rich-Edwards JW, Weiss ST, Gold DR, Kleinman K, Gillman MW. Maternal intake of vitamin D during pregnancy and risk of recurrent wheeze in children at 3 years of age. Am J Clin Nutr. 2007 Mar;85(3):788–795.

[41] Devereux G, Litonjua AA, Turner SW, Craig LC, McNeill G, Martindale S, Helms PJ, Seaton A, Weiss ST. Maternal vitamin D intake during pregnancy and early childhood wheezing. Am J Clin Nutr. 2007 Mar;85(3):853–859.

[42] Poon AH, Laprise C, Lemire M, Montpetit A, Sinnett D, Schurr E, Hudson TJ. Association of vitamin D receptor genetic variants with susceptibility to asthma and atopy. Am J Respir Crit Care Med. 2004 Nov 1;170(9):967–973.

[43] Raby BA, Lazarus R, Silverman EK, Lake S, Lange C, Wjst M, Weiss ST. Association of vitamin D receptor gene polymorphisms with childhood and adult asthma. Am J Respir Crit Care Med. 2004 Nov 15;170(10):1057–1065.

[44] Lee SJ, Kang BH, Choi BS. Vitamin D serum levels in children with allergic and vasomotor rhinitis. Korean J Pediatr. 2015 Sep;58(9):325–329.

[45] Della Giustina A, Landi M, Bellini F, Bosoni M, Ferrante G, Onorari M, Travaglini A, et al. Vitamin D, allergies and asthma: focus on pediatric patients. World Allergy Organ J. 2014 Dec 10;7(1):27.

[46] Navas-Nazario A, Li FY, Shabanova V, Weiss P, Cole DE, Carpenter TO, Bazzy-Asaad A. Effect of vitamin D-binding protein genotype on the development of asthma in children. Ann Allergy Asthma Immunol. 2014 Jun;112(6):519–524.

[47] Hollams EM, Hart PH, Holt BJ, Serralha M, Parsons F, de Klerk NH, Zhang G, Sly PD, Holt PG. Vitamin D and atopy and asthma phenotypes in children: a longitudinal cohort study. Eur Respir J. 2011 Dec;38(6):1320–1327.

[48] Esposito S, Lelii M. Vitamin D and respiratory tract infections in childhood. BMC Infect Dis. 2015 Oct 28;15:487.

[49] Najada AS, Habashneh MS, Khader M. The frequency of nutritional rickets among hospitalized infants and its relation to respiratory diseases. J Trop Pediatr. 2004 Dec; 50(6):364–368.

[50] Principi N, Bianchini S, Baggi E, Esposito S. Implications of maternal vitamin D deficiency for the fetus, the neonate and the young infant. Eur J Nutr. 2013 Apr;52(3): 859–867.

[51] Damera G, Fogle HW, Lim P, Goncharova EA, Zhao H, Banerjee A, Tliba O, Krymskaya VP, Panettieri RA Jr. Vitamin D inhibits growth of human airway smooth muscle cells through growth factor-induced phosphorylation of retinoblastoma protein and checkpoint kinase 1. Br J Pharmacol. 2009 Nov;158(6):1429–1441.

[52] Hansdottir S, Monick MM, Lovan N, Powers L, Gerke A, Hunninghake GW. Vitamin D decreases respiratory syncytial virus induction of NF-kappa B-linked chemokines and cytokines in airway epithelium while maintaining the antiviral state. J Immunol. 2010 Jan 15;184(2):965–974.

[53] Roth DE, Jones AB, Prosser C, Robinson JL, Vohra S. Vitamin D receptor polymorphisms and the risk of acute lower respiratory tract infection in early childhood. J Infect Dis. 2008 Mar 1;197(5):676–680.

[54] Charan J, Goyal JP, Saxena D, Yadav P. Vitamin D for prevention of respiratory tract infections: a systematic review and meta-analysis. J Pharmacol Pharmacother. 2012 Oct; 3(4):300–303.

[55] Rothers J, Wright AL, Stern DA, Halonen M, Camargo CA Jr. Cord blood 25-hydroxyvitamin D levels are associated with aeroallergen sensitization in children from Tucson, Arizona. J Allergy Clin Immunol. 2011 Nov;128(5):1093-9.e1-5.

[56] Mohamed WA, Al-Shehri MA. Cord blood 25-hydroxyvitamin D levels and the risk of acute lower respiratory tract infection in early childhood. J Trop Pediatr. 2013 Feb;59(1): 29–35.

[57] Devillier P, Dreyfus JF, Demoly P, Calderón MA. A meta-analysis of sublingual allergen immunotherapy and pharmacotherapy in pollen-induced seasonal allergic rhinoconjunctivitis. BMC Med. 2014 May 1;12:71.

[58] Hébert J, Blaiss M, Waserman S, Kim H, Creticos P, Maloney J, Kaur A, Li Z, Nelson H, Nolte H. The efficacy and safety of the Timothy grass allergy sublingual immunotherapy tablet in Canadian adults and children. Allergy Asthma Clin Immunol. 2014 Oct 30;10(1):53.

[59] Wahn U, Tabar A, Kuna P, Halken S, Montagut A, de Beaumont O, Le Gall M; SLIT Study Group. Efficacy and safety of 5-grass-pollen sublingual immunotherapy tablets in pediatric allergic rhinoconjunctivitis. J Allergy Clin Immunol. 2009 Jan;123(1):160–166.

[60] Blanco DB, Luckey D, Smart M, Pittelkow M, Kumar R, David CS, Mangalam AK. Modulation of immune response by ultra-violet light in HLA class-II transgenic mice. Jacobs J Allergy Immunol. 2014 Dec;1(1):pii:007.

[61] Martin A. Assessment and management of health claims in Europe 9 years after the adoption of the regulation. J Nutr Sci Vitaminol (Tokyo). 2015;61 Suppl:S50–S52.

[62] Płudowski P, Karczmarewicz E, Bayer M, Carter G, Chlebna-Sokół D, Czech-Kowalska J, Dębski R, et al. Practical guidelines for the supplementation of vitamin D and the treatment of deficits in Central Europe - recommended vitamin D intakes in the general population and groups at risk of vitamin D deficiency. Endokrynol Pol. 2013;64(4):319–327.

[63] Institute of Medicine, Food and Nutrition Board. Dietary Reference Intakes for Calcium and Vitamin D. Washington, DC: National Academy Press, 2010.

Chapter 8

Non-Bone Effects of Vitamin D in Children, Adolescents, and Young Adults

Mohsin Ali Cheema and Khalid Parvez Lone

Additional information is available at the end of the chapter

http://dx.doi.org/10.5772/65079

Abstract

Vitamin D, also known as "sunshine vitamin", has long been established as an essential component for the maintenance of adequate bone health. Large number of studies are available which demonstrate the various biochemical pathways of vitamin D in bone physiology and its important role in musculoskeletal health. In last five decades, data regarding the non-bone effects of vitamin D have started to emerge, and now many important non-bone physiological processes are explained by the biochemical pathways and functions of vitamin D. However, majority of the data regarding extra-skeletal effects of vitamin D are available regarding adult population. In this chapter, we try to focus on the role of vitamin D in aging and various diseases which are frequently seen in children, adolescents, and young adults such as cancer, type 1 diabetes mellitus, allergies, asthma, and various autoimmune diseases.

Keywords: vitamin D, autoimmune disease, aging, diabetes, cancer, children

1. Introduction

Vitamin D, also known as "sunshine vitamin", is considered essential for maintenance of bone health along with many other key roles in metabolic processes of the body. About 10,000–20,000 IU of vitamin D3 is produced in our skin from the sun's ultraviolet light (UVB) after full body exposure for 15 min [1]. Complexion or skin phototype, use of sunblock, smog, cloud cover, latitude, time of day (10:00 am to 02:00 pm), and season are some of the factors that affect our body's ability to produce vitamin D [2]. Latitudes from 0 to 35° north or south allow yearlong production of vitamin D in our body, but as latitude increases the amount of vitamin

D production will decrease [3]. The latitude of Pakistan is 32.0162°N which is suitable for yearlong production of vitamin D.

Recommended daily intake of vitamin D in the first year of life is 400 IU and 600 IU for everyone above 1 year of age [4]. Vitamin D is essential for proper bone growth and bone remodeling by the action of osteoblasts and osteoclasts [5]. Vitamin D insufficiency or deficiency can lead to osteoporosis by decreasing calcium absorption from intestine and kidney [6]. Harinarayan et al. have showed that normal homeostasis of bone has been seriously affected by decreased intake of calcium and vitamin D in diet [7]. In children, inadequate mineralization of the growing skeleton leading to rickets is an established fact [8]. Välimäki et al. discussed that vitamin D deficiency is common in young men, especially in winter, and this deficiency may have detrimental effects on the acquisition of maximal peak bone mass [9]. Gait disturbances and myopathy have also been associated with severe vitamin D deficiency [10]. Discovery of the vitamin D receptor (VDR) expressed in the cell nuclei of muscle cells and its association with muscle cell contractility has been documented by Bischoff-Ferrari et al [11]. Meta-analysis done by Papadimitropoulos et al. concluded that adequate levels of vitamin D decrease the rate of vertebral and non-vertebral fractures [12].

1.1. Prevalence of vitamin D deficiency

Studies from around the world have shown pandemic of vitamin D deficiency or insufficiency [13]. In last decade, various studies from Pakistan have found critically low levels of vitamin D in diverse groups of population. Work done by Siddiqui and Rai [14] in Hazara division of Pakistan showed that nutritional rickets resulting from vitamin D deficient diet is a predisposing factor for different childhood illnesses such as pneumonia, diarrhea, and delayed motor milestones. They attributed lack of sun exposure, malnutrition, and various antenatal factors as important causes for vitamin D deficiency. Similarly, Anwar et al. [15] found that 99.5% of women and 97.3% of neonates from urban population sample and 89% of women and 82% of neonates from rural population sample had below 50 nmol/l levels of vitamin D. Jamal et al. [16] and Qamar et al. [17] have also found insufficient/deficient vitamin D levels in more than 90% children included in their studies. A study done in our laboratory including 5- to 11-year-old children with intellectual disability also showed extremely low levels of vitamin D (12.08 ± 9.06 ng/ml) in 85% of study sample. These numbers are consistent with another population-based study done in Karachi by Iqbal et al. [18] who found that vitamin D was deficient in 73.7% and insufficient in 13.8% study subjects.

In last two decades, lot of studies came out from around the world reporting lower or deficient levels of vitamin D; however, few questions remained ambiguous: How much vitamin D is needed to achieve desirable health outcomes and how much of vitamin D is too much? In 2011, a new public health report on dietary intake requirements for calcium and vitamin D authored by the committee from the Institute of Medicine (IOM) gave their recommendations regarding above queries. The committee concluded that serum 25OHD levels of 16 ng/ml (40 nmol/l) cover the requirements of approximately half of the population, and levels of 20 ng/ml (50 nmol/l) cover the requirements of at least 97.5% of the population. The IOM committee also highlighted the fact that data for the upper cutoff limit were scarce and long-term effects of chronically high concentrations of vitamin D, and a margin of safety for public health recommendations was prudent. Thus, serum 25OHD levels above 50 ng/ml (125 nmol/l) should raise concerns among clinicians about potential adverse effects [4].

1.2. Vitamin D status and skin complexion

Complexion or skin phototype is another important factor that determines amount of vitamin D production in our body. Individuals with darker complexion have higher chances of developing vitamin D deficiency than fair-skinned people [19]. The deficiency in circulating levels of 25-hydroxyvitamin D in individuals with darker skin complexion is mediated largely by melanin which protects the skin from ultraviolet light rays. By blocking the sun's ultra violet rays, melanin greatly reduces the skin's ability to convert 7-dehydrocholesterol to cholecalciferol, the precursor of 25(OH)D, in the skin [20]. In 2011, Signorello et al. investigated the association between 94 single nucleotide polymorphisms (SNPs) in five vitamin D pathway genes (*GC, VDR, CYP2R1, CYP24A1, CYP27B1*) and serum 25-hydroxyvitamin D (25(OH)D) levels among 379 African American and 379 Caucasian participants, and they found that common variation in vitamin D pathway genes predicts circulating 25-hydroxyvitamin D levels among African Americans having darker complexion [21].

Vitamin D supplementation is considered as an important factor in improving the poor bone health of people with both fair and dark skin complexions. Studies have shown that people with both dark and white skin color appear to have similar capacities to synthesize vitamin D in the skin, but vitamin D synthesis is less efficient among blacks at usual levels of sun exposure as compared to white people [22]. Recently, a study done by Gallagher et al. demonstrated that despite having lower levels of vitamin D at baseline, the increase in serum vitamin D after supplementation in African American women was similar to that seen in the Caucasian women. Furthermore, 97.5% of both dark- and white-colored women on 800 IU/daily for 12 months reached a level of 20 ng/ml. They also concluded that because absorption and metabolism of oral vitamin D are similar in both dark and fair complexioned people, lower levels of serum 25OHD in African Americans must be due to lower production of vitamin D in skin [23].

1.3. Is Vitamin D synthesis gender specific?

Data regarding the association of gender with vitamin D synthesis are inconsistent and scarce. There are number of conflicting hypotheses which indicate that differences in the amount of adipose tissue in males and females and in their skins may lead to the variation in vitamin D synthesis. Number of studies has demonstrated decreased bioavailability of fat soluble vitamin D3 from cutaneous and dietary sources because of its deposition in body fat [24–27]. This may point toward decreased vitamin D levels in women because of their excessive adiposity as compared to men. On the other hand, it has also been hypothesized that the lighter color of female skin permits synthesis of relatively higher amounts of vitamin D3 during pregnancy and lactation [28]. On the contrary, data from a recent study in mice have demonstrated that androgens decrease the synthesis of vitamin D induced by ultraviolet rays in the skin of a male mice by an enzymatic mechanism [29].

1.4. Physiology of non-bone effects of vitamin D

Initially, it was thought that vitamin D has a restricted role in calcium homeostasis, but it is only now that the pleiotropic actions of vitamin D with their clinical significance are becoming apparent. Recently, 2776 genomic positions occupied by the vitamin D receptor (VDR) and 229 genes showing significant changes in expression in response to vitamin D have been identified [30]. VDR has been identified in wide variety of tissues other than small intestine, kidneys, and bone such as brain, heart, stomach, pancreas, activated T and B lymphocytes, skin and gonads [31]. One hydroxylase activity has also been identified in cultured cells from skin, colon, prostate, breast, lung, and brain [1, 32, 33]. Addition of vitamin D in our diet in the form of supplementation or food fortification can reduce the risk of asthma, influenza, respiratory tract infections, autoimmune diseases, diabetes, and cancer and improve bone and overall health [2]. This evidence points to the diverse functions of vitamin D and its implicated role in various diseases related to wide distribution of vitamin D receptors in many tissues.

In the following sections of this chapter, recent review of literature regarding physiological role of vitamin D in various non-bone effects is given.

2. Vitamin D and aging

It has been long known that aging decreases the capacity of human skin to produce vitamin D3 (7-dehydrocholesterol). This is evident from the classic study done by McLaughlin and Holick in 1985 in which they exposed the skin samples from various age groups to ultraviolet rays and compared the amount of previtamin D3 produced in the skin samples from 8- to 18-year-old subjects with the amount produced in the skin samples from 77- to 82-year-old subjects. The results revealed that aging can decrease the capacity of the skin to produce previtamin D3 by greater than twofold [34].

In humans, aging represents the accumulation of changes in a human being over time, encompassing physical, psychological, and social change. Telomeres, the DNA-protein structures located at the ends of chromosomes, have been proposed to act as a biomarker of aging [35]. A cross-sectional analysis of the Nurses' Health Study (NHS) data done by Liu et al. [36] demonstrated the association between vitamin D and telomere length in peripheral blood leukocytes by using plasma biomarkers of both 25(OH)D and 1,25(OH)2D. They found that higher plasma 25(OH)D levels are associated with longer telomeres, and this association may be modified by calcium intake. The difference in telomere length between those with high (vitamin D sufficient) and low (vitamin D insufficient) levels of vitamin D corresponded to 5 years of aging. The shortening of telomeres is thought to be caused by decreasing inflammatory mediators and cell proliferation [37]. Changes in the expression of VDR leading to vitamin D resistance with aging have also been elaborated in recent literature. Several studies have reported association of aging with decrease in the expression of VDR in bone, intestine, and muscle tissues [38–40]. Various factors that decrease with aging have been identified to influence VDR such as estrogen, growth hormones, and vitamin D itself [11, 41]. On the other hand, the expression of TNF alpha which increases with aging has been shown to downregu-

http://dx.doi.org/10.5772/65079

late the expression of VDR [42, 43]. An animal study has also demonstrated a decrease in number of vitamin D receptors in addition to decreased binding of active metabolite of vitamin D with VDR in aged rat intestinal subcellular fractions [44]. Association of premature aging phenotype with too high and too low vitamin D levels has also been demonstrated in mice [45]. A recent study found that individuals in the highest quartile of serum vitamin D had the longest lifespan compared to those in the lowest quartile [46]. Above evidence and already established associations of several common aging-associated diseases such as osteoporosis, hypertension, and diabetes with vitamin D deficiency point us to consider vitamin D levels as a biomarker for aging.

3. Cancer in children

Although cancer in children is rare, it is the leading cause of death by disease past infancy among children in the United States. The most common types of cancer diagnosed in children and adolescents are leukemia, brain and other central nervous system tumors, lymphoma, rhabdomyosarcoma, neuroblastoma, Wilms tumor, bone cancer, and gonadal (testicular and ovarian) germ cell tumors [47]. Pakistan is a developing country with a population of about 180 million people out of which 39% are under 15 years of age. Although malnutrition and communicable diseases are still the major killers, cancer is becoming an important cause of morbidity and mortality in children. Cancer among children is fortunately less common than in adult accounting for 3–5% of all cancers. Based on Karachi Cancer Registry, it is estimated that about 7000–7500 children get cancer every year in Pakistan [48]. A limited number of small studies have examined the vitamin D status of pediatric oncology patients, and the results indicate an increased prevalence of hypovitaminosis D [49, 50].

3.1. Role of vitamin D in cancer

Specific vitamin D receptors found in nearly all tissues produce 1,25(OH)2D in the presence of 25(OH)D. This 1,25(OH)2D is recognized to cause activation of VDR resulting in heterodimerization with the retinoid X receptor and binding to cognate vitamin D response elements (VDREs) in target genes involved in cellular differentiation, cell growth, apoptosis, inflammation, and immune modulation [51, 52]. Formation of 1 alpha hydroxylase enzyme in cancer cells has also been demonstrated [53, 54]. In last decade of twentieth century, number of studies came out that highlighted the association of decreased vitamin D levels in individuals living at high latitudes and increased risk of dying of colon, breast, prostate, and ovarian cancer [55–57].

Polymorphism of VDR gene has been associated with high risk of cancer. Women with mutations of VDR gene have higher risk of breast cancer [58]. Results of a hospital-based case-control study conducted by our department demonstrated that the *BsmI* polymorphism in the *VDR* gene may be associated with an increased breast cancer risk in Pakistani women negative for *BRCA1/2* germline mutations [59].

Vitamin D-liganded VDR displays antiproliferative activities in many tumor types, as do activated members of the p53 family, through the induction of cell cycle arrest, senescence, differentiation, and apoptosis [60]. Several metabolites of vitamin D which do not cause hypercalcemia show antitumor activity in a subset of cancer patients with high VDR expression and are associated with good prognosis [61]. Some transcription factors repress VDR gene expression in human colon cancer cells. In human colon cancers, elevated expression of transcription factors correlates with downregulation of VDR [62]. The malignant cells are capable of annulling the antiproliferative activity of vitamin D by increasing the expression of transcriptional factors.

Several preclinical trials have demonstrated the antiproliferative effects of active form of vitamin D in various tumor types. The mechanisms by which vitamin D can exert antitumor effects may include inhibition of tumor angiogenesis, induction of apoptosis, regulation of different signaling pathways in tumor cells, and arrest of tumor cells in G0/G1 phase of cell cycle. Preclinical data indicate that maximal antitumor effects are seen with pharmacological doses of 1α,25(OH)2D3 and can be safely achieved in animals using a high-dose, intermittent schedule of administration [60]. Some clinical trial data indicate that 1α,25(OH)2D3 is well tolerated in cancer patients within a proper dosing schedule. Data support the hypothesis that vitamin D compounds may have an important role in cancer therapy and prevention, and merit further investigation [63].

Despite the presence of wealthy data regarding vitamin D and cancer, only a handful of studies have assessed vitamin D status in pediatric patients with malignancy, and all included relatively small numbers of patients [50]. This may point to the need of continuing elaborative research to identify the role of vitamin D in cancer, especially in populations with younger age groups.

4. Diabetes and vitamin D

Vitamin D deficiency and diabetes have one major trait in common: both are pandemic. The International Diabetes Federation estimated the number of people with diabetes worldwide to be nearly 415 million out of which three-quarters (75%) live in low- and middle-income countries. In Pakistan, prevalence of diabetes is 8.1%, whereas number of new cases of type 1 diabetes per 100,000 children per year is 0.5 [64]. Studies from different populations throughout the globe have reported high rates of vitamin D deficiency in children with type 1 diabetes [65–70]. Also several studies suggest that vitamin D deficiency correlates with the severity and frequency of type 1 diabetes in children as well as that vitamin D supplementation may reduce the risk of developing type 1 diabetes in younger age groups [71–73]. An analytical cross-sectional study conducted in our department demonstrated that offsprings of type 2 diabetics were severely deficient in vitamin D and its levels were inversely correlated with most of the components of metabolic syndrome [74]. Researchers have also explored the geographical variation in childhood diabetes. In study done by Mohr et al. in 2008, incidence rates of type 1 diabetes in children aged <14 years during 1990–1994 in 51 regions worldwide were assessed by using multiple regression. This study found an association between low ultraviolet B irradiance and high incidence rates of type 1 childhood diabetes after controlling for per capita health expenditure. Incidence rates of type 1 diabetes approached zero in regions worldwide with high UVB irradiance, adding new support to the concept of a role of vitamin D in reducing the risk of the disease [75]. Similarly, data from sixth edition of diabetes atlas showed that incidence of type 1 diabetes is higher in countries that are located furthest from the equator as compared to countries located closest to equator. Staples et al. in their another study, which specifically examined the latitude gradient within Australia's territories, found childhood diabetes to be positively related to latitude. There was a strong threefold increase in prevalence of type 1 diabetes moving from the most northern territory to the most southern territory of Australia [76]. The above discussion positively points toward the association of incidence of childhood type 1 diabetes with latitudinal location.

Type 1 diabetes mellitus results from a cellular-mediated autoimmune destruction of the beta cells of the pancreas [77]. Studies on mice suggest that vitamin D inhibits IL-12 production and pancreatic infiltration of T helper cells along with the increase in number of CD4 positive and d CD25 positive regulatory T cells in pancreatic lymph nodes which may help in limiting the immunological progression and preventing the clinical onset of type 1 diabetes [78, 79]. There is evidence that vitamin D is important in the prevention of islet cell death and might be useful in improving the survival of islet cell grafts [80]. The presence of VDR and vitamin D-dependant calcium binding protein on beta cells of pancreas has also been reported. The effects of vitamin D on beta cells may be by its regulation of extracellular calcium and calcium flux through the beta cell or through calcium-independent pathways. Vitamin D deficiency may also impair insulin secretion through its associated increase in parathormone levels. It may reduce insulin resistance by its immunomodulatory and antiinflammatory effects [61]. Above evidence suggests that vitamin D may play a role in the prevention and treatment of type 1 diabetes mellitus in children; however, definitive and conclusive evidence regarding the potential role of vitamin D in alleviating the increasing menace of diabetes needs to be further studied through its action on systemic inflammation, insulin secretion, and resistance.

5. Allergies and asthma in children

Asthma and allergies are common chronic diseases in developed world [81–83]. CDC reports from national surveys in United States show the prevalence of asthma in children as 8.3% in 2013. The highest rates were observed among children aged 5–14 years, boys of less than 18 years, and people below 100% of poverty level [84]. Epidemiology of allergies and asthma is not well documented in Pakistan. A survey conducted in the city of Karachi among school children of age 3–16 years reported the prevalence of asthma at 15.8% and that of allergic rhinitis at 28.50% [85]. Another two-stage community-based representative cross-sectional survey conducted in Karachi from March 2012 to April 2013 reported overall prevalence of asthma among study participants as 10.2% [86].

5.1. Role of vitamin D in allergies and asthma in children

Many potential reasons have been reported in literature that could account for the pattern of increased burden of allergic diseases and asthma in developing as well as developed world. In last couple of decades, several studies have proposed pandemic of vitamin D deficiency as an important candidate that could explain a significant proportion of increased prevalence of allergic disease and asthma [87]. Potential mechanisms of how vitamin D can affect the risk of developing asthma and allergies include genetic pathways, role of vitamin D in prevention of bacterial infections, immune system effects, and effects on lung development and functions [88]. Significant associations between polymorphisms in the VDR gene and asthma have been reported by number of studies [89–91]. It has been proposed that asthmatics with bacterial and viral infections are at higher risk of more severe symptoms [92, 93]. Vitamin D induces the production of the antimicrobial polypeptide cathelicidin which has both antibacterial and antiviral effects [94, 95], and supplementation with vitamin D in asthmatics demonstrated decreased incidence of cold or influenza symptoms [96]. Studies done in mice have shown that vitamin D deficiency *in utero* leads to decreased lung volumes by affecting lung development [97]. Similarly, vitamin D also appears to affect *in utero* immune system development that begins to exert its effects in early life. Work done by Chi et al. [98] showed that cord blood vitamin D levels were inversely associated with the proportions of CD25(+), CD25(Bright), and CD25(+) FoxP3 cells to total CD4(+) T cells. While the clinical consequence of this inverse association remains unclear, it supports the notion that *in utero* vitamin D levels affect immune development and may influence immune regulation early in life. The number of studies has shown that maternal supplementation with vitamin D during pregnancy may eventually decrease the risk of allergies and wheezing in their children [99–101]. Other studies have investigated the association between vitamin D levels and asthma and allergies in the post-partum period. Several case-control studies from different parts of the world have found greater prevalence of vitamin D deficiency among asthmatic children than in controls [102–104]. In a recent work done in our department, an inverse relation of vitamin D levels with severity of asthma has been found (Lone Unpublished data).

Although more studies have shown a beneficial effect of vitamin D on asthma and allergies than studies showing negative results, definitive clinical trials are lacking and the optimal dose and level of vitamin D for decreasing the burden of asthma and allergies in children remain unknown.

6. Autoimmune diseases in children and vitamin D

There are at least 80 recognized human autoimmune diseases with new diseases frequently added to the list [105]. Autoimmune diseases are generally rare in children; however, when they occur, they can be challenging to diagnose and difficult to treat because most of the autoimmune diseases that are common in children have not cured yet. Besides type 1 diabetes mellitus, other autoimmune diseases which commonly occur in children include celiac disease, lupus (SLE), juvenile dermatomyositis, scleroderma, juvenile idiopathic arthritis (JIA), and multiple sclerosis (MS) [106].

The effects of vitamin D in the immune system translate into an enhancement of innate immunity associated with a multifaceted regulation of acquired immunity [107]. Vitamin D supplementation is considered as an appealing therapy in pediatric autoimmune diseases. The diverse effects of vitamin D may minimize disease-related comorbidities resulting from bone weakening and infections in addition to attenuation of immune hyperactivation that is characteristic of pediatric SLE [108]. The SLE Disease Activity Index scores were found to be significantly higher in children who had 25(OH)D level less than 20 ng/ml [109]. Work done by Robinson et al. [110] demonstrated that low serum 25(OH)D level in children with SLE is associated with proteinuria and urinary vitamin D binding protein. Animal models have also shown that vitamin D and calcium supplementation can inhibit lupus activity [111, 112]. An incidental finding of an association between vitamin D deficiency and SLE nephritis has also been reported [113], but studies evaluating this relationship are lacking. Juvenile dermatomyositis (JDM) is a photosensitive rheumatic disease, treated commonly with sun avoidance, corticosteroids, methotrexate, and hydroxychloroquine. JDM shares many similarities with pediatric SLE in photosensitivity and methods of treatment, but it is not associated with proteinuria, making this an ideal comparison group for SLE. Robinson et al. [110] also reported that vitamin D deficiency is associated with disease activity in children with JDM.

Multiple sclerosis is an autoimmune disease of the central nervous system characterized by inadequate recognition of autoepitopes in myelinated nerve fibers by cells of the acquired immune system, generating an inflammatory immune response mediated by lymphocytes and macrophages, resulting in localized areas of inflammation and demyelination [114]. Some studies have also demonstrated the association of vitamin D deficiency and MS and its role not only in the reduction of relapse rates, but also in the prevention of its development [115, 116]. In MS, allelic variation in the MHC class II region exerts the single strongest effect on genetic risk. Environmental factors act at a population level. Sunlight or vitamin D is a key environmental factor in etiology and might interact with inherited factors in the MHC class II region [61]. A single MHC vitamin D response element (VDRE) gene has been identified as *HLA-DRB1*15* haplotypes. In a subgroup of individuals genetically predisposed to multiple sclerosis, deficiency of vitamin D may cause non-activation of histocompatibility genes necessary for differentiating between self and foreign proteins [117].

Despite the compelling evidence of low levels of vitamin D and its association with various autoimmune diseases, recently studies are coming out that challenge the assumption that serum levels of 25OH vitamin D are a sensitive marker of the autoimmune disease state [118]. Now many commentators are also advocating clinicians to stop costly measurements of 25OH vitamin D in asymptomatic patients [119–121]. The consideration of vitamin D as a marker for autoimmune diseases requires further sensitive evidence from randomized controlled clinical trials. The results then may elaborate the status of vitamin D as a marker for autoimmune diseases in future. Although more epidemiologic studies are needed to better understand the theory of vitamin D deficiency and its association with various auto immune diseases in children, the compelling data pointing to a role for vitamin D in immune regulation suggest that special attention should be paid to these at-risk populations.

6.1. Role of sunlight in autoimmune diseases

The ultraviolet radiation in sunlight can induce the onset of, or exacerbate, the symptoms of certain autoimmune diseases. Work done by Fraser et al. suggested that ultraviolet light exposure triggers production of reactive oxygen species as normal by-products. If the cell does not quickly eliminate the reactive oxygen species, however, the buildup can cause cellular and DNA damage. The *GSTM1* gene normally codes for an enzyme, glutathione S-transferase, which rids the body of reactive oxygen species. Individuals who have the GSTM1 null genotype and are missing the enzyme may, therefore, be at an increased risk of DNA damage and can induce the onset of lupus [122]. Work done by Love et al. to study the distribution of myositis phenotypes and ultraviolet radiation exposure in the United States showed that ultraviolet radiation may modulate the clinical and immunological expression of autoimmune disease in women [123]. The above evidence points toward the stimulation and exacerbation of autoimmune process in specific disease, whereas, on the other hand, ultraviolet radiations can also prevent or reduce the symptoms of other autoimmune diseases by vitamin D formation in skin as elaborated in various parts of this chapter.

7. Conclusion

In last decade, the number of studies that have investigated the non-bone effects of vitamin D has increased tremendously. Many studies from around the globe have measured circulating 25-hydroxyvitamin D as a determinant of vitamin D status. However, several queries regarding the definitive place of 25OH vitamin D as a biomarker of vitamin D status and the exact level of 25OHD that determines optimal vitamin D status in children for majority of non-bone effects of vitamin D still remain elusive. Although the committee from Institute of Medicine has recommended that a 25OHD level of 50 nmol/l (20 ng/ml) should be considered sufficient, this recommendation was mainly based on studies of bone health and the committee acknowledged that studies in other disease states are sorely lacking [4]. There are number of authors who have dissenting opinion regarding the IOM recommendations of optimal vitamin D levels [124]. Furthermore, there are data that suggest that optimal circulating levels regarding adequate bone health as well as other non-bone effects are much higher than the current IOM recommendations [125]. Most of the studies done in children have only measured vitamin D level once at one point in time. It is known that vitamin D levels vary over seasons and likely over time. Future studies need to measure 25OHD at multiple time points in relation to the outcome of interest. Finally, there is a continuing need for further conclusive studies to define appropriate levels of vitamin D status and recommended daily allowance regarding adequate bone health and other non-bone effects of vitamin D, especially in children, adolescents, and young adults.

Author details

Mohsin Ali Cheema and Khalid Parvez Lone*

*Address all correspondence to: khalid.lone@gmail.com

Department of Physiology and Cell Biology, and Centre for Research in Endocrinology and Reproductive Sciences (CRERS), University of Health Sciences, Lahore, Pakistan

References

[1] Holick MF. Vitamin D deficiency. N Engl J Med. 2007;357(3):266–81.

[2] Watson L. (2012). 10 things you need to know about vitamin D. [online] Available at: <http://www.drlisawatson.com/10-things-you-need-to-know-about-vitamin-d> [Accessed 10 January 2016].

[3] Pettifor JM, Moodley GP, Hough FS, Koch H, Chen T, Lu Z, Holick MF. The effect of season and latitude on *in vitro* vitamin D formation by sunlight in South Africa. S Afr Med J. 1996;86(10):1270–2.

[4] Ross AC, Manson JE, Abrams SA, Aloia JF, Brannon PM, Clinton SK, Durazo-Arvizu RA, Gallagher JC, Gallo RL, Jones G, Kovacs CS, Mayne ST, Rosen CJ, Shapses SA. The 2011 report on dietary reference intakes for calcium and vitamin D from the Institute of Medicine: what clinicians need to know. J Clin Endocrinol Metab. 2011;96(1):53–8.

[5] Cranney A, Horsley T, O'Donnell S, Weiler H, Puil L, Ooi D, Atkinson S, Ward L, Moher D, Hanley D, Fang M, Yazdi F, Garritty C, Sampson M, Barrowman N, Tsertsvadze A, Mamaladze V. Effectiveness and safety of vitamin D in relation to bone health. Evid Rep Technol Assess (Full Rep). 2007;(158):1–235.

[6] Heaney RP. Advances in therapy for osteoporosis. Clin Med Res. 2003;1(2):93–9. Review.

[7] Harinarayan CV, Ramalakshmi T, Prasad UV, Sudhakar D. Vitamin D status in Andhra Pradesh: a population based study. Indian J Med Res. 2008;127(3):211–8.

[8] Goldring SR, Krane SM, Aviolo LV. Disorders of calcification: osteomalacia and rickets. In: LJD, editor. Endocrinology, 3rd edition. Philadelphia (PA): WB Saunders, pp. 1204–27, 1995.

[9] Välimäki VV, Alfthan H, Lehmuskallio E, Löyttyniemi E, Sahi T, Stenman UH, Suominen H, Välimäki MJ. Vitamin D status as a determinant of peak bone mass in young Finnish men. J Clin Endocrinol Metab. 2004;89(1):76–80.

[10] Glerup H, Mikkelsen K, Poulsen L, Hass E, Overbeck S, Andersen H, Charles P, Eriksen EF. Hypovitaminosis D myopathy without biochemical signs of osteomalacic bone involvement. Calcif Tissue Int. 2000;66(6):419–424.

[11] Bischoff-Ferrari HA, Borchers M, Gudat F, Dürmüller U, Stähelin HB, Dick W. Vitamin D receptor expression in human muscle tissue decreases with age. J Bone Miner Res. 2004;19(2):265–9.

[12] Papadimitropoulos E, Wells G, Shea B, Gillespie W, Weaver B, Zytaruk N, Cranney A, Adachi J, Tugwell P, Josse R, Greenwood C, Guyatt G, Osteoporosis Methodology Group and The Osteoporosis Research Advisory Group. Meta-analyses of therapies for postmenopausal osteoporosis. VIII: meta-analysis of the efficacy of vitamin D treatment in preventing osteoporosis in postmenopausal women. Endocr Rev. 2002;23(4):560–9.

[13] Holick MF. Resurrection of vitamin D deficiency and rickets. J Clin Invest. 2006;116:2062–72.

[14] Siddiqui TS, Rai MI. Presentation and predisposing factors of nutritional rickets in children of Hazara Division. J Ayub Med Coll Abbottabad. 2005;17(3):29–32.

[15] Anwar S, Iqbal MP, Azam I, Habib A, Bhutta S, Soofi SB, Bhutta ZA. Urban and rural comparison of vitamin D status in Pakistani pregnant women and neonates. J Obstet Gynaecol. 2016;36(3):318–23. doi:10.3109/01443615.2015.1050647. Epub 2015 Oct 14.

[16] Jamal A, Khanani MR, Billo G, Asghar A, Jafri Z. Rickets in a slum of Karachi. Pak J Med Sci. 1996;12:247–50.

[17] Qamar S, Akbani S, Shamim S, Khan G. Vitamin D levels in children with growing pains. J Coll Phys Surg Pak. 2011;21(5):284–7.

[18] Iqbal R, Jafri L, Haroon A, Habib Khan A. Illuminating the dark side—vitamin D status in different localities of Karachi. J Coll Phys Surg Pak. 2013;23(8):604–6. doi:08.2013/JCPSP.604606.

[19] Signorello LB, Williams SM, Zheng W, Smith JR, Long J, Cai Q, Hargreaves MK, Hollis BW, Blot WJ. Blood vitamin D levels in relation to genetic estimation of African ancestry. Cancer Epidemiol Biomarkers Prev. 2010;19(9):2325–31. doi:10.1158/1055-9965.EPI-10-0482. Epub 2010 Jul 20.

[20] Shoenfeld N, Amital H, Shoenfeld Y. The effect of melanism and vitamin D synthesis on the incidence of autoimmune disease. Nat Clin Pract Rheumatol. 2009;5(2):99–105. doi:10.1038/ncprheum0989.

[21] Signorello LB, Shi J, Cai Q, Zheng W, Williams SM, Long J, Cohen SS, Li G, Hollis BW, Smith JR, Blot WJ. Common variation in vitamin D pathway genes predicts circulating 25-hydroxyvitamin D levels among African Americans. PLoS One. 2011;6(12):e28623. doi:10.1371/journal.pone.0028623. Epub 2011 Dec 21.

[22] Dawson-Hughes B. Racial/ethnic considerations in making recommendations for vitamin D for adult and elderly men and women. Am J Clin Nutr. 2004;80(6 Suppl): 1763S–6S.

[23] Gallagher JC, Peacock M, Yalamanchili V, Smith LM. Effects of vitamin D supplementation in older African American women. J Clin Endocrinol Metab. 2013;98(3):1137–46. doi:10.1210/jc.2012-3106. Epub 2013 Feb 5.

[24] Wortsman J, Matsuoka LY, Chen TC, Lu Z, Holick MF. Decreased bioavailability of vitamin D in obesity. Am J Clin Nutr. 2000;72(3):690–3.

[25] Arthritis Rheum. 2009 Aug;60(8):2499–504. doi: 10.1002/art.24702. Ultraviolet radiation intensity predicts the relative distribution of dermatomyositis and anti-Mi-2 autoantibodies in women. Love LA, Weinberg CR, McConnaughey DR, Oddis CV, Medsger TA Jr, Reveille JD, Arnett FC, Targoff IN, Miller FW.

[26] Lagunova Z, Porojnicu AC, Lindberg F, Hexeberg S, Moan J. The dependency of vitamin D status on body mass index, gender, age and season. Anticancer Res. 2009;29(9):3713–20.

[27] Verdoia M, Schaffer A, Barbieri L, Di Giovine G, Marino P, Suryapranata H, De Luca G; Novara Atherosclerosis Study Group (NAS). Impact of gender difference on vitamin D status and its relationship with the extent of coronary artery disease. Nutr Metab Cardiovasc Dis. 2015;25(5):464–70. doi:10.1016/j.numecd.2015.01.009. Epub 2015 Feb 7.

[28] Jablonski NG, Chaplin G. The evolution of human skin coloration. J Hum Evol. 2000;39(1):57–106.

[29] Xue Y, Ying L, Horst RL, Watson G, Goltzman D. Androgens attenuate Vitamin D production induced by UVB irradiation of the skin of male mice by an enzymatic mechanism. J Invest Dermatol. 2015;135(12):3125–32. doi:10.1038/jid.2015.297. Epub 2015 Jul 27.

[30] Ramagopalan SV, Heger A, Berlanga AJ, Maugeri NJ, Lincoln MR, Burrell A, Handun netthi L, Handel AE, Disanto G, Orton SM, Watson CT, Morahan JM, Giovannoni G, Ponting CP, Ebers GC, Knight JC. A ChIP-seq defined genome-wide map of vitamin D receptor binding: associations with disease and evolution. Genome Res. 2010;20(10): 1352–60. doi:10.1101/gr.107920.110. Epub 2010 Aug 24.

[31] Holick MF. Vitamin D: a millennium perspective. J Cell Biochem. 2003;88(2):296–307.

[32] Colston K, Colston MJ, Feldman D. 1,25-Dihydroxyvitamin D3 and malignant melanoma: the presence of receptors and inhibition of cell growth in culture. Endocrinology. 1981;108(3):1083–6.

[33] Cross HS, Bareis P, Hofer H, Bischof MG, Bajna E, Kriwanek S, Bonner E, Peterlik M. 25-Hydroxyvitamin D(3)-1alpha-hydroxylase and vitamin D receptor gene expression in human colonic mucosa is elevated during early cancerogenesis. Steroids. 2001;66(3–5):287–92.

[34] MacLaughlin J, Holick MF. Aging decreases the capacity of human skin to produce vitamin D3. J Clin Invest. 1985;76(4):1536–8.

[35] Mather KA, Jorm AF, Parslow RA, Christensen H. Is telomere length a biomarker of aging? A review. J Gerontol A Biol Sci Med Sci. 2011;66(2):202–13. doi:10.1093/gerona/glq180. Epub 2010 Oct 28.

[36] Liu JJ, Prescott J, Giovannucci E, Hankinson SE, Rosner B, Han J, De Vivo I. Plasma vitamin D biomarkers and leukocyte telomere length. Am J Epidemiol. 2013;177(12): 1411–7. doi:10.1093/aje/kws435. Epub 2013 May 9.

[37] Tuohimaa P. Vitamin D and aging. J Steroid Biochem Mol Biol. 2009;114(1–2):78–84.

[38] Duque G, El Abdaimi K, Macoritto M, Miller MM, Kremer R. Estrogens (E2) regulate expression and response of 1,25-dihydroxyvitamin D3 receptors in bone cells: changes with aging and hormone deprivation. Biochem Biophys Res Commun. 2002;299(3):446–54.

[39] Walters JR, Balesaria S, Chavele KM, Taylor V, Berry JL, Khair U, Barley NF, van Heel DA, Field J, Hayat JO, Bhattacharjee A, Jeffery R, Poulsom R. Calcium channel TRPV6 expression in human duodenum: different relationships to the vitamin D system and aging in men and women. J Bone Miner Res. 2006;21(11):1770–7.

[40] Welsh J, Wietzke JA, Zinser GM, Smyczek S, Romu S, Tribble E, Welsh JC, Byrne B, Narvaez CJ. Impact of the Vitamin D3 receptor on growth-regulatory pathways in mammary gland and breast cancer. J Steroid Biochem Mol Biol. 2002;83(1–5):85–92.

[41] Klaus G, Weber L, Rodríguez J, Fernández P, Klein T, Grulich-Henn J, Hügel U, Ritz E, Mehls P. Interaction of IGF-I and 1 alpha, 25(OH)2D3 on receptor expression and growth stimulation in rat growth plate chondrocytes. Kidney Int. 1998;53(5):1152–61.

[42] Andress D. Nonclassical aspects of differential vitamin D receptor activation: implications for survival in patients with chronic kidney disease. Drugs. 2007;67(14):1999–2012

[43] Fernandez-Martin JL, Kurian S, Farmer P, Nanes MS. Tumor necrosis factor activates a nuclear inhibitor of vitamin D and retinoid-X receptors. Mol Cell Endocrinol. 1998;141(1–2):65–72.

[44] González Pardo V, Boland R, de Boland AR. Vitamin D receptor levels and binding are reduced in aged rat intestinal subcellular fractions. Biogerontology. 2008;9(2):109–18. Epub 2007 Dec 1.

[45] Keisala T, Minasyan A, Lou YR, Zou J, Kalueff AV, Pyykkö I, Tuohimaa P. Premature aging in vitamin D receptor mutant mice. J Steroid Biochem Mol Biol. 2009;115(3–5): 91–7. doi:10.1016/j.jsbmb.2009.03.007. Epub 2009 Mar 31.

[46] Smit E, Crespo CJ, Michael Y, Ramirez-Marrero FA, Brodowicz GR, Bartlett S, Andersen RE. The effect of vitamin D and frailty on mortality among non-institutionalized US older adults. Eur J Clin Nutr. 2012;66(9):1024–8. doi:10.1038/ejcn.2012.67. Epub 2012 Jun 13.

[47] Online available at: http://www.cancer.gov/types/childhood-cancers/child-adolescent-cancers-fact-sheet

[48] Online available at: http://ccfpakistan.org/2011/05/23/childhood-cancer/

[49] Helou M, Ning Y, Yang S, Irvine P, Bachmann LM, Godder K, Massey G. Vitamin D deficiency in children with cancer. J Pediatr Hematol Oncol. 2014;36(3):212–7. doi: 10.1097/MPH.0b013e31829f3754.

[50] Modan-Moses D, Pinhas-Hamiel O, Munitz-Shenkar D, Temam V, Kanety H, Toren A. Vitamin D status in pediatric patients with a history of malignancy. Pediatr Res. 2012;72(6):620–4. doi:10.1038/pr.2012.131. Epub 2012 Oct 4.

[51] Lin R, White JH. The pleiotropic actions of vitamin D. Bioessays. 2004;26(1):21–8. pmid: 14696037 doi:10.1002/bies.10368

[52] Jones G, Strugnell SA, DeLuca HF. Current understanding of the molecular actions of vitamin D. Physiol Rev. 1998;78(4):1193–231. pmid:9790574

[53] Tangpricha V, Flanagan JN, Whitlatch LW, Tseng CC, Chen TC, Holt PR, Lipkin MS, Holick MF. 25-hydroxyvitamin D-1alpha-hydroxylase in normal and malignant colon tissue. Lancet. 2001;357(9269):1673–4.

[54] Mawer EB, Hayes ME, Heys SE, Davies M, White A, Stewart MF, Smith GN. Constitutive synthesis of 1,25-dihydroxyvitamin D3 by a human small cell lung cancer cell line. J Clin Endocrinol Metab. 1994;79(2):554–60.

[55] Garland CF, Garland FC, Gorham ED. Can colon cancer incidence and death rates be reduced with calcium and vitamin D? Am J Clin Nutr. 1991;54(1 Suppl):193S–201S.

[56] Garland CF, Comstock GW, Garland FC, Helsing KJ, Shaw EK, Gorham ED. Serum 25-hydroxyvitamin D and colon cancer: eight-year prospective study. Lancet. 1989;2(8673): 1176–8.

[57] Hanchette CL, Schwartz GG. Geographic patterns of prostate cancer mortality. Evidence for a protective effect of ultraviolet radiation. Cancer. 1992;70(12):2861–9.

[58] Guy M, Lowe LC, Bretherton-Watt D, Mansi JL, Peckitt C, Bliss J, Wilson RG, Thomas V, Colston KW. Vitamin D receptor gene polymorphisms and breast cancer risk. Clin Cancer Res. 2004;10(16):5472–81.

[59] Rashid MU, Muzaffar M, Khan FA, Kabisch M, Muhammad N, Faiz S, Loya A, Hamann U. Association between the BsmI polymorphism in the vitamin D receptor gene and breast cancer risk: results from a Pakistani Case-Control Study. PLoS One. 2015;10(10):e0141562. doi:10.1371/journal.pone.0141562. eCollection 2015.

[60] Feldman D, Krishnan AV, Swami S, Giovannucci E, Feldman BJ. The role of vitamin D in reducing cancer risk and progression. Nat Rev Cancer. 2014;14(5):342–57. doi: 10.1038/nrc3691. pmid:24705652. Epub 2014 Apr 4.

[61] Visweswaran RK, Lekha H. Extraskeletal effects and manifestations of Vitamin D deficiency. Indian J Endocrinol Metab. 2013;17(4):602–10. doi:10.4103/2230-8210.113750.

[62] Pálmer HG, Larriba MJ, García JM, Ordóñez-Morán P, Peña C, Peiró S, Puig I, Rodríguez R, de la Fuente R, Bernad A, Pollán M, Bonilla F, Gamallo C, de Herreros AG, Muñoz A. The transcription factor SNAIL represses vitamin D receptor expression and responsiveness in human colon cancer. Nat Med. 2004;10(9):917–9.

[63] Deeb KK, Trump DL, Johnson CS. Vitamin D signalling pathways in cancer: potential for anticancer therapeutics. Nat Rev Cancer. 2007;7(9):684–700.

[64] Online available at: http://www.diabetesatlas.org/

[65] Pozzilli P, Manfrini S, Crino A, Picardi A, Leomanni C, Cherubini V, Valente L, Khazrai M, Visalli N; IMDIAB group. Low levels of 25-hydroxyvitamin D3 and 1,25-dihydroxyvitamin D3 in patients with newly diagnosed type 1 diabetes. Horm Metab Res. 2005;37:680–3. [PubMed].

[66] Littorin B, Blom P, Scholin A, Arnqvist HJ, Blohme G, Bolinder J, Ekbom-Schnell A, Eriksson JW, Gudbjörnsdottir S, Nyström L, Ostman J, Sundkvist G. Lower levels of plasma 25-hydroxyvitamin D among young adults at diagnosis of autoimmune T1D compared with control subjects: results from the nationwide Diabetes Incidence Study in Sweden (DISS) Diabetologia. 2006;49:2847–52. [PubMed].

[67] Greer RM, Rogers MA, Bowling FG, Buntain HM, Harris M, Leong GM, Cotterill AM. Australian children and adolescents with T1D have low vitamin D levels. Med J Aust. 2007;187:59–60. [PubMed].

[68] Svoren BM, Volkening LK, Wood JR, Laffel LM. Significant vitamin D deficiency in youth with T1D mellitus. J Pediatr. 2009;154:132–4. [PMC free article] [PubMed].

[69] Borkar VV, Devidayal, Verma S, Bhalla AK. Low levels of vitamin D in North Indian children with newly diagnosed type 1 diabetes. Pediatr Diabetes. 2010;11:345–50. [PubMed].

[70] Bener A, Alsaied A, Al-Ali M, Al-Kubaisi A, Basha B, Abraham A, Guiter G, Mian M. High prevalence of vitamin D deficiency in T1D mellitus and healthy children. Acta Diabetol. 2009;46:183–9. [PubMed].

[71] Zipitis CS, Akobeng AK. Vitamin D supplementation in early childhood and risk of type 1 diabetes: a systematic review and meta-analysis. Arch Dis Child. 2008;93:512–7.

[72] Hyppönen E, Läärä E, Reunanen A, Järvelin MR, Virtanen SM. Intake of vitamin D and risk of type 1 diabetes: a birth-cohort study. Lancet. 2001;358:1500–3.

[73] Stene LC, Joner G. Use of cod liver oil during the first year of life is associated with lower risk of childhood onset type 1 diabetes: a large population-based, case control study. Am J Clin Nutr. 2003;78:1128–34.

[74] Roomi MA, Lone KP, Madassar A. Vitamin D and cardiometabolic risk factors in adult non-diabetic offspring of type 2 diabetic parents. J Pak Med Assoc. 2014;64(11):1229–34.

[75] Mohr SB, Garland CF, Gorham ED, Garland FC. The association between ultraviolet B irradiance, vitamin D status and incidence rates of type 1 diabetes in 51 regions worldwide. Diabetologia. 2008;51(8):1391–8. doi:10.1007/s00125-008-1061-5. Epub 2008 Jun 12.

[76] Staples JA, Ponsonby AL, Lim LL, McMichael AJ. Ecologic analysis of some immune-related disorders, including type 1 diabetes, in Australia: latitude, regional ultraviolet radiation, and disease prevalence. Environ Health Perspect. 2003;111(4):518–23.

[77] American Diabetes Association. Diagnosis and classification of diabetes mellitus. Diabetes Care. 2009;32(Suppl 1):S62–7.

[78] Gregori S, Giarratana N, Smiroldo S, Uskokovic M, Adorini L. A 1alpha,25-dihydroxyvitamin D(3) analog enhances regulatory T-cells and arrests autoimmune diabetes in NOD mice. Diabetes. 2002;51(5):1367–74.

[79] Mathieu C1, Waer M, Casteels K, Laureys J, Bouillon R. Prevention of type I diabetes in NOD mice by nonhypercalcemic doses of a new structural analog of 1,25-dihydroxyvitamin D3, KH1060. Endocrinology. 1995;136(3):866–72.

[80] Aljabri KS, Bokhari SA, and Khan MJ. Glycemic changes after vitamin D supplementation in patients with type 1 diabetes mellitus and vitamin D deficiency. Ann Saudi Med. 2010;30(6):454–8.

[81] Masoli M, Fabian D, Holt S, Beasley R. The global burden of asthma: executive summary of the GINA Dissemination Committee report. Allergy. 2004;59:469–78.

[82] Mannino DM, Homa DM, Akinbami LJ, et al. Surveillance for asthma: United States, 1980–1999. MMWR Surveill Summ. 2002;51:1–13.

[83] Devereux G. The increase in allergic disease: environment and susceptibility. Proceedings of a symposium held at the Royal Society of Edinburgh, 4th June 2002. Clin Exp Allergy. 2003;33:394–406.

[84] (CDC) CfDCaP. Vital signs: asthma prevalence, disease characteristics, and self-management education: United States, 2001–2009. Morb Mortal Wkly Rep. 2011;60:547–52.

[85] Hasnain SM, Khan M, Saleem A, Waqar MA. Prevalence of asthma and allergic rhinitis among school children of Karachi, Pakistan, 2007. J Asthma. 2009;46(1):86–90. doi: 10.1080/02770900802513023.

[86] Khan AA, Tanzil S, Jamali T, Shahid A, Naeem S, Sahito A, Siddiqui FA, Nafees AA, Fatmi Z. Burden of asthma among children in a developing megacity: childhood

asthma study, Pakistan. J Asthma. 2014;51(9):891–9. doi:10.3109/02770903.2014.930882. Epub 2014 Jul 3.

[87] Litonjua AA. Vitamin D deficiency as a risk factor for childhood allergic disease and asthma. Curr Opin Allergy Clin Immunol. 2012;12(2):179–85. doi:10.1097/ACI.0b013e3283507927.

[88] Litonjua AA. Childhood asthma may be a consequence of vitamin D deficiency. Curr Opin Allergy Clin Immunol. 2009;9(3):202–7. doi:10.1097/ACI.0b013e32832b36cd.

[89] Raby BA, Lazarus R, Silverman EK, Lake S, Lange C, Wjst M, Weiss ST. Association of vitamin D receptor gene polymorphisms with childhood and adult asthma. Am J Respir Crit Care Med. 2004;170(10):1057–65. Epub 2004 Jul 28.

[90] Poon AH, Laprise C, Lemire M, Montpetit A, Sinnett D, Schurr E, Hudson TJ. Association of vitamin D receptor genetic variants with susceptibility to asthma and atopy. Am J Respir Crit Care Med. 2004;170(9):967–73. Epub 2004 Jul 28.

[91] Wjst M, Altmüller J, Faus-Kessler T, Braig C, Bahnweg M, André E. Asthma families show transmission disequilibrium of gene variants in the vitamin D metabolism and signalling pathway. Respir Res. 2006;7:60.

[92] Ramsey CD, Celedón JC. The hygiene hypothesis and asthma. Curr Opin Pulm Med. 2005;11(1):14–20

[93] Corne JM, Marshall C, Smith S, Schreiber J, Sanderson G, Holgate ST, Johnston SL. Frequency, severity, and duration of rhinovirus infections in asthmatic and non-asthmatic individuals: a longitudinal cohort study. Lancet. 2002;359(9309):831–4.

[94] Liu PT, Stenger S, Tang DH, Modlin RL. Cutting edge: vitamin D-mediated human antimicrobial activity against Mycobacterium tuberculosis is dependent on the induction of cathelicidin. J Immunol. 2007;179(4):2060–3

[95] Herr C, Shaykhiev R, Bals R. The role of cathelicidin and defensins in pulmonary inflammatory diseases. Expert Opin Biol Ther. 2007;7(9):1449–61.

[96] Ginde AA, Mansbach JM, Camargo CA Jr. Association between serum 25-hydroxyvitamin D level and upper respiratory tract infection in the Third National Health and Nutrition Examination Survey. Arch Intern Med. 2009;169(4):384–90. doi:10.1001/archinternmed.2008.560.

[97] Zosky GR, Berry LJ, Elliot JG, James AL, Gorman S, Hart PH Vitamin D deficiency causes deficits in lung function and alters lung structure. Am J Respir Crit Care Med. 2011;183(10):1336–43.

[98] Chi A, Wildfire J, McLoughlin R, Wood RA, Bloomberg GR, Kattan M, Gergen P, Gold DR, Witter F, Chen T, Holick M, Visness C, Gern J, O'Connor GT. Umbilical cord plasma 25-hydroxyvitamin D concentration and immune function at birth: the Urban Envi-

ronment and Childhood Asthma study. Clin Exp Allergy. 2011;41(6):842–50. doi: 10.1111/j.1365-2222.2011.03712.x. Epub 2011 Apr 11.

[99] Miyake Y, Sasaki S, Tanaka K, Hirota Y. Dairy food, calcium and vitamin D intake in pregnancy, and wheeze and eczema in infants. Eur Respir J. 2010;35(6):1228–34.

[100] Rothers J, Wright AL, Stern DA, Halonen M, Camargo CA Jr. Cord blood 25-hydroxyvitamin D levels are associated with aeroallergen sensitization in children from Tucson, Arizona. J Allergy Clin Immunol. 2011;128(5):1093–9.e1–5.

[101] Camargo CA Jr, Ingham T, Wickens K, Thadhani R, Silvers KM, Epton MJ, Town GI, Pattemore PK, Espinola JA, Crane J, New Zealand Asthma and Allergy Cohort Study Group. Cord-blood 25-hydroxyvitamin D levels and risk of respiratory infection, wheezing, and asthma. Pediatrics. 2011;127(1):e180–7.

[102] Freishtat RJ, Iqbal SF, Pillai DK, Klein CJ, Ryan LM, Benton AS, Teach SJ. High prevalence of vitamin D deficiency among inner-city African American youth with asthma in Washington, DC. J Pediatr. 2010;156(6):948–52.

[103] Bener A, Ehlayel MS, Tulic MK, Hamid Q. Vitamin D deficiency as a strong predictor of asthma in children. Int Arch Allergy Immunol. 2012;157(2):168–75.

[104] Alyasin S, Momen T, Kashef S, Alipour A, Amin R. The relationship between serum 25 hydroxy vitamin D levels and asthma in children. Allergy Asthma Immunol Res. 2011;3(4):251–5.

[105] Available at: http://autoimmune.pathology.jhmi.edu/faqs.cfm. [Accessed 23 March 23 2016].

[106] Available at: http://www.childrenshospital.org/conditions-and-treatments/conditions/autoimmune-diseases 15th April, 2016

[107] Adorini A, Penna G. Control of autoimmune diseases by the vitamin D endocrine system. Nat Clin Pract Rheumatol. 2008;4:404–12

[108] Bultink IE, Lems WF, Kostense PJ, Dijkmans BA, Voskuyl AE. Prevalence of and risk factors for low bone mineral density and vertebral fractures in patients with systemic lupus erythematosus. Arthritis Rheum. 2005;52(7):2044–50.

[109] Wright TB, Shults J, Leonard MB, Zemel BS, Burnham JM. Hypovitaminosis D is associated with greater body mass index and disease activity in pediatric systemic lupus erythematosus. J Pediatr. 2009;155(2):260–5.

[110] Robinson AB, Thierry-Palmer M, Gibson KL, Rabinovich CE. Disease activity, proteinuria, and vitamin D status in children with systemic lupus erythematosus and juvenile dermatomyositis. J Pediatr. 2012;160(2):297–302. doi:10.1016/j.jpeds.2011.08.011. Epub 2011 Sep 15.

[111] Abe J, Nakamura K, Takita Y, Nakano T, Irie H, Nishii Y. Prevention of immunological disorders in MRL/l mice by a new synthetic analogue of vitamin D3: 22-oxa-1 alpha, 25-dihydroxyvitamin D3. J Nutr Sci Vitaminol (Tokyo). 1990;36(1):21–31

[112] Lemire JM, Ince A, Takashima M. 1,25-Dihydroxyvitamin D3 attenuates the expression of experimental murine lupus of MRL/l mice. Autoimmunity. 1992;12(2):143–8.

[113] Kamen DL, Cooper GS, Bouali H, Shaftman SR, Hollis BW, Gilkeson GS. Vitamin D deficiency in systemic lupus erythematosus. Autoimmun Rev. 2006;5(2):114–7

[114] Nagpal S, Na S, Rathnachalam R. Noncalcemic actions of vitamin D receptor ligands. Endocr Rev. 2005;26(5):662–87.

[115] Munger KL, Zhang SM, O'Reilly E, Hernán MA, Olek MJ, Willett WC. Vitamin D intake and incidence of multiple sclerosis. Neurology. 2004;62:60–5.

[116] Goldberg P, Fleming MC, Picard EH. Multiple sclerosis: decreased relapse rate through dietary supplementation with calcium, magnesium and vitamin D. Med Hypotheses. 1986;21:193–200.

[117] Ramagopalan SV, Maugeri NJ, Handunnetthi L, Lincoln MR, Orton SM, Dyment DA, Deluca GC, Herrera BM, Chao MJ, Sadovnick AD, Ebers GC, Knight JC. Expression of the multiple sclerosis-associated MHC class II Allele HLA-DRB1*1501 is regulated by vitamin D. PLoS Genet. 2009;5(2):e1000369.

[118] Blaney GP, Albert PJ, Proal AD. Vitamin D metabolites as clinical markers in autoimmune and chronic disease. Ann N Y Acad Sci. 2009;1173:384–90. doi:10.1111/j.1749-6632.2009.04875.x.

[119] Welsh P, Sattar N. Vitamin D and chronic disease prevention. BMJ. 2014;348:g2280. doi: 10.1136/bmj.g2280.

[120] Sattar N, Welsh P, Panarelli M, Forouhi NG. Increasing requests for vitamin D measurement: costly, confusing, and without credibility. Lancet. 2012;379(9811):95–6. doi: 10.1016/S0140-6736(11)61816-3.

[121] Fletcher RH. Review: vitamin D3 supplementation may reduce mortality in adults; vitamin D2 does not. Ann Intern Med. 2014;161(2):JC5. doi:10.7326/0003-4819-161-2-201407150-02005.

[122] Fraser PA, Ding WZ, Mohseni M, Treadwell EL, Dooley MA, St Clair EW, Gilkeson GS, Cooper GS. Glutathione S-transferase M null homozygosity and risk of systemic lupus erythematosus associated with sun exposure: a possible gene-environment interaction for autoimmunity. J Rheumatol. 2003;30(2):276–82.

[123] Love LA, Weinberg CR, McConnaughey DR, Oddis CV, Medsger TA Jr, Reveille JD, Arnett FC, Targoff IN, Miller FW. Ultraviolet radiation intensity predicts the relative distribution of dermatomyositis and anti-Mi-2 autoantibodies in women. Arthritis Rheum. 2009;60(8):2499–504. doi:10.1002/art.24702.

[124] Heaney RP, Holick MF. Why the IOM recommendations for vitamin D are deficient. J Bone Miner Res. 2011;26(3):455–7. doi:10.1002/jbmr.328.

[125] Heaney RP, Armas LA, Shary JR, Bell NH, Binkley N, Hollis BW. 25-Hydroxylation of vitamin D3: relation to circulating vitamin D3 under various input conditions. Am J Clin Nutr. 2008;87(6):1738–42.

Section 3

Vitamin D on Musculoskeletal and Neurological System

Chapter 9

Role of Vitamin D in Myogenesis

Akira Wagatsuma

Additional information is available at the end of the chapter

http://dx.doi.org/10.5772/64514

Abstract

The secosteroid, 1α,25-dihydroxyvitamin D_3 [1α,25$(OH)_2D_3$] plays a crucial role in regulating bone formation, remodeling and repair. Beyond its well-established role in skeletogenesis, gene-targeting studies support a physiological role for 1α,25$(OH)_2D_3$ in muscle development. There is evidence for expression of vitamin D receptor/vitamin D synthesizing enzyme/transport protein, local production of 1α,25$(OH)_2D_3$ and uptake of 25(OH)D_3, implying the existence of vitamin D-endocrine system in myogenic cells. Recently, much interest has been devoted to the effects of 1α,25$(OH)_2D_3$ on myogenesis. Simply stated, 1α,25$(OH)_2D_3$ has potent antiproliferative activity on myoblasts and inhibits myoblast differentiation. Intriguingly, recent studies suggest that 1α,25$(OH)_2D_3$ may stimulate protein synthesis in myotubes and have a role in self-renewal of muscle stem cells. 1α,25$(OH)_2D_3$ regulates myogenesis probably through its genomic or nongenomic actions. Understanding how vitamin D signaling contributes to muscle homeostasis may provide a valuable insight into an effective intervention strategy for muscle disorders. In this review, we summarize the current knowledge about a possible role of vitamin D in myogenesis.

Keywords: skeletal muscle, proliferation, differentiation, vitamin D, myogenesis

1. Introduction

The biologically active metabolite of vitamin D, 1α,25-dihydroxyvitamin D_3 [1α,25$(OH)_2D_3$] functions by binding to vitamin D receptor (VDR) [1]. Thus, accurate identification of VDR in tissues is critical to understand the physiological and pathological significance of vitamin D [2]. The VDR has shown to be expressed in a wide variety of tissues, including bone, bronchus, intestine, kidney, mammary gland, pancreas, parathyroid, pituitary gland, prostate gland, spleen, testis and thymus [2]. However, in skeletal muscle, some controversies have existed [3–9]. Despite conflict, gene-targeting studies suggest a physiological role for 1α,25(OH)$_2$D$_3$ in muscle development. VDR null-mutant mice exhibit impaired muscle development [10, 11]. Intriguingly, there is evidence for expression of vitamin D receptor/vitamin D synthesizing enzyme/transport protein [6, 9, 12–16], local conversion of 25(OH)D$_3$ to 1α,25(OH)$_2$D$_3$ [6, 15] and uptake of 25(OH)D$_3$ [17], implying the existence of vitamin D-endocrine system in myogenic cells.

Recently, much interest has been devoted to the effects of 1α,25(OH)$_2$D$_3$ on myogenesis. It has been generally supposed that 1α,25(OH)$_2$D$_3$ exerts its effects through binding to VDR, inducing genomic or nongenomic actions. A genomic action occurs through nuclear transportation of 1α,25(OH)$_2$D$_3$-VDR complex [1]. VDR heterodimerizes with 9-cis-retinoic acid receptor (RXR), which modulates gene expression via binding to specific target gene promoter regions, known as vitamin D response elements (VDREs) [1]. The very large number of genes (estimated at approximately 3% of the mouse or human genome) is regulated, directly and/or indirectly, by vitamin D-endocrine system [18]. A nongenomic response to 1α,25(OH)$_2$D$_3$ is characterized by a rapid (the seconds to minutes range) activation of signaling cascades and an insensitivity to inhibitors of transcription and protein synthesis [19]. Although a consensus cannot be developed about how the nongenomic actions are initiated [1], several data suggest that nongenomic actions begin at the plasma membrane and occur through a putative plasma membrane-associated receptor [VDR(mem)] present in a caveolae [20] and a 1α,25(OH)$_2$D$_3$-membrane-associated rapid response steroid (1α,25D$_3$-MARRS) binding protein [21]. Although relative contribution of genomic and nongenomic action to myogenesis is currently unknown, 1α,25(OH)$_2$D$_3$ appears to have antiproliferative effect on myoblasts [9, 12–15] and inhibits myoblast differentiation [9, 16, 18]. Some hypotheses are currently proposed that 1α, 25(OH)$_2$D$_3$ may have anabolic effects on protein metabolism in myotubes [12, 15] and may be involved in self-renewal of muscle stem cells [9].

Better characterization of effects of vitamin D in myogenic cells will help to provide a valuable insight into an effective intervention strategy for muscle disorders such as sarcopenia, myopathy and neuromuscular diseases. Here, we will summarize the current evidence for a role of vitamin D in myogenesis.

2. VDR expression in skeletal muscle and myogenic cells

2.1. Animal studies

Although a number of studies have been published on the expression of VDR in skeletal muscle by immunohistochemistry, the specificity of various commercially available VDR antibodies used is called into question by Wang et al. [22]. They systematically characterize these antibodies in terms of their specificity and immunosensitivity using negative control samples from VDR knockout mice. For example, despite widely used for immunohistochemical studies, the rat monoclonal antibody (9A7, Affinity BioReagents) that recognizes an epitope between amino acid residues 89 and 105 of human VDR has been shown to bind the VDR but nonspecifically cross-react with other unidentified proteins from mice [22]. They identify the mouse monoclonal VDR antibody against the C-terminus of human VDR (D-6, Santa Cruz Biotechnology), which possesses high specificity, high sensitivity and versatility [22]. They show that VDR protein is not detected in muscle fibers from 6- to 7-week-old C57BL/6 mice [8]. In contrast, Girgis et al. [6] demonstrated that VDR protein is localized to nuclei of muscle fibers from 3-week-old C57BL/6 mice using the same antibody. These data suggest that VDR expression level may be dependent on the age of animal because its expression is progressively decreased with advancing age [6]. However, intriguingly, VDR has shown to be re-expressed in the regenerating muscle fibers from 12-week-old C57BL/6 mice [13]. Immunohistochemical analysis using a rabbit polyclonal antibody demonstrated that VDR is clearly detected in the central nuclei of newly formed regenerating muscle fibers but not in noninjured muscle fibers, suggesting that vitamin D signal via VDR may be involved in muscle regeneration irrespective of age. Besides immunohistochemical evidence, Srikuea et al. [13] provide strong evidence for the presence of VDR in mouse C2C12 cells, by combining PCR-based cloning and DNA sequencing. The full-length VDR mRNA transcript could be isolated from myoblasts and myotubes.

2.2. Human studies

Costa et al. [23] initially identify $1\alpha,25(OH)_2D_3$ binding protein in cloned human skeletal muscle cells derived from patients undergoing orthopedic surgery by a radiolabelled-ligand binding assay. The cloned human myogenic cells have a binding protein compatible with classical $1\alpha,25(OH)_2D_3$ receptors and functionally responded to $1\alpha,25(OH)_2D_3$ at physiological concentrations [23]. Bischoff et al. [3, 4] report the nuclear localization of VDR in muscle fibers from middle-aged and older female patients with osteoarthritis/osteoporosis or undergoing surgery using a rat monoclonal antibody (9A7). Ceglia et al. [5] investigate VDR expression in skeletal muscles from healthy postmenopausal women aged 65–85 years using a mouse monoclonal VDR antibody (H4537, R&D systems) whose accuracy is confirmed by immunoblot analysis using a mouse monoclonal VDR antibody (D-6). They report that VDR is detected in the nuclei of muscle fibers without relation to muscle fiber subtype. Ryan et al. [7] demonstrate that VDR is detectable in human muscle fibers by immunohistochemistry and immunoblot analysis using a polyclonal antibody against the human VDR. However, Olsson et al. [9] show that VDR is undetectable in muscle fibers from healthy individuals of age 20–27 years by immunoblot analysis using a mouse monoclonal VDR antibody (D-6). Overall, we cannot rule out the possibility of VDR expression in both animal and human muscle fibers at low levels. However, conflicting data should be carefully discussed because the detection of VDR expressed in muscle fibers may be dependent on the methodology for protein extraction and primary antibody. Indeed, Girgis et al. [6] note that the importance of specific conditions in the detection of VDR proteins expressed at low levels in skeletal muscle.

3. Vitamin D metabolism

3.1. Vitamin D metabolism in the liver and kidney

The process of $1\alpha,25(OH)_2D_3$ synthesis is highly regulated. Vitamin D, in the form of vitamin D_3, is nonenzymatically synthesized from 7-dehydrocholesterol in the skin through the action of ultraviolet irradiation [24]. Alternatively, vitamin D, in the form of either vitamin D_2 or vitamin D_3, can also be taken in the diet [24]. A biologically active form, $1\alpha,25(OH)_2D_3$, is synthesized from vitamin D_3 through two hydroxylation steps [24]. Vitamin D_3 is converted to 25-hydroxyvitamin D_3 [$25(OH)D_3$] in the liver by 25-hydroxylases (encoded by the gene *CYP27A1*) [24]. The generated $25(OH)D_3$ is further hydroxylated to $1\alpha,25(OH)_2D_3$ by 25-hydroxyvitamin D_3 1α-hydroxylase (encoded by the gene *CYP27B1*) in the kidney [24]. The synthesis of $1\alpha,25(OH)_2D_3$ from $25(OH)D_3$ is stimulated by parathyroid hormone (PTH) and suppressed by calcium, inorganic phosphate and $1\alpha,25(OH)_2D_3$ itself [1]. As just described, it is generally recognized that vitamin D is metabolized sequentially in the liver and kidney. However, CYP27B1 seems to be more widely distributed in several extrarenal tissues than previously expected, such as skin, placenta, colon, pancreas, vasculature and brain [25–27]. Several lines of evidence suggest that myogenic cells have the ability to internalize circulating $25(OH)D_3$ into the cell cytoplasm and subsequently locally convert it to $1\alpha,25(OH)_2D_3$. In the following section, we will briefly summarize our understanding of the uptake of $25(OH)D_3$ and local synthesis of $1\alpha,25(OH)_2D_3$ occurred in myogenic cells.

3.2. Uptake of $25(OH)D_3$ in myogenic cells

Vitamin D metabolites are lipophilic molecules with low aqueous solubility that must be transported in the circulation bound to plasma proteins [28]. The most important of these carrier proteins is the vitamin D-binding protein (DBP) [29]. Almost all circulating vitamin D metabolites are bound to DBP, to a lesser extent, to albumin and lipoproteins [28], providing a major impact on their pharmacokinetics. DBP-bound vitamin D metabolites restrict access to target cells [29] and, therefore, are less susceptible to hepatic metabolism and subsequent biliary excretion, leading to a longer circulating half-life [28]. $25(OH)D_3$ does not simply diffuse into the cells. In the kidney proximal tubule, uptake of the $25(OH)D_3$-DBP complex by epithelial cells occurs depending on receptor-mediated endocytosis via the multiligand megalin/cubilin tandem receptor [30]. Both receptors are expressed primarily in polarized epithelial cells [28]. Once internalized by epithelial cells, DBP is degraded in lysosomes, releasing $25(OH)D_3$ for activation to $1\alpha,25(OH)_2D_3$ by CYP27B1 [30]. Intriguingly, it seems likely that muscle fibers endocytose $25(OH)D_3$ through a similar mechanism. In support of this idea, megalin and cubilin are expressed in muscle fibers [17]. Uptake of ^{3}H-labeled $25(OH)D_3$ into C2C12 myotubes was drastically reduced by megalin inhibitor [17]. These data suggest that $25(OH)D_3$ may be internalized through megalin/cubilin-mediated endocytosis in myogenic cells.

3.3. Conversion of $25(OH)D_3$ to $1\alpha,25(OH)_2D_3$ in myogenic cells

Several studies demonstrate that myogenic cells express key components of the vitamin D-endocrine system, including VDR, CYP27B1, CYP24A1 and DBP [6, 12–16]. It is supposed therefore that localized, muscle-specific, conversion of $25(OH)D_3$ to $1\alpha,25(OH)_2D_3$ might drive many of the recognized effects of vitamin D. Srikuea et al. [13] report that C2C12 cells express the full-length CYP27B1 mRNA transcript and CYP27B1 protein is detected in the cytoplasm of myoblasts, exhibiting partially overlapping with the mitochondria to which CYP27B1 has been reported to be typically localized [31]. Girgis et al. [6, 15] confirm this possibility using a luciferase reporter assay system. In this system, luciferase activity results from $1\alpha,25(OH)_2D_3$ binding to GAL-4-VDR and subsequent activation of the UASTK-luciferase gene through its GAL4 promoter. The assay system show a dose-dependent increase in luciferase activity after 24 hours treatment with $25(OH)D_3$, indicating the intracellular conversion of $25(OH)D_3$ to $1\alpha,25(OH)_2D_3$ by CYP27B1 and the subsequent activation of luciferase expression through $1\alpha,25(OH)_2D_3$-bound GAL4-VDR [6, 15]. Although the functional importance of CYP27B1 in myogenic cells has not been fully understood, these data provide evidence that this enzyme is biologically active and mediates to convert $25(OH)D_3$ to $1\alpha,25(OH)_2D_3$.

4. Vitamin D signaling

4.1. Genomic pathway

Microarray analysis in squamous cell carcinoma treated with $1\alpha,25(OH)_2D_3$ identifies many genes including cell adhesion, cytoskeleton, extracellular matrix, growth factors/receptors, signal transduction, transcription factors, cell cycle and channels/transporters [32–34]. Profiling data provide insight into a much broader range of action for $1\alpha,25(OH)_2D_3$. With respect to several genes related to myogenesis, we will describe them in more detail. A consensus DR3-type VDRE has been identified in the dysferlin (DYSF) promoter [34], which plays a crucial role in membrane repair in skeletal muscle [35]. $1\alpha,25(OH)_2D_3$ increases dysferlin expression in human myotubes from carriers of one mutation in the DYSF gene probably through the binding of VDR to promoter of the DYSF gene [36]. FOXO1 has also the same element in the promoter [34]. Foxo1 is a member of the Foxo subfamily of forkhead/winged helix family of transcription factors, governs muscle growth, metabolism and myoblast differentiation. Foxo1 physically and functionally interacts with Notch by promoting corepressor clearance from DNA-binding protein, CSL [CBF1/RBPjk/Su(H)/Lag-1], leading to inhibition of myoblast differentiation through activation of Notch target genes [37]. Integrin β3 has a consensus sequence (−–756/−770) in the immediate 5′-flanking region [38]. The integrins are heterodimeric cell surface receptors, formed by the combination of 18 α-subunits and 8 β-subunits, mediate adhesion to the extracellular matrix [39]. Deficiency of integrin β3 in myoblast decreases Rac1-GTPase activity, downregulates myogenin expression, disrupts focal adhesion formation and interrupts actin organization, resulting in impaired myoblast migration and fusion [40]. Similarly, a β3-integrin-neutralizing antibody blocks myotube formation [40]. Consistent with in vitro studies, integrin β3 null-mutant mice reveals a defective muscle regeneration [40]. Id (inhibitor of differentiation) gene is also known as target of $1\alpha,25(OH)_2D_3$ [41]. $1\alpha,25(OH)_2D_3$ exerts its negative effect on Id1 gene transcription via the 57 bp upstream response sequence (−1146/−1090) [41]. Id proteins (Id1, Id2, Id3 and Id4) dimerize and neutralize the transcriptional activity of basic helix-loop-helix proteins [42]. Id inhibits MyoD activity either by forming transcriptionally inactive complexes of MyoD-Id or by forming heterodimers with E-proteins and effectively blocking the formation of active MyoD/E-protein complexes [43]. VDR null-mutant mice show no differences in expression levels of Id1 and Id2 in skeletal muscle [44]. Finally, expression of insulin-like growth factor binding proteins (IGFBPs) is known to be under the control of multiple VDREs [45]. The IGF system includes not only three ligands (insulin, IGF-I and IGF-II), three receptors [the insulin receptor (IR), the IGF-I receptor (IGF-IR) and the mannose-6-phosphate IGF-II receptor (M6P/IGF-IIR)] but also six IGFBPs. IGFs through its receptor signal can stimulate both myoblast proliferation and differentiation, which are two mutually exclusive biological events during myogenesis [46]. In the circulation, IGFBPs act as carrier proteins for the IGFs and regulate IGFs turnover, transport and half-life [47]. At the local level, they function as modulators of IGF activity [48]. Of six human IGFBP genes, IGFBP-1, -3 and -5 have been shown to be primary $1\alpha,25(OH)_2D_3$-target genes [45]. IGFBP-5 is the major IGFBP secreted by skeletal muscle and appears to mediate IGF-II expression via an autoregulatory loop mechanism [46]. In belief, the induction of IGFBP-5 occurs earlier than the induction of IGF-II in early stages of myogenesis. IGFBP-5 located on the cell surface binds to autocrine IGF-II and potentiates its interaction with IGF-IR, leading to the enhanced activation of the IGF-IR-PI3K-Akt signaling pathway. Consequently, IGF-II expression is further increased, promoting myoblast differentiation [49]. Besides the regulation of IGF bioavailability, IGFBP-5 directly regulates the transcriptional response of osteoblasts to $1\alpha,25(OH)_2D_3$. IGFBP-5 interacts with VDR and prevents VDR:RXRα heterodimerization, probably leading to impair vitamin D-stimulated transcription and cell cycle arrest [49].

Recently, Ryan et al. [7] report that 1947 mRNAs are differentially expressed in human myogenic cells following treatment of $1\alpha,25(OH)_2D_3$. The most significantly increased or repressed mRNA are CYP24A1 (>25,000-fold) or calpain 11 (0.1-fold), respectively. Messenger RNAs encode proteins involved in muscle relaxation (Parvalbumin), protein synthesis (Ig-like, fibronectin type III domain-containing 1), cytoskeletal dynamics [Rho Guanine Nucleotide Exchange Factor (GEF) 16], RNA and nucleotide binding (Tudor domain-containing protein 10), cellular energy metabolism (insulin-like growth factor 3), apoptosis (Fas apoptotic inhibitory molecule 2) and nucleosome function (Histone cluster 1, H3j) are significantly upregulated, whereas messenger RNAs encode proteins involved in cell migration (Podocan) and cellular proliferation (WAP four-disulfide core domain 1) are significantly downregulated [7]. Very little is currently known about the direct $1\alpha,25(OH)_2D_3$-target genes in myogenic cells, it is to be hoped that future research will clarify this issue.

4.2. Nongenomic pathway

After short-term treating (1–10 min) chick myoblasts with 1α,25$(OH)_2D_3$, translocation of VDR from the nucleus to the plasma membrane rapidly occurs [50]. Microtubule-depolymerizing agents block the translocation [50]. The translocation depends on intact caveolae that are specialized plasmalemmal microdomains [51]. Methyl-beta-cyclodextrin, which disrupts the caveolae structure, abolishes 1α,25$(OH)_2D_3$-dependent VDR translocation to the plasma membrane [52]. In addition, chemically-induced disruption of caveolae and small-interfering RNA (siRNA)-mediated silencing of caveolin-1 suppress 1α,25$(OH)_2D_3$-dependent activation of proto-oncogene, c-Src [52]. Caveolin-1 colocalizes with c-Src near the plasma membrane under basal conditions [52]. After treating with 1α,25$(OH)_2D_3$, the colocalization is disrupted and they are redistributed into cytoplasm and nucleus [52]. Therefore, it can be hypothesized that (1) interaction caveolin-1/c-Src inactivates the kinase under basal conditions and (2) when 1α,25$(OH)_2D_3$ stimulates VDR translocation to the plasma membrane, it dissociates the caveolin-1/c-Src complex, allowing c-Src activation [52]. The initial activation of c-Src by 1α, 25$(OH)_2D_3$ is assumed to be a gateway to the nongenomic actions in myogenic cells [53, 54]. 1α,25$(OH)_2D_3$ can potentially activate multiple signaling pathways, including cyclic adenosine monophosphate (cAMP)/protein kinase A (PKA), PKC, calmodulin/calmodulin-dependent kinase, protein kinase B (PKB/Akt) and multiple mitogen-activated protein kinases (MAPKs), including extracellular signal-regulated kinase 1 and 2 (ERK1/2), p38 and c-Jun NH_2-terminal 1 and 2 [53]. For example, short hairpin RNA (shRNA)-mediated silencing of VDR in C2C12 myoblasts reduces activation of c-Src, ERK1/2, p38, and Akt induced by 1α,25$(OH)_2D_3$ [55, 56], suggesting that 1α,25$(OH)_2D_3$ exerts nongenomic actions through VDR in myogenic cells. It should be noted, however, that their relative contribution to myogenesis remains to be established. For further details on nongenomic pathways in myogenic cells, excellent reviews are available [53, 54].

The rapid nongenomic effect observed in myogenic cells is likely to be mediated, at least in part, by VDR-independent mechanisms. Indeed, studies using myoblasts lacking VDR demonstrates that ERK1/2 and Akt phosphorylation by 1α,25$(OH)_2D_3$ is only partially suppressed [55, 56]. A putative membrane-associated receptor for 1α,25$(OH)_2D_3$, 1,25D_3-MARRS has been purified from chick intestinal basal lateral membranes [21], cloned and sequenced [57]. A full-length cDNA of 1,25D_3-MARRS is identical to that previously described as glucose-regulated thiol-disulfide oxidoreductase protein precursor, ERp57, as it has also been referred to as GRP58, ERp60/61, PDI-Q2 and PDIA3 [58]. Studies using neutralizing antibodies [59, 60], ribozyme technology [58] and *Cre/loxP* technology [61, 62] demonstrate that 1,25D_3-MARRS mediated at least some nongenomic actions of 1α,25$(OH)_2D_3$. A rapid activation of PLC and PKC by 1α,25$(OH)_2D_3$, which is normally observed in chondrocytes derived from the growth zone, is not reduced in cells from VDR null-mutant mice [59]. However, these responses are blocked by rabbit polyclonal antibody against the N-terminal sequence of the 1,25D_3-MARRS [59]. Although no experimental data are currently available, 1α,25$(OH)_2D_3$, possibly through 1,25D_3-MARRS, may act directly on the muscle cell membrane.

5. Role of vitamin D signaling in muscle development

5.1. VDR null-mutant mice

Gene knockout mice provide an excellent possibility to investigate complex regulatory systems that cannot be modeled in cell culture systems in vitro. In two independent laboratories [63, 64], VDR null-mutant mice are generated by gene targeting. Yoshizawa et al. [63] disrupt exon 2 of the VDR gene, which encodes the first zinc finger motif in the DNA-binding domain essential for the biological functions of VDR, whereas Li et al. [64] ablate a VDR fragment spanning exons 3–5, which encode the second zinc finger motif in the DNA-binding domain. Both VDR null-mutant mice are phenotypically normal at birth and grow properly until weaning, but thereafter show various pathological conditions including hypocalcemia, hypophosphatemia, secondary hyperparathyroidism and osteomalacia, as a typical feature of human vitamin D-dependent rickets type II [63, 64]. This phenotype is progressively exacerbated until VDR null-mutant mice are fed a rescue diet containing the high-calcium, -phosphorus and -lactose [65].

5.2. Muscle morphology and gene expression in VDR null-mutant mice

At the age of 3 week, no significant differences are observed between VDR null-mutant mice [63] and wild-type mice in body weight or serum concentrations of calcium, phosphate, alkaline phosphatase and vitamin D metabolites [10]. At the cellular level, the skeletal muscle from VDR null-mutant mice has already begun to atrophy (approximately −20%, regardless of muscle fiber types) compared with the wild-type mice [10], implying that muscle atrophy observed in 3-week-old VDR null-mutant mice occurs independent of secondary systemic metabolic changes. By 8 weeks of age, the morphological changes further progress probably because of the absence of VDR or the systemic metabolic changes that had not been present at 3-week [10]. The morphological abnormalities are observed in VDR null-mutant mice fed a rescue diet [10], suggesting that the absence of VDR is likely to be a major cause rather than the secondary systemic metabolic changes. Finally, neither degenerative nor necrotic changes are observed in skeletal muscle from VDR null-mutant mice [10]. Similar results are obtained with biceps femoris, medial gastrocnemius, anterior tibial and soleus muscles [10]. Therefore, the skeletal muscle abnormalities in VDR null-mutant mice may occur diffusely spread throughout the body. At the gene level, Myf5, myogenin, E2A, embryonic and neonatal myosin heavy chain (MyHC) genes, which should have already been downregulated in wild-type mice, are still expressed in VDR null-mutant mice [10]. These data suggest two possibilities as to impaired muscle development. One possibility is that the absence of VDR affects myogenesis during embryonic stage. In favor of the first possibility, it is noteworthy that VDR is primarily expressed in chick skeletal muscle at embryonic stage rather than at adult stage and VDR-binding activity gradually decreased between embryonic and perinatal stages to levels that is equal to low activity observed in adult stage [66]. Consistent with VDR null-mutant mice, newborn rats from vitamin D-deficient mothers have smaller muscle fibers compared to newborn rats from vitamin D-adequate mothers [67]. When VDR expression is suppressed by siRNA in myoblasts, they failed to differentiate into myotubes [16].

Taken together, impaired muscle development may be a consequence of defective myogenesis during embryonic and fetal stages. On the other hand, if overall myoblast differentiation occurred normally as the authors noted [10], another possibility should be taken into account. Vitamin D-deficient rats show a decrease in type II muscle fiber size concomitantly with an increase in protein degradation and decrease in protein synthesis [68]. Vitamin D deficiency leads to upregulation of muscle-specific E3 ubiquitin ligases, muscle atrophy F-box (MAFbx)/Atrogin-1 and/or muscle ring-finger protein 1 (MuRF1) [11, 68] that are transcriptionally increased before and at the onset of muscle wasting [69] and subsequently activated ubiquitin-proteasome pathway [68]. MAFbx/Atrogin-1 targets MyoD for degradation in several models of skeletal muscle atrophy [70], whereas MuRF1 is involved in degradation of MyHC protein in dexamethasone-treated skeletal muscle [71]. More recently, some ubiquitin-proteasome-related genes are shown to be upregulated in skeletal muscle of newborn rat from vitamin D-deficient mothers compared with that of newborn rat from vitamin D-adequate mothers [67]. The newborn rats show smaller muscle fiber size than the control rats [67]. Therefore, impaired muscle development may result from an imbalance between protein synthesis and degradation. To more effectively address the issue, tissue-specific deletion of VDR at early or late stage of muscle development or conditional postnatal deletion may help to clarify the role of VDR.

6. Effects of vitamin D on myogenesis in vitro

6.1. Myoblast proliferation

Some studies show that 1α,25$(OH)_2D_3$ stimulates myoblast proliferation. Giuliani et al. [72] report that 1α,25$(OH)_2D_3$ (0.13 nM) increases cell density of chick myoblasts. Drittanti et al. [73] show the biphasic effects of 1α,25$(OH)_2D_3$ (0.1 nM) that exhibited a mitogenic effect during the stage of myoblast proliferation. In contrast, most studies suggest that 1α,25$(OH)_2D_3$ or 25$(OH)D_3$ has antiproliferative effect on myogenic cells [9, 12–16]. 1α,25$(OH)_2D_3$ (1–100 nM) inhibits proliferation of C2C12 myoblasts in a dose-dependent manner [11–13] without inducing necrotic and apoptotic cell death [9, 15]. To elucidate the mechanism by which 1α,25$(OH)_2D_3$ exerts its antiproliferative effect, Okuno et al. [14] perform cell cycle analysis using flow cytometry. 1α,25$(OH)_2D_3$ arrests the cells in the G0/G1 phase concomitantly with induction of cyclin-dependent kinase (CDK) inhibitors, $p21^{WAF1/CIP1}$ that facilitates cell cycle withdrawal [74] and $p27^{Kip1}$ that inhibits a wide range of CDKs essential for cell cycle progression [75]. Girgis et al. [4] also report the increased expression of genes involved in G0/G1 arrest including Rb (retinoblastoma protein) and ATM (ataxia telangiectasia mutated) and decreased expression of genes involved in G1/S transition, such as c-myc and cyclin D1. Hypophosphorylated Rb protein is decreased [15] that is active form, blocks entry into S-phase by inhibiting the E2F transcriptional program [76, 77]. Overall, these data support the antiproliferative role of 1α,25$(OH)_2D_3$ in myogenic cells.

6.2. Myoblast differentiation

Some studies [78, 79] report that 1α,25(OH)$_2$D$_3$ (0.1 or 1 nM) has inhibitory effects on DNA synthesis in differentiating chick myoblasts, with an increase in MyHC expression, an increase in myofibrillar and microsomal protein synthesis and an elevation of creatine kinase activity. Garcia et al. [12] report that prolonged treatment of C2C12 myoblasts with 1α,25(OH)$_2$D$_3$ (100 nM) enhanced myoblast differentiation by inhibiting cell proliferation and modulating the expression of promyogenic and antimyogenic growth factors. 1α,25(OH)$_2$D$_3$ downregulates insulin-like growth factor-I (IGF-I) and myostatin expression and upregulates IGF-II and follistatin expression [12]. Follistatin antagonizes myostatin-mediated inhibition of myogenesis [80]. Intriguingly, inhibition of myostatin is characterized by increased expression of IGF-1 and IGF-II [81–86], which are known to be potent stimulus of myogenesis [45, 87]. Therefore, it can be hypothesized that 1α,25(OH)$_2$D$_3$ may contribute to myogenesis by inducing IGF-II expression through modulation of myostatin-follistatin system. It should be noted, however, that only small thin myotubes with few nuclei are observed until day 10 [12]. This may not recapitulate normal C2C12 myoblast differentiation as previously reported [88].

In general, C2C12 myoblasts normally proliferate and are mononucleated when kept subconfluently in high-mitogen medium (e.g., 10–20% fetal bovine serum). To initiate cell cycle exit and myogenic differentiation, by switching from high-mitogen medium to low-mitogen medium (e.g., 2% horse serum), they fuse and differentiate into postmitotic, elongated and multinucleated myotubes. Using this C2C12 myoblast differentiation system, Buitrago et al. [89] show that 1α,25(OH)$_2$D$_3$ (1 nM) enhanced the expression of MyHC and myogenin at 72 h after treatment. Okuno et al. [14] examine the effects of 1α,25(OH)$_2$D$_3$ (1–100 nM) on differentiating and differentiated stage of C2C12 myoblasts. In differentiating phase, 1α,25(OH)$_2$D$_3$ treatment downregulates the expression of neonatal MyHC and myogenin and inhibits myotube formation in a dose-dependent manner [14]. The expression of fast-type MyHC isoforms is increased when fully differentiated myotubes are treated with 1α,25(OH)$_2$D$_3$ [14, 16]. Girgis et al. [15] investigate the prolonged treatment of 1α,25(OH)$_2$D$_3$ (100 nM) on C2C12 myoblast differentiation. When myoblast is treated with 1α,25(OH)$_2$D$_3$ throughout proliferative, differentiating and differentiated stages, myotube formation is delayed by day 10 concomitantly with downregulation of Myf5 and myogenin [15]. However, myotubes treated with 1α,25(OH)$_2$D$_3$ exhibited larger cell size than nontreated myotubes [15]. These results suggest that 1α,25(OH)$_2$D$_3$ may biphasically act in the process of early and late myoblast differentiation. They show that 1α,25(OH)$_2$D$_3$-mediated hypertrophic effect on myotubes is accompanied with downregulation of myostatin [15]. Several studies have provided evidence that myostatin acts as a negative regulator of the Akt/mammalian target of rapamycin (mTOR) signaling pathway [90–93], which plays a key role in the regulation of protein synthesis [94]. For example, Trendelenburg et al. [92] show that myostatin reduces Akt/mTOR signaling complex 1 (TORC1)/p70 S6 kinase (p70S6K) signaling, inhibiting myoblast differentiation and reducing myotube size. Intriguingly, 1α,25(OH)$_2$D$_3$ sensitizes the Akt/mTOR signaling pathway to the stimulating effect of leucine and insulin, resulting in a further activation of protein synthesis in C2C12 myotubes [95]. Recently, however, Olsson et al. [9] report that myostatin expression remains unchanged in response to 1α,25(OH)$_2$D$_3$ in human muscle precursor cells. These conflicting results may be due to the variable methods of analysis and to the species differences. Therefore, further studies are required to clarify whether 1α,25(OH)$_2$D$_3$ affects myoblast differentiation by modulating myostatin/Akt/mTOR signaling. Olsson et al. [9] report that 1α,25(OH)$_2$D$_3$ inhibits myotubes formation with concomitant downregulation of myogenic regulatory factors, myocyte enhancer factor 2 (MEF2) transcription factors and muscle structural proteins including MyHCs, myosin light chains (MyLCs), troponin and titin. They suggest the possibility that 1α,25(OH)$_2$D$_3$ may play a role in the promotion of self-renewal and maintenance of the satellite stem cell pool, through the modulation of the FOXO and Notch signaling pathways [9]. The proposed concept regarding the direct effects of 1α,25(OH)$_2$D$_3$ on myogenesis is intriguing because FOXO3 promotes quiescence in muscle stem cells by activating Notch signaling [96], which is required to maintain quiescent state in myogenic stem cells [97, 98]. Notch signaling inhibits myoblast differentiation [99]. For example, Notch signaling activated Hes1, which inhibits MyoD expression [100] and the ability of MEF2C to cooperate with MyoD and myogenin to activate myogenesis [101]. Further studies will be required to fully elucidate the molecular mechanisms of 1α,25(OH)$_2$D$_3$ actions on myogenesis.

7. Conclusion

This review highlighted the role of 1α,25(OH)$_2$D$_3$ in myogenesis. Although VDR null-mutant mice exhibit impaired muscle development, the precise mechanisms remains to be elucidated. However, they provide insight into the physiological roles of vitamin D in muscle develop-

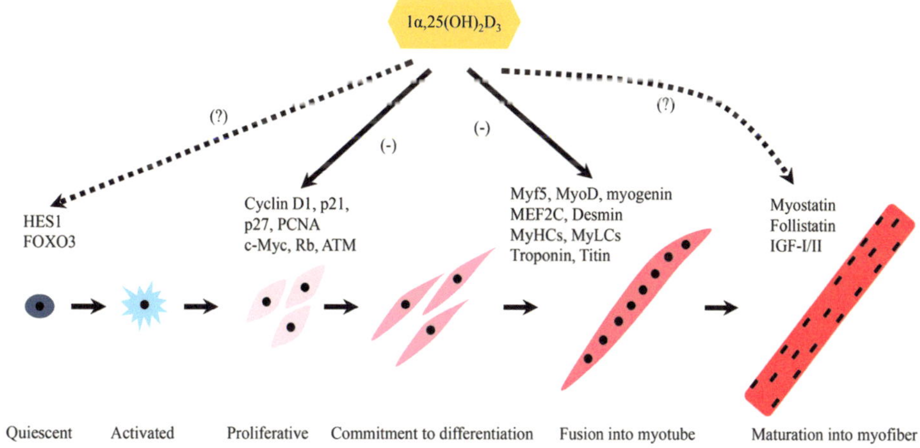

Figure 1. Proposed effects of 1α,25(OH)$_2$D$_3$ on myogenesis based upon in vitro data. Illustration of myogenesis was cited from Zammit et al. [102] with minor modifications. 1α,25(OH)$_2$D$_3$ modulates expression of key components of the vitamin D-endocrine system in myoblasts/myotubes. Potential effects of 1α,25(OH)$_2$D$_3$ are shown (dashed lines). 1α, 25(OH)$_2$D$_3$ has antiproliferative effect and inhibits myoblast differentiation.

ment. Myogenic cells appear to retain vitamin D-endocrine system. $1\alpha,25(OH)_2D_3$ regulates myogenesis probably through genomic and nongenomic actions. **Figure 1** shows proposed effects of $1\alpha,25(OH)_2D_3$ on myogenesis. Simply stated, $1\alpha,25(OH)_2D_3$ has potent antiproliferative activity on myoblasts and inhibits myoblast differentiation. Further studies will be required to clarify whether $1\alpha,25(OH)_2D_3$ enhances protein synthesis in myotubes and promotes self-renewal of muscle stem cells. Although the effects of $1\alpha,25(OH)_2D_3$ on myogenesis have gradually come to be known, several questions remain unanswered and clouded. For example, what are the genes specifically induced by $1\alpha,25(OH)_2D_3$? How does multiple signaling pathways activated by $1\alpha,25(OH)_2D_3$ contribute to myogenesis? What are the molecular events underlying the cross-talk that occurs between genomic and nongenomic pathways and how, ultimately, are these pathways regulated? Given the beneficial role of $1\alpha,25(OH)_2D_3$ in myogenesis, answering these questions will be critical if we want to develop vitamin D therapy to treat muscle disorders.

Conflict of interests

The authors declare that there is no conflict of interests regarding the publication of this paper.

Acknowledgements

This research was supported by MEXT (the Ministry of Education, Culture, Sports, Science and Technology) (Grant in Aid for Scientific Research (C), 25350882), Japan.

Author details

Akira Wagatsuma

Address all correspondence to: akira_wagatsuma@ipc.i.u-tokyo.ac.jp

Graduate School of Information Science and Technology, The University of Tokyo, Tokyo, Japan

References

[1] Deeb KK, Trump DL, Johnson CS (2007). Vitamin D signalling pathways in cancer: potential for anticancer therapeutics. Nat Rev Cancer 7: 684–700.

[2] Wang Y, Zhu J, DeLuca HF (2012). Where is the vitamin D receptor? Arch Biochem Biophys 523: 123–133.

[3] Bischoff HA, Borchers M, Gudat F, Duermueller U, Theiler R, Stähelin HB, Dick W (2001). In situ detection of 1,25-dihydroxyvitamin D3 receptor in human skeletal muscle tissue. Histochem J 33: 19–24.

[4] Bischoff-Ferrari HA, Borchers M, Gudat F, Dürmüller U, Stähelin HB, Dick W (2004). Vitamin D receptor expression in human muscle tissue decreases with age. J Bone Miner Res 19: 265–269.

[5] Ceglia L, da Silva Morais M, Park LK, Morris E, Harris SS, Bischoff-Ferrari HA, Fielding RA, Dawson-Hughes B (2010). Multi-step immunofluorescent analysis of vitamin D receptor loci and myosin heavy chain isoforms in human skeletal muscle. J Mol Histol 41: 137–142.

[6] Girgis CM, Mokbel N, Cha KM, Houweling PJ, Abboud M, Fraser DR, Mason RS, Clifton-Bligh RJ, Gunton JE (2014). The Vitamin D Receptor (VDR) is expressed in skeletal muscle of male mice and modulates 25-hydroxyvitamin D (25OHD) uptake in myofibers. Endocrinology 155: 3227–3237.

[7] Ryan ZC, Craig TA, Folmes CD, Wang X, Lanza IR, Schaible NS, Salisbury JL, Nair KS, Terzic A, Sieck GC, Kumar R (2016). 1α,25-Dihydroxyvitamin D3 regulates mitochondrial oxygen consumption and dynamics in human skeletal muscle cells. J Biol Chem 291:1514–1528.

[8] Wang Y, DeLuca HF (2011). Is the vitamin D receptor found in muscle? Endocrinology 152: 354–363.

[9] Olsson K, Saini A, Strömberg A, Alam S, Lilja M, Rullman E, Gustafsson T (2016). Evidence for vitamin D receptor expression and direct effects of 1α,25(OH)2D3 in human skeletal muscle precursor cells. Endocrinology 157: 98–111.

[10] Endo I, Inoue D, Mitsui T, Umaki Y, Akaike M, Yoshizawa T, Kato S, Matsumoto T (2003). Deletion of vitamin D receptor gene in mice results in abnormal skeletal muscle development with deregulated expression of myoregulatory transcription factors. Endocrinology 144: 5138–5144.

[11] Girgis CM, Cha KM, Houweling PJ, Rao R, Mokbel N, Lin M, Clifton-Bligh RJ, Gunton JE (2015). Vitamin D receptor ablation and vitamin D deficiency result in reduced grip strength, altered muscle fibers, and increased myostatin in mice. Calcif Tissue Int 97: 602–610.

[12] Garcia LA, King KK, Ferrini MG, Norris KC, Artaza JN (2011). 1,25(OH)2 Vitamin D3 stimulates myogenic differentiation by inhibiting cell proliferation and modulating the expression of promyogenic growth factors and myostatin in C2C12 skeletal muscle cells. Endocrinology 152: 2976–2986.

[13] Srikuea R, Zhang X, Park-Sarge OK, Esser KA (2012). VDR and CYP27B1 are expressed in C2C12 cells and regenerating skeletal muscle: potential role in suppression of myoblast proliferation. Am J Physiol Cell Physiol 303: C396-C405.

[14] Okuno H, Kishimoto KN, Hatori M, Itoi E (2012). 1α,25-Dihydroxyvitamin D_3 enhances fast-myosin heavy chain expression in differentiated C2C12 myoblasts. Cell Biol Int 36: 441–447.

[15] Girgis CM, Clifton-Bligh RJ, Mokbel N, Cheng K, Gunton JE (2014). Vitamin D signaling regulates proliferation, differentiation, and myotube size in C2C12 skeletal muscle cells. Endocrinology 155: 347–357.

[16] Tanaka M, Kishimoto KN, Okuno H, Saito H, Itoi E (2014). Vitamin D receptor gene silencing effects on differentiation of myogenic cell lines. Muscle Nerve 49: 700–708.

[17] Abboud M, Puglisi DA, Davies BN, Rybchyn M, Whitehead NP, Brock KE, Cole L, Gordon-Thomson C, Fraser DR, Mason RS (2013). Evidence for a specific uptake and retention mechanism for 25-hydroxyvitamin D (25OHD) in skeletal muscle cells. Endocrinology 154: 3022–3030.

[18] Bouillon R, Carmeliet G, Verlinden L, van Etten E, Verstuyf A, Luderer HF, Lieben L, Mathieu C, Demay M (2008). Vitamin D and human health: lessons from vitamin D receptor null mice. Endocr Rev 29: 726–776.

[19] Lösel R, Wehling M (2003). Nongenomic actions of steroid hormones. Nat Rev Mol Cell Biol 4: 46–56.

[20] Huhtakangas JA, Olivera CJ, Bishop JE, Zanello LP, Norman AW (2004). The vitamin D receptor is present in caveolae-enriched plasma membranes and binds 1 alpha, 25(OH)2-vitamin D3 in vivo and in vitro. Mol Endocrinol 18: 2660–2671.

[21] Nemere I, Dormanen MC, Hammond MW, Okamura WH, Norman AW (1994). Identification of a specific binding protein for 1 alpha,25-dihydroxyvitamin D3 in basal-lateral membranes of chick intestinal epithelium and relationship to transcaltachia. J Biol Chem 269: 23750–23756.

[22] Wang Y, Becklund BR, DeLuca HF (2010). Identification of a highly specific and versatile vitamin D receptor antibody. Arch Biochem Biophys 494: 166–177.

[23] Costa EM, Blau HM, Feldman D (1986). 1,25-Dihydroxyvitamin D3 receptors and hormonal responses in cloned human skeletal muscle cells. Endocrinology 119: 2214–2220.

[24] Jones G, Strugnell SA, DeLuca HF (1998). Current understanding of the molecular actions of vitamin D. Physiol Rev 78: 1193–1231.

[25] Zehnder D, Bland R, Williams MC, McNinch RW, Howie AJ, Stewart PM, Hewison M (2001). Extrarenal expression of 25-hydroxyvitamin d(3)-1 alpha-hydroxylase. J Clin Endocrinol Metab 86: 888–894.

http://dx.doi.org/10.5772/64514

[26] Zehnder D, Bland R, Chana RS, Wheeler DC, Howie AJ, Williams MC, Stewart PM, Hewison M (2002). Synthesis of 1,25-dihydroxyvitamin D(3) by human endothelial cells is regulated by inflammatory cytokines: a novel autocrine determinant of vascular cell adhesion. J Am Soc Nephrol 13: 621–629.

[27] Zehnder D, Evans KN, Kilby MD, Bulmer JN, Innes BA, Stewart PM, Hewison M (2002). The ontogeny of 25-hydroxyvitamin D(3) 1alpha-hydroxylase expression in human placenta and decidua. Am J Pathol 161: 105–114.

[28] Dusso AS, Brown AJ, Slatopolsky E (2005). Vitamin D. Am J Physiol Renal Physiol 289: F8-F28.

[29] Cooke NE, Haddad JG (1989). Vitamin D binding protein (Gc-globulin). Endocr Rev 10: 294–307.

[30] Christensen EI, Birn H (2002). Megalin and cubilin: multifunctional endocytic receptors. Nat Rev Mol Cell Biol 3: 256–266.

[31] Nakamura Y, Eto TA, Taniguchi T, Miyamoto K, Nagatomo J, Shiotsuki H, Sueta H, Higashi S, Okuda KI, Setoguchi T (1997). Purification and characterization of 25-hydroxyvitamin D3 1alpha-hydroxylase from rat kidney mitochondria. FEBS Lett 419: 45–48.

[32] Akutsu N, Lin R, Bastien Y, Bestawros A, Enepekides DJ, Black MJ, White JH (2001). Regulation of gene expression by 1alpha,25-dihydroxyvitamin D3 and its analog EB1089 under growth-inhibitory conditions in squamous carcinoma cells. Mol Endocrinol 15: 1127–1139.

[33] Lin R, Nagai Y, Sladek R, Bastien Y, Ho J, Petrecca K, Sotiropoulou G, Diamandis EP, Hudson TJ, White JH (2002). Expression profiling in squamous carcinoma cells reveals pleiotropic effects of vitamin D3 analog EB1089 signaling on cell proliferation, differentiation, and immune system regulation. Mol Endocrinol 16: 1243–1256.

[34] Wang TT, Tavera-Mendoza LE, Laperriere D, Libby E, MacLeod NB, Nagai Y, Bourdeau V, Konstorum A, Lallemant B, Zhang R, Mader S, White JH (2005). Large-scale in silico and microarray-based identification of direct 1,25-dihydroxyvitamin D3 target genes. Mol Endocrinol 19: 2685–2695.

[35] Han R, Campbell KP (2007). Dysferlin and muscle membrane repair. Curr Opin Cell Biol 19: 409–416.

[36] De Luna N, Díaz-Manera J, Paradas C, Iturriaga C, Rojas-García R, Araque J, Genebriera M, Gich I, Illa I, Gallardo E (2012). 1α,25(OH)(2)-Vitamin D3 increases dysferlin expression in vitro and in a human clinical trial. Mol Ther 20 (10): 1988–1997.

[37] Kitamura T, Kitamura YI, Funahashi Y, Shawber CJ, Castrillon DH, Kollipara R, DePinho RA, Kitajewski J, Accili D (2007). A Foxo/Notch pathway controls myogenic differentiation and fiber type specification. J Clin Invest 117: 2477–2485.

[38] Cao X, Ross FP, Zhang L, MacDonald PN, Chappel J, Teitelbaum SL (1993). Cloning of the promoter for the avian integrin beta 3 subunit gene and its regulation by 1,25-dihydroxyvitamin D3. J Biol Chem 268: 27371–27380.

[39] Desgrosellier JS, Cheresh DA (2010). Integrins in cancer: biological implications and therapeutic opportunities. Nat Rev Cancer 10: 9–22.

[40] Liu H, Niu A, Chen SE, Li YP (2011). Beta3-integrin mediates satellite cell differentiation in regenerating mouse muscle. FASEB J 25: 1914–1921.

[41] Ezura Y, Tournay O, Nifuji A, Noda M (1997). Identification of a novel suppressive vitamin D response sequence in the 5'-flanking region of the murine Id1 gene. J Biol Chem 272: 29865–29872.

[42] Lasorella A, Benezra R, Iavarone A (2014). The ID proteins: master regulators of cancer stem cells and tumour aggressiveness. Nat Rev Cancer 14: 77–91.

[43] Jen Y, Weintraub H, Benezra R (1992). Overexpression of Id protein inhibits the muscle differentiation program: in vivo association of Id with E2A proteins. Genes Dev 6: 1466–1479.

[44] Matilainen M, Malinen M, Saavalainen K, Carlberg C (2005). Regulation of multiple insulin-like growth factor binding protein genes by 1alpha,25-dihydroxyvitamin D3. Nucleic Acids Res 33: 5521–5532.

[45] Duan C, Ren H, Gao S (2010). Insulin-like growth factors (IGFs), IGF receptors, and IGF-binding proteins: roles in skeletal muscle growth and differentiation. Gen Comp Endocrinol 167: 344–351.

[46] Jones JI, Clemmons DR (1995). Insulin-like growth factors and their binding proteins: biological actions. Endocr Rev 16: 3–34.

[47] Zapf J (1995). Physiological role of the insulin-like growth factor binding proteins. Eur J Endocrinol 132: 645–654.

[48] Ren H, Yin P, Duan C (2008). IGFBP-5 regulates muscle cell differentiation by binding to IGF-II and switching on the IGF-II auto-regulation loop. J Cell Biol 182: 979–991.

[49] Schedlich LJ, Muthukaruppan A, O'Han MK, Baxter RC (2007). Insulin-like growth factor binding protein-5 interacts with the vitamin D receptor and modulates the vitamin D response in osteoblasts. Mol Endocrinol 21: 2378–2390.

[50] Capiati D, Benassati S, Boland RL (2002). 1,25(OH)2-Vitamin D3 induces translocation of the vitamin D receptor (VDR) to the plasma membrane in skeletal muscle cells. J Cell Biochem 86: 128–135.

[51] Parton RG, del Pozo MA (2013). Caveolae as plasma membrane sensors, protectors and organizers. Nat Rev Mol Cell Biol 14: 98–112.

[52] Buitrago C, Boland R (2010). Caveolae and caveolin-1 are implicated in 1alpha,25(OH)2-vitamin D3-dependent modulation of Src, MAPK cascades and VDR localization in skeletal muscle cells. J Steroid Biochem Mol Biol 121: 169–175.

[53] Dirks-Naylor AJ, Lennon-Edwards S (2011). The effects of vitamin D on skeletal muscle function and cellular signaling. J Steroid Biochem Mol Biol 125: 159–168.

[54] Girgis CM, Clifton-Bligh RJ, Hamrick MW, Holick MF, Gunton JE (2013). The roles of vitamin D in skeletal muscle: form, function, and metabolism. Endocr Rev 34: 33–83.

[55] Boland R, De Boland AR, Buitrago C, Morelli S, Santillán G, Vazquez G, Capiati, D, Baldi C (2002). Non-genomic stimulation of tyrosine phosphorylation cascades by 1,25(OH)(2)D(3) by VDR-dependent and -independent mechanisms in muscle cells. Steroids 67: 477–482.

[56] Buitrago C, Pardo VG, Boland R (2013). Role of VDR in 1α,25-dihydroxyvitamin D3-dependent non-genomic activation of MAPKs, Src and Akt in skeletal muscle cells. J Steroid Biochem Mol Biol 136: 125–130.

[57] Nemere I, Safford SE, Rohe B, DeSouza MM, Farach-Carson MC (2004). Identification and characterization of 1,25D3-membrane-associated rapid response, steroid (1,25D3-MARRS) binding protein. J Steroid Biochem Mol Biol 89–90: 281–285.

[58] Nemere I, Farach-Carson MC, Rohe B, Sterling TM, Norman AW, Boyan BD, Safford SE (2004). Ribozyme knockdown functionally links a 1,25(OH)2D3 membrane binding protein (1,25D3-MARRS) and phosphate uptake in intestinal cells. Proc Natl Acad Sci U S A 101: 7392–7397.

[59] Boyan BD, Sylvia VL, McKinney N, Schwartz Z (2003). Membrane actions of vitamin D metabolites 1alpha,25(OH)2D3 and 24R,25(OH)2D3 are retained in growth plate cartilage cells from vitamin D receptor knockout mice. J Cell Biochem 90: 1207–1223.

[60] Nemere I (2005). The 1,25D3-MARRS protein: contribution to steroid stimulated calcium uptake in chicks and rats. Steroids 70: 455–457

[61] Nemere I, Garbi N, Hämmerling GJ, Khanal RC (2010). Intestinal cell calcium uptake and the targeted knockout of the 1,25D3-MARRS (membrane-associated, rapid response steroid-binding) receptor/PDIA3/Erp57. J Biol Chem 285: 31859–31866.

[62] Nemere I, Garcia-Garbi N, Hämmerling GJ, Winger Q (2012). Intestinal cell phosphate uptake and the targeted knockout of the 1,25D3-MARRS receptor/PDIA3/ERp57. Endocrinology 153: 1609–1615.

[63] Yoshizawa T, Handa Y, Uematsu Y, Takeda S, Sekine K, Yoshihara Y, Kawakami T, Arioka K, Sato H, Uchiyama Y, Masushige S, Fukamizu A, Matsumoto T, Kato S (1997). Mice lacking the vitamin D receptor exhibit impaired bone formation, uterine hypoplasia and growth retardation after weaning. Nat Genet 16: 391–396.

[64] Li YC, Pirro AE, Amling M, Delling G, Baron R, Bronson R, Demay MB (1997). Targeted ablation of the vitamin D receptor: an animal model of vitamin D-dependent rickets type II with alopecia. Proc Natl Acad Sci U S A 94: 9831–9835.

[65] Li YC, Amling M, Pirro AE, Priemel M, Meuse J, Baron R, Delling G, Demay MB (1998). Normalization of mineral ion homeostasis by dietary means prevents hyperparathyroidism, rickets, and osteomalacia, but not alopecia in vitamin D receptor-ablated mice. Endocrinology 139: 4391–4396.

[66] Haussler MR, Whitfield GK, Kaneko I, Haussler CA, Hsieh D, Hsieh JC, Jurutka PW (2013). Molecular mechanisms of vitamin D action. Calcif Tissue Int 92: 77–98.

[67] Max D, Brandsch C, Schumann S, Kühne H, Frommhagen M, Schutkowski A, Hirche F, Staege MS, Stangl GI (2014). Maternal vitamin D deficiency causes smaller muscle fibers and altered transcript levels of genes involved in protein degradation, myogenesis, and cytoskeleton organization in the newborn rat. Mol Nutr Food Res 58: 343–352.

[68] Bhat M, Kalam R, Qadri SS, Madabushi S, Ismail A (2013). Vitamin D deficiency-induced muscle wasting occurs through the ubiquitin proteasome pathway and is partially corrected by calcium in male rats. Endocrinology 154: 4018–4029.

[69] Bodine SC, Latres E, Baumhueter S, Lai VK, Nunez L, Clarke BA, Poueymirou WT, Panaro FJ, Na E, Dharmarajan K, Pan ZQ, Valenzuela DM, DeChiara TM, Stitt TN, Yancopoulos GD, Glass DJ (2001). Identification of ubiquitin ligases required for skeletal muscle atrophy. Science 294: 1704–1708.

[70] Lagirand-Cantaloube J, Cornille K, Csibi A, Batonnet-Pichon S, Leibovitch MP, Leibovitch SA (2009). Inhibition of atrogin-1/MAFbx mediated MyoD proteolysis prevents skeletal muscle atrophy in vivo. PLoS One 4: e4973.

[71] Clarke BA, Drujan D, Willis MS, Murphy LO, Corpina RA, Burova E, Rakhilin SV, Stitt TN, Patterson C, Latres E, Glass DJ (2007). The E3 Ligase MuRF1 degrades myosin heavy chain protein in dexamethasone-treated skeletal muscle. Cell Metab 6: 376–385.

[72] Giuliani DL, Boland RL (1984). Effects of vitamin D3 metabolites on calcium fluxes in intact chicken skeletal muscle and myoblasts cultured in vitro. Calcif Tissue Int 36: 200–205.

[73] Drittanti L, de Boland AR, Boland R (1989). Modulation of DNA synthesis in cultured muscle cells by 1,25-dihydroxyvitamin D-3. Biochim Biophys Acta 1014: 112–119.

[74] Dotto GP (2000). p21(WAF1/Cip1): more than a break to the cell cycle? Biochim Biophys Acta 1471: M43-M56.

[75] Sherr CJ, Roberts JM (1999). CDK inhibitors: positive and negative regulators of G1-phase progression. Genes Dev 13: 1501–1512.

[76] Nevins JR (1992). E2F: a link between the Rb tumor suppressor protein and viral oncoproteins. Science 258: 424–429.

[77] Sherr CJ (1996). Cancer cell cycles. Science 274: 1672–1677.

[78] Drittanti LN, Boland RL, de Boland AR (1989). Induction of specific proteins in cultured skeletal muscle cells by 1,25-dihydroxyvitamin D-3. Biochim Biophys Acta 1012: 16–23.

[79] Capiati DA, Téllez-Iñón MT, Boland RL (1999). Participation of protein kinase C alpha in 1,25-dihydroxy-vitamin D3 regulation of chick myoblast proliferation and differentiation. Mol Cell Endocrinol 153: 39–45.

[80] Amthor H, Nicholas G, McKinnell I, Kemp CF, Sharma M, Kambadur R, Patel K (2004). Follistatin complexes myostatin and antagonises myostatin-mediated inhibition of myogenesis. Dev Biol 270: 19–30.

[81] Kocamis H, Gahr SA, Batelli L, Hubbs AF, Killefer J (2002). IGF-I, IGF-II, and IGF-receptor-1 transcript and IGF-II protein expression in myostatin knockout mice tissues. Muscle Nerve 26: 55–63.

[82] Gilson H, Schakman O, Combaret L, Lause P, Grobet L, Attaix D, Ketelslegers JM, Thissen JP (2007). Myostatin gene deletion prevents glucocorticoid-induced muscle atrophy. Endocrinology 148: 452–460.

[83] Marshall A, Salerno MS, Thomas M, Davies T, Berry C, Dyer K, Bracegirdle J, Watson T, Dziadek M, Kambadur R, Bower R, Sharma M (2008). Mighty is a novel promyogenic factor in skeletal myogenesis. Exp Cell Res 314: 1013–1029.

[84] Gilson H, Schakman O, Kalista S, Lause P, Tsuchida K, Thissen JP (2009). Follistatin induces muscle hypertrophy through satellite cell proliferation and inhibition of both myostatin and activin. Am J Physiol Endocrinol Metab 297: E157-E164.

[85] Miyake M, Hayashi S, Taketa Y, Iwasaki S, Watanabe K, Ohwada S, Aso H, Yamaguchi T (2010) Myostatin down-regulates the IGF-2 expression via ALK-Smad signaling during myogenesis in cattle. Anim Sci J 81: 223–229.

[86] Williams NG, Interlichia JP, Jackson MF, Hwang D, Cohen P, Rodgers BD (2011). Endocrine actions of myostatin: systemic regulation of the IGF and IGF binding protein axis. Endocrinology 152: 172–180.

[87] Zanou N, Gailly P (2013). Skeletal muscle hypertrophy and regeneration: interplay between the myogenic regulatory factors (MRFs) and insulin-like growth factors (IGFs) pathways. Cell Mol Life Sci 70: 4117–4130.

[88] Shen X, Collier JM, Hlaing M, Zhang L, Delshad EH, Bristow J, Bernstein HS (2003). Genome-wide examination of myoblast cell cycle withdrawal during differentiation. Dev Dyn 226: 128–138.

[89] Buitrago CG, Arango NS, Boland RL (2012). 1α,25(OH)2D3-dependent modulation of Akt in proliferating and differentiating C2C12 skeletal muscle cells. J Cell Biochem 113: 1170–1181.

[90] Amirouche A, Durieux AC, Banzet S, Koulmann N, Bonnefoy R, Mouret C, Bigard X, Peinnequin A, Freyssenet D (2009). Down-regulation of Akt/mammalian target of

rapamycin signaling pathway in response to myostatin overexpression in skeletal muscle. Endocrinology 150: 286–294.

[91] Sartori R, Milan G, Patron M, Mammucari C, Blaauw B, Abraham R, Sandri M (2009). Smad2 and 3 transcription factors control muscle mass in adulthood. Am J Physiol Cell Physiol 296: C1248-C1257.

[92] Trendelenburg AU, Meyer A, Rohner D, Boyle J, Hatakeyama S, Glass DJ (2009). Myostatin reduces Akt/TORC1/p70S6K signaling, inhibiting myoblast differentiation and myotube size. Am J Physiol Cell Physiol 296: C1258-C1270.

[93] Lipina C, Kendall H, McPherron AC, Taylor PM, Hundal HS (2010). Mechanisms involved in the enhancement of mammalian target of rapamycin signalling and hypertrophy in skeletal muscle of myostatin-deficient mice. FEBS Lett 584: 2403–2408.

[94] Sarbassov DD, Ali SM, Sabatini DM (2005). Growing roles for the mTOR pathway. Curr Opin Cell Biol 17: 596–603.

[95] Salles J, Chanet A, Giraudet C, Patrac V, Pierre P, Jourdan M, Luiking YC, Verlaan S, Migné C, Boirie Y, Walrand S (2013). 1,25(OH)2-Vitamin D3 enhances the stimulating effect of leucine and insulin on protein synthesis rate through Akt/PKB and mTOR mediated pathways in murine C2C12 skeletal myotubes. Mol Nutr Food Res 57: 2137–4216.

[96] Gopinath SD, Webb AE, Brunet A, Rando TA (2014). FOXO3 promotes quiescence in adult muscle stem cells during the process of self-renewal. Stem Cell Rep 2: 414–426.

[97] Bjornson CR, Cheung TH, Liu L, Tripathi PV, Steeper KM, Rando TA (2012). Notch signaling is necessary to maintain quiescence in adult muscle stem cells. Stem Cells 30: 232–342.

[98] Mourikis P, Sambasivan R, Castel D, Rocheteau P, Bizzarro V, Tajbakhsh S (2012). A critical requirement for notch signaling in maintenance of the quiescent skeletal muscle stem cell state. Stem Cells 30: 243–252.

[99] Buas MF, Kadesch T (2010). Regulation of skeletal myogenesis by Notch. Exp Cell Res 316: 3028–3033.

[100] Kopan R, Nye JS, Weintraub H (1994). The intracellular domain of mouse Notch: a constitutively activated repressor of myogenesis directed at the basic helix-loop-helix region of MyoD. Development 120: 2385–2396.

[101] Wilson-Rawls J, Molkentin JD, Black BL, Olson EN (1999). Activated notch inhibits myogenic activity of the MADS-Box transcription factor myocyte enhancer factor 2C. Mol Cell Biol 19: 2853–2862.

[102] Zammit PS, Partridge TA, Yablonka-Reuveni Z (2006). The skeletal muscle satellite cell: the stem cell that came in from the cold. J Histochem Cytochem 54: 1177–1191.

Chapter 10

Skeletal and Extraskeletal Benefits of Vitamin D

Enrique Casado and Marta Larrosa

Additional information is available at the end of the chapter

http://dx.doi.org/10.5772/65101

Abstract

Vitamin D exerts its principal actions on bone metabolism, so it has important benefits on the skeleton. Serum 25(OH)D is directly related to bone mineral density (BMD), so subjects with lower levels have lower BMD and higher prevalence of osteoporosis and fractures, mainly hip and non-vertebral fractures. But, vitamin D has also many other beneficial effects, and its deficit has been associated with a great variety of diseases, such as asthma, cancer, diabetes, hypertension and other cardiovascular diseases, some inflammatory and autoimmune diseases, infections and some liver diseases. It is also remarkable its direct effect on muscle strength, so patients with vitamin D deficiency have higher risk of falls. Supplementation with vitamin D in patients with low 25(OH)D levels has shown a favourable effect not only on bone and muscle, reducing the risk of fracture, but also on inflammation, cell proliferation or immune system, reducing the risk of other diseases and complications. However, observational studies are needed with larger numbers of patients and well-designed randomized clinical trials, with baseline vitamin D determination and accurate monitoring to establish a cause-effect relationship between vitamin D deficiency and some diseases.

Keywords: vitamin D, benefits, skeletal, extraskeletal, osteoporosis

1. Introduction

The main vitamin D metabolite (1,25-dihydroxyvitamin D or 1,25$(OH)_2$D) exerts its biological actions through binding to Vitamin D receptor (VDR). VDR is a ligand-induced nuclear receptor that regulates the expression of over 900 genes throughout the genome. 1,25$(OH)_2$D dissociates from serum vitamin D-binding protein (VDBP) and enters the cell and binds to and activates the VDR, leading to the promotion and modulation of the expression of the targeted genes [1].

VDR is found not only in the organs responsible for calcium homeostasis (bone, kidney, intestine and parathyroid) but also in many other tissues such as immune system cells, muscle and myocardium, which explains the extraskeletal effects of this vitamin (**Figure 1**) [1].

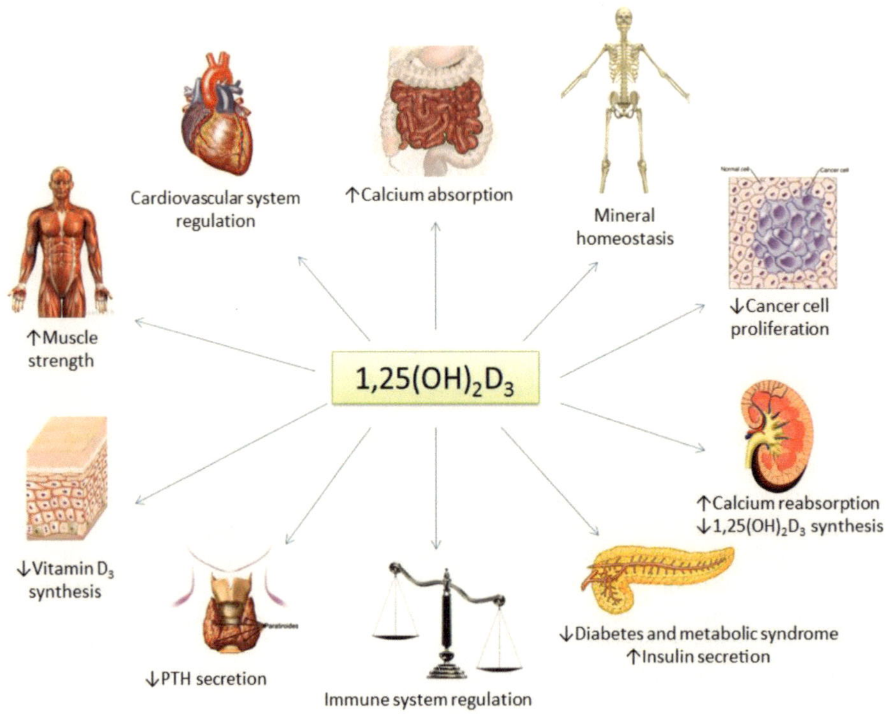

Figure 1. Principal targets and actions of vitamin D.

Serum determination of $1,25(OH)_2D$ does not have much interest, as it has a very short half-life and its levels are highly regulated by some hormones such as parathyroid hormone (PTH), so subjects with vitamin D deficiency can have normal $1,25(OH)_2D$ levels. Therefore, the best assessment of vitamin D status is provided by serum 25(OH)D levels and thus should be the only vitamin D assay typically performed, since its half-life is about 2 weeks.

2. Skeletal benefits of vitamin D

Vitamin D exerts its principal action on bone metabolism regulating the intestinal absorption of calcium and bone remodeling [2]. This vitamin is essential for the normal development of the skeleton in utero and during childhood and adolescence and also to maintain bone health in adults. Optimal levels of vitamin D are needed to achieve a proper balance of calcium and phosphorus for a normal bone mineralization. A longstanding vitamin D deficiency has been associated with growth retardation and rickets in children, and osteoporosis or, in the most severe cases, osteomalacia in adults [3].

http://dx.doi.org/10.5772/65101

Serum 25(OH)D in both sexes and in all races is directly related to bone mineral density (BMD), so that people with lower levels have lower BMD and higher prevalence of osteoporosis and fractures [4]. This is, in part, because a sustained decrease in 25(OH)D induces a higher secretion of PTH (secondary hyperparathyroidism), with an increase in osteoclast differentiation and bone resorption over bone formation.

Levels of 25(OH)D are inversely associated with the risk of non-vertebral and hip fracture [5] (**Figure 2**), so patients with lower levels of vitamin D have higher risk of fracture.

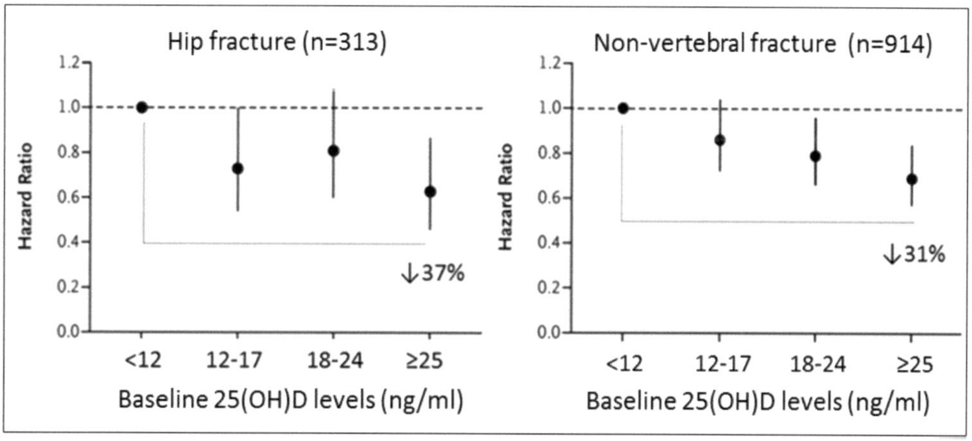

Figure 2. Risk of fracture, according to quartile of baseline 25-hydroxyvitamin D level [5].

Larrosa et al. described an association between vitamin D deficiency and the severity of hip fracture [6]. Patients with more severe femoral neck or intertrochanteric fractures (Garden III IV and Kyle III–IV) have higher prevalence of vitamin D deficiency (74%) and lower levels of 25(OH)D (20 ± 15 ng/ml) than patients with less severe fractures (57% and 26 ± 21 ng/ml).

The supplementation with vitamin D has shown to improve the skeletal health reducing the risk of fracture.

A meta-analysis conducted by Bischoff-Ferrari et al. [7] revealed that supplementation with 700–800 IU/day of vitamin D reduced up to 23% the risk of non-vertebral fractures and 26% the risk of hip fracture, while doses <400 IU had no effect. The minimum levels of 25(OH)D to reduce the risk of hip fracture are 29.6 ng/ml. These data have served as the basis for the International Osteoporosis Foundation (IOF) to set as "optimal" 25(OH)D ≥ 30 ng/ml.

More recently, the same group conducted another meta-analysis involving more than 30,000 subjects and showing that only supplementation with high doses of vitamin D (800–2000 IU/day) had an effect in the reduction in the risk of fracture (30% for hip and 14% for non-vertebral) [5], in population over 65 years regardless of whether they were institutionalized or not.

It is controversial whether the association of calcium with vitamin D contributes to a more efficacy. In fact, in the last meta-analysis [5], patients supplemented with less calcium (<1000 mg/day) had lower fracture risk than those receiving higher doses (≥1000 mg/day). Furthermore, in another meta-analysis, the authors found that supplementation with calcium alone (without vitamin D) could even increase the risk of hip fracture. However, they recommend taking this data with caution as participants included in some trials had been receiving more dietary calcium than usual in the normal population. So, it cannot be ruled out that supplementation with calcium is most beneficial in patients with poor dietary intake.

Despite this controversy, there is sufficient evidence of benefit of calcium supplementation (preferably from diet) and vitamin D in osteoporosis patients, especially when receiving antiresorptive treatment, so that clinical practice guidelines recommend the use of combined calcium (1000–1200 mg/day) and vitamin D (800–1000 IU/day), and if not are achieved with diet should be in the form of supplements [8].

3. Extraskeletal benefits of vitamin D

In addition to its main benefits on bone health, vitamin D has many other beneficial effects and its deficit has been associated with a great variety of diseases.

3.1. Vitamin D and asthma

Vitamin D regulates certain key genes for lung growth during embryonic development, and a deficiency can lead to a change in the structure and lung function.

It has been reported that the forced vital capacity in children of both sexes is strongly associated with maternal 25(OH)D and that vitamin D deficiency in mothers is associated with an increased incidence of asthma in children [9].

The anti-inflammatory effect of vitamin D (inhibits interleukin-6 and Tumor Necrosis Factor), its immunomodulation of both the innate and adaptive immune systems and its potential antimicrobial action also contribute to explain the protective effect of vitamin D in asthma [10].

Vitamin D also enhances the response of asthmatic patients to treatment and could have a synergistic effect with steroids [11].

Some authors have even been suggested that vitamin D deficiency may be responsible in part for the increased prevalence of asthma worldwide and that 25(OH)D levels could be a potentially modifiable marker of severe asthma [12].

3.2. Vitamin D and cancer

The antitumour activity of vitamin D has been shown in a variety of malignancies, and these seem to be the main mechanisms by which this action is exerted:

– Inhibition of cell proliferation. Vitamin D inhibits the phosphorylation of some proteins, leading to inhibition of a series of genes responsible for the progression of cellular cycle [13].

– Apoptosis. Vitamin D induces cell death in some tumours through inhibition of antiapoptotic factors and/or stimulation of proapoptotic factors, depending on the cell type [14].

http://dx.doi.org/10.5772/65101

– Inhibition of angiogenesis. Vitamin D inhibits proliferation of endothelial cells and some angiogenic factors, such as transforming growth factor alpha (TGFα), the epidermal growth factor (EGF) and vascular endothelial growth factor (VEGF) [15]. Vitamin D also reduces the migration and invasiveness of tumour cells.

3.2.1. *Breast cancer*

Breast cancer is probably the most studied cancer in regard to vitamin D status. In a cohort of American women with breast cancer, authors found that women with optimal levels of 25(OH)D were up to 63% lower risk to present a breast cancer [16]. However, in prospective studies, it has not succeeded in showing any relationship between 25(OH)D and breast cancer [17]. Peppone et al. [18] in a retrospective study of 224 women with non-metastatic breast cancer concluded that vitamin D deficiency is associated with poor prognosis and quality of life. With these data, it seems reasonable to determine 25(OH)D levels in women with breast cancer and supplement those with vitamin D deficiency.

3.2.2. *Prostate cancer*

When we analyse the association between vitamin D and prostate cancer, the literature is more controversial. A meta-analysis published in 2014 that included 21 studies concluded that higher concentrations of 25(OH)D were associated with an increased risk of developing prostate cancer [19]. However, a more recent study has described a 30% higher risk in patients with 25(OH)D in the lowest quartile than patients in the highest quartile [20].

The use of calcitriol combined with conventional treatment in patients with prostate cancer has been proved to be effective, both in vitro and in vivo, so vitamin D enhances the antitumour effect of the conventional treatment [21]. A clinical trial of 63 patients showed that patients with prostate cancer treated with 4000 IU of vitamin D3 for one year had a lower tumour progression than controls (38 versus 63%) [22]. However, to date, there is no such randomized clinical trial with a greater number of patients that can confirm these results.

3.2.3. *Colon and rectal cancer*

Colon and rectal cancer have also been associated with vitamin D deficiency, especially rectal cancer. Several meta-analysis have described an inverse association between 25(OH)D and the incidence of colorectal cancer [23]. Patients with 25(OH)D in the highest quartile have 33% lower risk of developing this cancer than patients with 25(OH)D in the lowest quartile [24].

At the moment, there is no evidence that vitamin D influences in the progression of colorectal cancer, although it seems to have a protective effect on mortality [25].

3.3. Vitamin D and cardiovascular disease

Vitamin D deficiency has been associated with inflammation and endothelial and platelet dysfunction, which favours the risk of cardiovascular complications [26].

$1,25(OH)_2D$ has a direct action on myocardial cells, smooth muscle fibres and vascular endothelial cells stimulating the calcium ATPase activity, promoting the calcium transfer to the intracellular space [1]. Vitamin D also regulates blood pressure by decreasing gene expression of renin and aldosterone synthesis. All these explain why vitamin D deficiency has been associated with cardiovascular disease, including hypertension, ischemic heart disease and heart failure [27].

Patients with hypertension and vitamin D deficiency have two-fold increased risk of cardiovascular complications [28] and up to 52% more risk of stroke [29]. In these patients, supplementation with vitamin D could reduce blood pressure by decreasing the activity of renin and angiotensin II values, although this effect has not been demonstrated in all studies.

A multicenter study [30] showed that 96% of patients with myocardial infarction (MI) had 25(OH)D < 30 ng/ml. On the other hand, it is reported that 25(OH)D is an independent predictor of cardiovascular complications in patients with MI and that the risk is 40% higher in patients with levels <7.3 ng/ml [31].

However, one of the effects of vitamin D is the increase in phosphate levels and a high amount of phosphorus increases vascular calcification and, consequently, may increase morbidity and mortalidad [32]. It is, therefore, important to keep vitamin D levels in a safe threshold, where the benefits outweigh the risks.

3.4. Vitamin D and diabetes

Vitamin D also has receptors in pancreatic cells and exerts a regulatory action on glucose metabolism.

A meta-analysis revealed that vitamin D supplementation in children reduces the risk of developing type 1 diabetes (odds ratio: 0.71) [33] and also seems to prevent the development of type 2 diabetes, as evidenced by a meta-analysis published in 2012. In this study, vitamin D deficiency was associated with a 43% higher incidence of diabetes and 62% more progression of prediabetes to diabetes [34].

3.5. Vitamin D in inflammatory and autoimmune diseases

Vitamin D has been linked to many other autoimmune diseases, beyond type 1 diabetes.

Some studies have reported a higher prevalence of some autoimmune diseases at higher latitudes, suggesting that sun exposure and, therefore, the production of vitamin D may play a role in the pathogenesis of some diseases such as type 1 diabetes, multiple sclerosis and Crohn disease [33].

In patients with vitamin D deficiency has been described a higher incidence of Crohn's disease, so, in addition to its immunomodulatory effect, vitamin D promotes the function of the intestinal barrier and stimulates the synthesis of antimicrobial peptides, all of them protective factors for the development of inflammatory bowel disease (IBD) [35]. A prospective study found that vitamin D deficiency could be a risk factor for developing IBD, and higher 25(OH)D seemed to be associated with a significant reduction in the incidence of Crohn's disease, but not ulcerative colitis [36].

Supplementation with vitamin D does not seem to improve significantly the clinical course of Crohn's disease, but it can decrease relapse in some patients.

The risk of developing multiple sclerosis is also associated with lower levels of 25(OH)D. When we analyse the association between 25(OH)D and flare-up in patients with multiple sclerosis before and after supplementation with 3000 IU/day of vitamin D, a strong negative association was found between the incidence of flares and 25(OH)D levels ($p < 0.0001$) [37].

Vitamin D seems to have a protective role in rheumatoid arthritis (RA). In a prospective cohort of nearly 30,000 women aged 55–69 years, authors found that the incidence of RA was 33% lower in patients receiving more vitamin D [38].

Low vitamin D levels have also been associated with an increased incidence and/or relapses of other autoimmune diseases such as systemic lupus erythematosus, partly due to VDR gene polymorphisms [39].

3.6. Vitamin D and infections

Vitamin D has a protective role against infections by stimulating the production of cathelicidin (antimicrobial peptide) and modulating the production of cytokines and the inflammatory cascade during the infection [40], so vitamin D deficiency has also been associated with an increased risk of infections, especially those of the respiratory tract, including tuberculosis [41].

3.7. Vitamin D and liver disease

It has also been suggested that vitamin D may play a role in the development of some chronic liver diseases such as non-alcoholic fatty liver disease, cholestatic disease and autoimmune liver disease; and that supplementation with vitamin D could improve the patient response to antiviral therapy in hepatitis C [42].

3.8. Vitamin D and falls

Vitamin D has a direct effect on muscle strength through its action on specific receptors on the muscle.

Vitamin D deficiency has been associated with type II muscle fibres atrophy, which leads to impaired muscle function and disability, increasing the risk of falls, especially in the elderly [43].

Supplementation with vitamin D at doses of 700–1000 IU/day reduces the risk of falls up to 34%, contributing to the antifracture effect [44]. This reduction in the risk of falls seems apparent already in the first 6 months of treatment with vitamin D.

Levels of 25(OH)D > 24 ng/ml are needed to achieve a significant reduction in the risk of falls (relative risk [RR]: 0.77; confidence interval [CI]: 95%, 0.65–0.90), while lower values do not seem sufficient (RR: 1.35; 95% CI: 0.98–1.84) [45].

4. Summary and conclusions

Besides the well-known benefits on the skeleton, especially as regards on reducing the risk of non-vertebral and hip fracture, vitamin D also has other favourable effects on many organs. Its immunomodulatory, anti-inflammatory, antitumour and antimicrobial effects, as well as their effects on glucose metabolism, cardiovascular system and muscle, are remarkable.

Although the association of vitamin D deficiency with many diseases, as well as the benefits of supplementation, seems clear, it is not easy to establish a cause-effect relationship. Observational studies with larger numbers of patients and well-designed randomized clinical trials are needed with accurate determination and close monitoring of vitamin D.

Conflict of interests

The author declares no conflict of interest.

Author details

Enrique Casado* and Marta Larrosa

*Address all correspondence to: ecasado@tauli.cat

Rheumatology Department, University Hospital Parc Taulí (UAB), Sabadell, Spain

References

[1] Wolden-Kirk H, Gysemans C, Verstuyf A, Mathieu C. Extraskeletal effects of vitamin D. Endocrinol Metab Clin North Am. 2012;41:571–94.

[2] Lips P, Van Schoor NM. The effect of vitamin D on bone and osteoporosis. Best Pract Res Clin Endocrinol Metab. 2011;25:585–91.

[3] Bischoff-Ferrari HA, Kiel DP, Dawson-Hughes B, Orav JE, Li R, Spiegelman D, et al. Dietary calcium and serum 25-hydroxyvitamin D status in relation to BMD among U.S. adults. J Bone Miner Res. 2009;24:935–42.

[4] Cauley JA, LaCroix AZ, Wu L, Horwitz M, Danielson ME, Bauer DC, et al. Serum 25-hydroxyvitamin D concentrations and risk for hip fractures. Ann Int Med. 2008;149:242–50.

[5] Bischoff-Ferrari HA, Willett WC, Orav EJ, Lips P, Meunier PJ, Lyons RA, et al. A pooled analysis of vitamin D dose requirements for fracture prevention. N Engl J Med. 2012;367:40–9.

[6] Larrosa M, Gómez A, Casado E, Moreno M, Vázquez I, Orellana C, et al. Hypovitaminosis D as a risk factor of hip fracture severity. Osteoporos Int. 2012;23: 607–14.

[7] Bischoff-Ferrari HA, Dawson-Hughes B, Baron JA, Burckhardt P, Li R, Spiegelman D, et al. Calcium intake and hip fracture risk in men and women: a meta-analysis of prospective cohort studies and randomized controlled trials. Am J Clin Nutr. 2007;86:1780–90.

[8] González-Macías J, Del Pino-Montes J, Olmos JM, Nogués X; en nombre de la Comisión de Redacción de las Guías de Osteoporosis de la SEIOMM. Clinical practice guidelines for posmenopausal, glucocorticoid-induced and male osteoporosis. Spanish Society for Research on Bone and Mineral Metabolism (3rd updated version 2014). Rev Clin Esp 2015; 215 (9): 515–26.

[9] Zosky GR, Hart PH, Whitehouse AJ, Kusel MM, Ang W, Foong RE, et al. Vitamin D deficiency at 16 to 20 weeks' gestation is associated with impaired lung function and asthma at 6 years of age. Ann Am Thorac Soc. 2014;11:571–7.

[10] Kerley CP, Elnazir B, Faul J, Cormican L. Vitamin D as an adjunctive therapy in asthma. Part 1: a review of potential mechanisms. Pulm Pharmacol Ther. 2015;32:60–74.

[11] Litonjua AA. Vitamin D and corticosteroids in asthma: synergy, interaction and potential therapeutic effects. Expert Rev Respir Med. 2013;7:101–4.

[12] Forno E, Celedon JC. Predicting asthma exacerbations in children. Curr Opin Pulm Med. 2012;18:63–9.

[13] Jensen SS, Madsen MW, Lukas J, Binderup L, Bartek J. Inhibitory effects of 1alpha,25-dihydroxyvitamin D(3) on the G(1) S phase controlling machinery. Mol Endocrinol. 2001;15:1370–80.

[14] Kizildag S, Ates H, Kizildag S. Treatment of K562 cells with 1,25-dihydroxyvitamin D(3) induces distinct alterations in the expression of apoptosis-related genes BCL2, BAX, BCL(XL), and p21. Ann Hematol. 2009;89:1–7.

[15] Ben-Shoshan M, Amir S, Dang DT, Dang LH, Weisman Y, Mabjeesh NJ. 1alpha,25-dihydroxyvitamin D3 (Calcitriol) inhibits hypoxia-inducible factor-1/vascular endothelial growth factor pathway in human cancer cells. Mol Cancer Ther. 2007;6:1433–9.

[16] Yao S, Sucheston LE, Millen AE, Johnson CS, Trump DL, Nesline MK, et al. Pretreatment serum concentrations of 25-hydroxyvitamin D and breast cancer prognostic characteristics: a case-control and a case-series study. PLoS One. 2011;6:e17251.

[17] Yin L, Grandi N, Raum E, Haug U, Arndt V, Brenner H. Meta-analysis: serum vitamin D and breast cancer risk. Eur J Cancer. 2010;46:2196–205.

[18] Peppone LJ, Huston AJ, Reid ME, Rosier RN, Zakharia Y, Trump DL, et al. The effect of various vitamin D supplementation regimens in breast cancer patients. Breast Cancer Res Treat. 2011;127:171–7.

[19] Xu Y, Shao X, Yao Y, Xu L, Chang L, Jiang Z, et al. Positive association between circulating 25-hydroxyvitamin D levels and prostate cancer risk: new findings from an updated meta-analysis. J Cancer Res Clin Oncol. 2014;140:1465–77.

[20] Deschasaux M, Souberbielle JC, Latino-Martel P, Sutton A, Charnaux N, Druesne-Pecollo N, et al. A prospective study of plasma 25-hydroxyvitamin D concentration and prostate cancer risk. Br J Nutr. 2016;115:305–14.

[21] Ben-Eltriki M, Deb S, Guns ES. Calcitriol in combination therapy for prostate cancer: pharmacokinetic and pharmacodynamic interactions. J Cancer. 2016;7:391–407.

[22] Marshall DT, Savage SJ, Garrett-Mayer E, Keane TE, Hollis BW, Horst RL, et al. Vitamin D3 supplementation at 4000 international units per day for one year results in a decrease of positive cores at repeat biopsy in subjects with low-risk prostate cancer under active surveillance. J Clin Endocrinol Metab. 2012;97:2315–24.

[23] Gandini S, Boniol M, Haukka J, Byrnes G, Cox B, Sneyd MJ, et al. Meta-analysis of observational studies of serum 25-hydroxyvitamin D levels and colorectal, breast and prostate cancer and colorectal adenoma. Int J Cancer. 2011;128:1414–24.

[24] Ma Y, Zhang P, Wang F, Yang J, Liu Z, Qin H. Association between vitamin D and risk of colorectal cancer: a systematic review of prospective studies. J Clin Oncol. 2011;29:3775–82.

[25] Jacobs ET, Kohler LN, Kunihiro AG, Jurutka PW. Vitamin D and colorectal, breast, and prostate cancers: a review of the epidemiological evidence. J Cancer. 2016;7:232–40.

[26] Kunadian V, Ford GA, Bawamia B, Qiu W, Manson JE. Vitamin D deficiency and coronary artery disease: a review of the evidence. Am Heart J. 2014;167:283–91

[27] Vacek JL, Vanga SR, Good M, Lai SM, Lakkireddy D, Howard PA. Vitamin D deficiency and supplementation and relation to cardiovascular health. Am J Cardiol. 2012; 109:359–63.

[28] Wang TJ, Pencina MJ, Booth SL et al. Vitamin D deficiency and risk of cardiovascular disease. Circulation. 2008; 4:503–511.

[29] Sun Q, Pan A, Hu FB, Manson JE, Rexrode KM. 25- Hydroxyvitamin D levels and the risk of stroke: a prospective study and meta-analysis. Stroke. 2012; 43:1470–7.

[30] Lee JH, Gadi R, Spertus JA et al. Prevalence of vitamin d deficiency in patients with acute myocardial infarction. Am J Cardiol. 2011; 107: 1636–8.

[31] Ng LL, Sandhu JK, Squire IB et al. Vitamin D and prognosis in acute myocardial infarction. Int J Cardiol. 2013; 168:2341–6.

[32] Body JJ, Bergmann P, Boonen S, Devogelaer JP, Gielen E, Goemaere S et al. Extraskeletal benefits and risks of calcium, vitamin D and anti-osteoporosis medications. Osteoporos Int. 2012; 23 (Suppl 1): S1–S23.

[33] Antico A, Tampoia M, Tozzoli R, Bizzaro N. Can supplementation with vitamin D reduce the risk or modify the course of autoimmune diseases? A systematic review of the literature. Autoimmun Rev. 2012; 12:127–36

[34] Deleskog A, Hilding A, Brismar AK, Hamsten A, Efendic S, Ostenson CG. Low serum 25-hydroxyvitamin D level predicts progression to type 2 diabetes in individuals with prediabetes but not with normal glucose tolerance. Diabetologia. 2012; 55:1668–78.

[35] Meeker S, Seamons A, Maggio-Price L, Paik J. Protective links between vitamin D, inflammatory bowel disease and colon cancer. World J Gastroenterol. 2016; 22:933–48.

[36] Ananthakrishnan AN, Khalili H, Higuchi LM, Bao Y, Korzenik JR, Giovannucci EL et al. Higher predicted vitamin D status is associated with reduced risk of Crohn's disease. Gastroenterology. 2012; 142:482–9.

[37] Pierrot-Deseilligny C, Rivaud-Pechoux S, Clerson P, de Paz R, Souberbielle JC. Relationship between 25-OH-D serum level and relapse rate in multiple sclerosis patients before and after vitamin D supplementation. Ther Adv Neurol Disord. 2012; 5:187–198.

[38] Merlino LA, Curtis J, Mikuls TR, Cerhan JR, Criswell LA, Saag KG. Vitamin D intake is inversely associated with rheumatoid arthritis: Results from the Iowa Women's Health Study. Arthritis Rheum. 2004; 50:72–77.

[39] Monticielo OA, Teixeira TM, Chies JA, Brenol JC, Xavier RM. Vitamin D and polymorphisms of VDR gene in patients with systemic lupus erythematosus. Clin Rheumatol. 2012; 31:1411–21

[40] Gunville CF, Mourani PM, Ginde AA. The role of vitamin D in prevention and treatment of infection. Inflamm Allergy Drug Targets. 2013; 12:239–245

[41] de Sa Del Fiol F, Barberato-Filho S, Lopes LC, de Cassia Bergamaschi C. Vitamin D and respiratory infections. J Infect Dev Ctries. 2015; 9:355–61.

[42] Lim LY, Chalasani N. Vitamin d deficiency in patients with chronic liver disease and cirrhosis. Curr Gastroenterol Rep. 2012; 14:67–73.

[43] Halfon M, Phan O, Teta D. Vitamin D: a review on its effects on muscle strength, the risk of fall, and frailty. Biomed Res Int. 2015; 2015:953241.

[44] Bischoff-Ferrari HA. Relevance of vitamin D in muscle health. Rev Endocr Metab Disord. 2012; 13:71–7

[45] Bischoff-Ferrari HA, Dawson-Hughes B, Staehelin HB, Orav JE, Stuck AE, Theiler R et al. Fall prevention with supplemental and active forms of vitamin D: a meta-analysis of randomised controlled trials. BMJ. 2009; 339: b3692.

Chapter 11

Optimising Vitamin D Levels after Hip Fractures

Jenson Mak

Additional information is available at the end of the chapter

http://dx.doi.org/10.5772/66546

Abstract

Older people presenting with hip fractures requiring surgery have a high prevalence of hypovitaminosis D, which is an important modifiable risk factor for falls and fractures. Inadequate sun exposure is the main reason for vitamin D deficiency in older people. Vitamin D supplements, with or without calcium, have been shown to reduce falls and fracture risk in this population. A small number of randomised controlled trials (RCTs) have shown that increased 25-OHD levels with a loading dose of vitamin D may improve falls and fractures. It is not previously known whether oral vitamin D replenishment using a loading dose is effective, and if it is, what is the interplay this is with patient characteristics, in particular lower limb mobility and 25-OHD levels. The results of a recent multisite randomised controlled trial (REVITAHIP) provide early evidence of the benefits of an early loading-dose oral vitamin D replenishment on functional mobility, falls, fractures, grip strength, health-related quality of life and mortality.

Keywords: Fragility fractures, femoral fractures, cholecalciferol, aged care, osteoporosis, rehabilitation, trauma surgery

1. Introduction

Hip fractures and related disabilities are important public health issues for older people around the world. Despite the age-adjusted hip fracture rates reducing in countries such as Australia, Italy, France, Japan and the United States, the actual numbers of fractures are increasing steadily due to the increasing proportion of the elderly population [1]. Outcomes for people who survive hip fracture are of concern, with more than one-quarter dying within a two-year period [2] and most not recovering their previous level of function. For example, around half of the people surviving a hip fracture require long-term help with routine activities and cannot walk unaided, one quarter require full-time nursing-home care [3] and 16% reporting depressive symptoms [4]. Given that people require assistance to recover from a hip fracture, personal and societal costs are often incurred following surgery due to the need for rehabilitation, outpatient visits for follow-up treatment, temporary residential aged care facility placement if required and assistance with activities of daily living at home during the recovery period. Given these factors, the quest in improving function after a hip fracture has the potential to be of enormous benefit to elderly people by reducing disability and improving quality of life. This can then reduce direct treatment costs and costs of long-term community or residential aged care services. The ability to mobilise is the key activity underlying functioning and quality of life.

Low vitamin D levels are commonly associated with hip fracture in older people [4] and occur because of multiple factors such as decreased sun exposure with reduced skin production of vitamin D and low dietary D2/D3 intake. More importantly, the relationship between vitamin D and frailty is postulated to be largely mediated via the development of sarcopenia, a condition characterised by a combination of the reduction in muscle mass, plus either muscle strength or performance.

Whilst an independent relationship has not been established on vitamin D and frailty in observational studies, there is a dearth of interventional studies yielding a positive effect on frailty using supplemental vitamin D, mainly via improvements in the physical performance parameters [5]. Further, vitamin D replacement (with calcium) has been used successfully to prevent fractures as well as falls among older people [6], and those living in the community [7]. However, in the absence of preventive treatment, hypovitaminosis D following hip fracture may result in proximal muscle weakness, pain, reduced dynamic balance and performance speed [8] affecting mobilisation during the acute postoperative and rehabilitation periods. Further, there had been concerned from one RCT with the safety data of 'megadose' vitamin D increasing the risk of falls and fractures in the first 3 months of treatment [9].

A pilot study examining the efficacy and effectiveness of moderate dose oral vitamin D therapy in maintaining 25-hydroxyvitamin D (25-OHD) levels following a hip fracture found that vitamin D levels significantly fall in the first 14 days following hip fracture despite regular oral vitamin D treatment (1000 IU daily) [10]. Whilst 1000 IU of cholecalciferol is an accepted dosage for maintenance therapy, this study found that the dose only raised levels by a small amount and took a longer timeframe to do so compared with loading doses.

As a further concern, mild but especially severe hypovitaminosis D can contribute to and exacerbate symptomatic hypocalcaemia occurring rarely but significantly following intravenous bisphosphonate (zoledronic acid), a first-line recommended treatment following hip fracture [11]. Whilst rare, hypocalcaemia may be life threatening and require immediate resuscitation and evaluation, often requiring hospitalisation to prevent additional morbidity and mortality risk from seizures, tetany, refractory hypotension or arrhythmias. Therefore, optimisation of vitamin D levels after hip fractures in older people can improve osteoporosis by providing an optimal environment for medications to act on the bone matrix, as well as improving non-skeletal factors such as ensuring adequate muscle function and controlling pain. The ability to simultaneously improve skeletal and non-skeletal factors using vitamin D replenishment techniques, therefore, has enormous potential to improve mobility and functional during the post-acute and rehabilitation periods, as well as to prevent falls and further fractures in the medium and long term.

The REVITAHIP multisite randomised controlled trial was conducted over three sites in Sydney and the Central Coast of Australia, evaluating the effect of an initial loading dose of vitamin D to improve rehabilitation outcomes following hip fractures. The study examined the impact of a loading dose of the vitamin D on physical performance measures designed to measure mobility-related disability. The results of this trial in relationship to other available trials on the effectiveness of optimisation strategies of vitamin D levels following hip fracture and its implication on public health will be discussed in this chapter.

2. Hypovitaminosis D, hip fracture surgery and old people

2.1. Vitamin D synthesis (in older people)

Vitamin D is recognised to have a wide role as a prohormone, with many cells in the body having receptors for its active form. The key source of vitamin D in humans is its synthesis from sunlight, and dietary sources are less significant. The daily recommended dietary allowance (RDA) of vitamin D is 600 IU (International Units) for children and adults less than 70 years of age and 800 IU for those aged over 70 years [12].

Vitamin D from diet and the skin is metabolised to 25-hydroxyvitamin D (25-OHD) in the liver, and it is in this form that 25-OHD is used to determine one's vitamin D status. Then in the kidneys, 25-OHD is metabolised by the enzyme 25-OHD-1α-hydroxylase (CYP27B1) to its active form 1,25-dihydroxyvitamin D (1,25 OHD). The renal production of 1,25-OHD is regulated tightly by plasma parathyroid hormone (PTH) levels, as well as serum calcium and phosphorus levels [6, 7]. Furthermore, there is a strong negative feedback regulation by 1,25-OHD itself in its own production (**Figure 1**). Inadequate sun exposure is the main reason for vitamin D deficiency in older people [13].

Humans obtain vitamin D from exposure to sunlight, from their diet and from dietary supplements (as above). Whilst theoretically a diet rich in oily fish helps to improve 25-OHD levels and prevents vitamin D deficiency, in reality this is challenging to achieve, in particular for older people and those with dietary restrictions. Solar ultraviolet B radiation (wavelength, 290–315 nm) traverses the skin and converts 7-dehydrocholesterol to previtamin D3, which is rapidly converted to vitamin D3. Because any excess previtamin D3 or vitamin D3 is destroyed by sunlight, there is also reversible conversion to inactive sterols in the skin, and excessive exposure to sunlight does not cause vitamin D3 intoxication [14]. Despite this, practical attempts for supervised 'sunlight therapy' in residential aged care have only produced mild 25-OHD improves, did not reach optimal levels and depended on the season of exposure [15]. On a pragmatic level, excessive ultraviolet exposure is the main cause of skin cancer, including cutaneous malignant melanoma, basal cell carcinoma and squamous cell carcinoma [16], and the general advice is 'avoid too much sun. Use sun protection. Do not use sunbeams'.

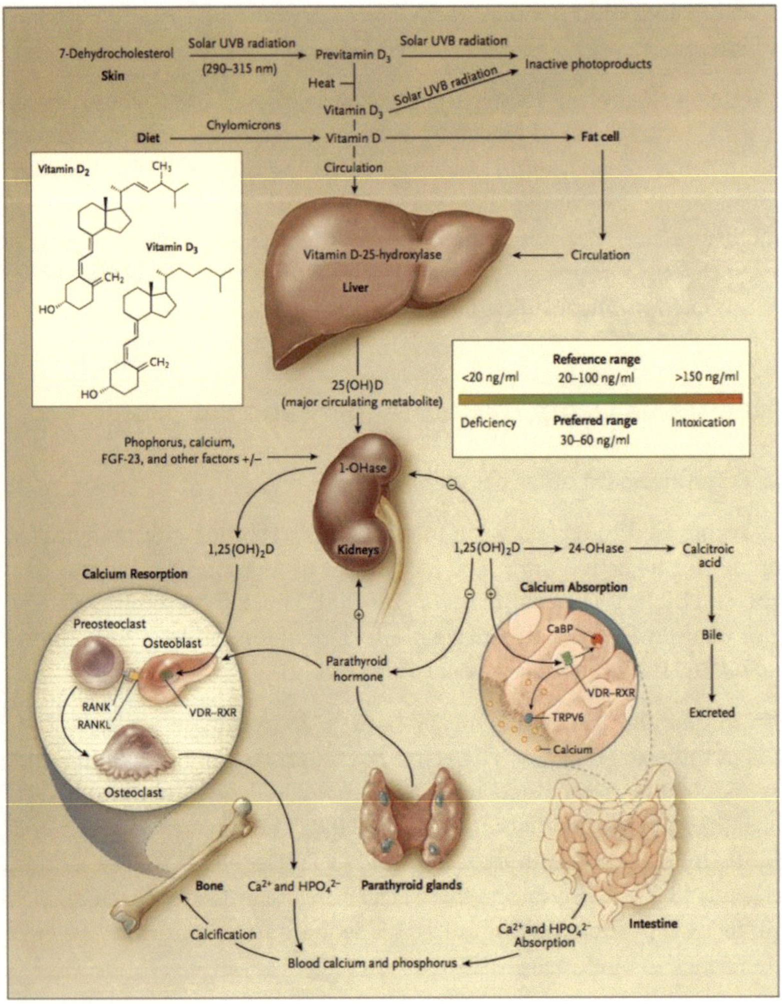

Figure 1. Synthesis and metabolism of vitamin D in the regulation of calcium, phosphorus and bone metabolism (from Holick [8]).

Age is the most important factor in vitamin D synthesis. The capacity of older people to synthesise vitamin D after UVB exposure has been investigated in laboratory and clinical studies. MacLaughlin and Holick [17] showed decreasing concentrations of 7-dehydrocholesterol with age and when irradiated, ageing skin had decreased capacity to produce previtamin D. With the ageing process, the ability to synthesise vitamin D from sunlight is reduced. Further, the ability of vitamin D to be activated to 1,25-OHD in the kidney also decreases with age. Finally, elderly people (especially those who stay mainly at home) are less likely to engage in regular exercise in the outdoor setting.

2.2. Definition of vitamin D deficiency

The optimal level of vitamin D has been subject to considerable debate, but a serum 25-OHD level of ≥50 nmol/L at the end of winter (with a 10–20 nmol/L higher level at the end of summer to allow for the winter decrease) is required for optimal musculoskeletal health. 25-OHD levels are inversely associated with PTH levels until the former reach 30–40 ng/mL (75–100 nmol/L), at which point PTH levels begin to level off (at their nadir) [18]. The Working Group recommended that although higher levels are likely to play a role in other diseases, 'there is insufficient evidence from randomised controlled trials to recommend higher targets' [19]. The Working Group also defined vitamin D status according to levels of serum 25-OHD:

- Vitamin D adequacy: ≥50 nmol/L (at end of winter).
- Mild vitamin D deficiency: 30–49 nmol/L.
- Moderate vitamin D deficiency: 12.5–29 nmol/L.
- Severe vitamin D deficiency: <12.5 nmol/L.

2.3. Prevalence of vitamin D deficiency following hip fracture surgery

The reported prevalence of vitamin D deficiency following hip fracture, dependent on racial groups and gender, ranges from 65.8 to 96.7% [20, 21] and, in up to 32% of cases, is complicated by secondary hyperparathyroidism (PTH > 5.25 pmol/L in the presence of hypovitaminosis D). Furthermore, soon after hip fracture surgery (at 2 weeks), 25-OHD levels can decrease after hip fracture despite standard oral vitamin D treatment (of 1000 IU daily) compared with higher replacement doses (**Figure 2** [10]).

2.4. Correction of vitamin D deficiency in older people following hip fracture surgery

There have been several studies looking at the optimal method in replenishing 25-OHD levels following hip fracture surgery. A moderate oral dose (4000 IU daily) replacement approach can significantly improve and maintain 25-OHD levels within 2 weeks after a hip fracture, with a mean vitamin D increase in 22.4 ± 18.3 nmol/L and up to 88.9% of participants with optimal 25-OHD levels [10]. de Jong et al. [22] found that substitution with 50,000 IU oral cholecalciferol daily for 7 days increased 25-OHD levels rapidly, safely and consistently. Finally,

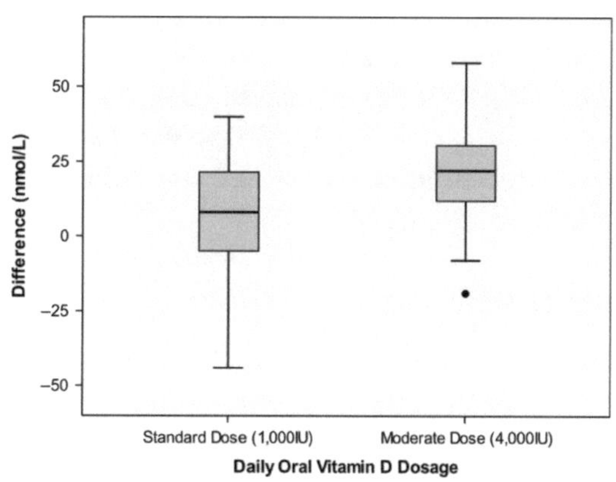

Figure 2. Differences in 25-hydroxyvitamin D levels 14 days after hip fracture according to daily oral vitamin D protocol (n = 66); 37.5% of participants on 1000 IU vitamin D reduced 25-OHD levels, from Mak et al. [10].

Lyles et al. [23], in the Health Outcomes and Reduced Incidence with Zoledronic Acid Once Yearly (HORIZON) Recurrent Fracture Trial, successfully utilised the strategy of a loading dose of either vitamin D3 or D2 (at a dose of 50,000–125,000 IU given orally or intramuscularly) 14 days before first infusion of a study drug. This was followed by daily supplementation with oral calcium (1000–1500 mg) and vitamin D (800–1200 IU). Whilst these studies highlight the various strategies that can utilised to replenish vitamin D (oral/intramuscular), loading dose followed by regular daily or interval dosing, there has been a sparsity of outcomes data including lower limb function, falls and fractures and mortality.

3. REVITAHIP Study

The Replenishment of Vitamin D in Patients with Hip Fracture (REVITAHIP) Trial was a multicenter, randomised, double-blind and placebo-controlled trial involving patients with recent hip fracture [24]. Patients were randomly assigned to receive either an oral loading dose of cholecalciferol (at a dose of 250,000 IU vitamin D3) or placebo, followed by daily supplementation with oral vitamin D (800 IU) and calcium (600 mg). Deviation from this protocol occurred for any participants with an initial serum 25-hydroxyvitamin D level of 10 nmol/L or less due to the known risks. Such cases received a 14-day loading dose of vitamin D3 (at a dose of 4000 IU given orally), continuing on as per other patients for the remainder of the trial. Patients were monitored for up to 26 weeks with telephone interviews or clinic/home visits at 2, 4, 12 (telephone interview only) and 26 weeks. All study procedures were approved by the local institutional review board at each participating site (HREC Number 10/HARBR/14).

3.1. Results of the REVITAHIP Study

The REVITAHIP cohort included 220 participants, with a mean age of 83.9 years (SD 7.1) and 77.1% women. The REVITAHIP study [25, 26] found that in these patients following a hip fracture (within 7 days, median 3.3), a loading dose of 250,000 IU vitamin D3 compared with placebo (the 'Active' method) resulted in higher 25-OHD levels and a greater percentage with target 'sufficient' 25-OHD levels (>50 nmol/L [31]), with no significant differences in gait velocity at 4 weeks. A significantly reduced incidence of falls was also noted in the active group compared with placebo over the study period. This is surprising because in the study cohort the baseline level of 25-OHD is higher than in other studies of patients with hip fracture, and the differences in 25-OHD at weeks 2 and 4, between the active and placebo groups, whilst significant, were not large. Further, there was a significant reduction in falls and non-significant reduction in mortality (**Figure 3**).

3.2. Limitations to the REVITAHIP trial

There were several limitations to this research. The REVITAHIP study patients were slightly younger with less unhealthy than are patients with hip fracture in the general population, as suggested by data regarding 1-year mortality. Despite this, patients in our study ranged widely in age (up to 101 years), with a significant percentage having cognitive impairment. Perhaps due to inclusion and exclusion criteria, there was under-recruitment of people with

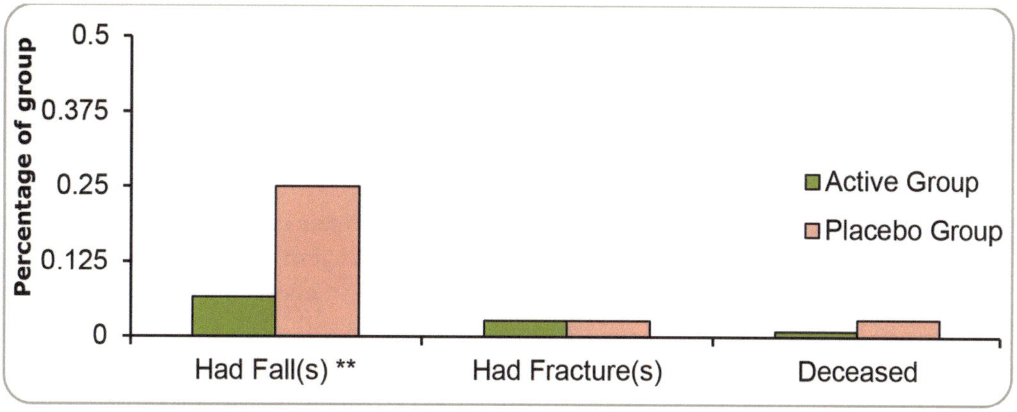

Figure 3. Fall, fractures and deaths over 4 weeks by intervention group in the REVITAHIP trial ($p = 0.024$).

severe pre-existing disability compared to the usual population with hip fractures: (1) low number of participants from residential aged care facility (10%) and (2) participants with fewer total comorbidities. The REVITAHIP participants did have relatively good function with moderate independence prior to their hip fracture, most likely due to the inclusion and exclusion criteria. However, the study team argued that this could be considered a strength, as it alerts clinicians to potential problems (e.g. in ADLs, mobility, high psychotropic medication use) even in a relatively well-functioning cohort.

The REVITAHIP participants had higher levels of baseline 25-OHD compared to previous cohorts (e.g. the ICHIBAN cohort [26] had a lower baseline 25-OHD levels than the REVITAHIP (44.1 ± 23.2 vs 53.1 ± 24.2 nmol/L)), likely owing to a less frail participant population. Another limitation was that the planned sample size of 450 could not be achieved due to less efficient than expected participant recruitment. Nevertheless, as 25-OHD levels in the REVITAHIP trial were higher than usual levels, it may have underestimated its effects on falls, fractures and death. It was also noted that the falls reduction effects at week 4 were higher than expected from other studies.

Furthermore, due to technical difficulties with obtain accurate fall diaries over a longer term, the data for fall rates of participants at week 12 and 26 could not be utilised and were therefore not able to be included in the presentation of the data. This is mainly due to the limited funds available resulting in limited to no administrative assistance with the trial. Documentation of various patient outcomes would have been improved with such help. Consequently, results of health services and community service utilisation and the economics analyses were not reported in the trial as data were incomplete, and analyses were not able to be done.

Finally, there was no direct comparison in this study between a loading dose and a high daily dose of vitamin D, so absolute conclusions cannot be drawn on optimal dosing regimen. The advantages of the loading dose were good compliance and likely rapid increase in 25OHD, both of which are likely to be helpful in this setting. But there are some disadvantages of loading doses, as opposed to daily doses of vitamin D [27], so the dosing issue is unclear.

Nevertheless, the study appears to indicate that a higher D status in this group of patients is beneficial for overall outcomes.

4. Implications of research findings

It has been generally established that vitamin D, usually with calcium supplement, decreases falls and fractures in older people from aged care facilities [28] but has not been proven convincingly in post-menopausal women in large meta-analyses. Following a hip fracture, vitamin D levels are universally low, and its levels may drop as early as 2 weeks following hip fracture [10]. Poor adherence to vitamin D therapy has been shown to compromise the efficacy of this treatment for fracture reduction and to increase medical costs [29]; such findings have been particularly notable in frail older adults [30]. Vitamin D deficiency is frequently observed in older patients and is associated with an increased risk of hypocalcaemia when intravenous bisphosphonates are administered before a normal vitamin D level has been achieved [8].

The REVITAHIP study was the first to show that treatment with 250,000 IU cholecalciferol within 7 days after hip fracture surgery (followed by regular maintenance vitamin D) is associated with higher percentage of replete 25-OHD and may reduce the rate of falls. However, the positive results occurred despite the group of participants having a higher baseline 25-OHD (with a resultant smaller than expected gain in 25-OHD levels), and the recruitment numbers were smaller than initially planned for a sample size to detect a difference in gait velocity. Thus, in practical terms, the use of the Active REVITAHIP vitamin D replenishment method would appear to be an efficient (through ease of administration) and effective means of optimising vitamin D levels for the older patients following hip fracture surgery. Indeed, several meta-analyses of studies of vitamin D supplementation, usually with calcium supplementation, have shown an overall benefit on falls and fracture reductions [31]. The intervention also had an optimal safety profile (one case of hypercalcaemia) and was relatively inexpensive to administer.

5. Recommendations of research findings

Currently, Australia is unique in that it does not have a 50,000 IU oral vitamin D tablet, readily available compared to other countries. Whilst (1) a liquid form of cholecalciferol exists (OsteVit-D) which potentially can be administered in higher doses; (2) preparations such as Fosamax Plus D (containing a weekly dose of 5600 IU daily); and (3) higher-dose formations of vitamin D manufactured from compounding pharmacists, the former (1) may be liable to accurate dosage; (2) may not be practical to administer whilst hip fractured patients are recovering in the post-acute period; and (3) the universal reliability of compounding pharmacist cannot be guaranteed. The approval of higher-dose vitamin D such as from the Active REVITAHIP method should be made readily available through Australian Government subsidies, and vitamin D manufacturers from overseas be supported to import such products into Australia.

For most other countries, higher dosage vitamin D formulations are available both in oral and injectable formats.

Finally, the author advocates that the outcomes of the Active REVITAHIP method of vitamin D replenishment be disseminated, and adaptation into local guidelines be considered.

6. Conclusion

The studies reviewed in this chapter provides an overall positive view on vitamin D replacement (which is deemed necessary) and showed that treatment with 250,000 IU cholecalciferol after hip fracture surgery in older patients (followed by regular daily maintenance vitamin D) is associated with higher percentage of replete 25-OHD and reduced rate of falls in the short term. Recognition of these factors may also improve the delivery of post-hip fracture surgical and rehabilitation care in this often vulnerable population. The author recommends further confirmation of the results of this latest research using a larger number of participants (with a higher diversity in ethnic populations who may be likely to have higher prevalence of hypovitaminosis D) in future studies.

Author details

Jenson Mak

Address all correspondence to: jenson.mak@gmail.com

Rehabilitation Studies Unit, University of Sydney, Australia

School of Medicine and Public Health, University of Newcastle, Australia

References

[1] Kanis JA, Odén A, McCloskey EV, Johansson H, Wahl DA, Cooper C; IOF Working Group on Epidemiology and Quality of Life. A systematic review of hip fracture incidence and probability of fracture worldwide. Osteoporos Int. 2012;23(9):2239–56.

[2] Burns A, Younger J, Morris J, Baldwin R, Tarrier N, Pendleton N, Cohen P, Horan M, Banerjee S. Outcomes following hip fracture surgery: a 2-year prospective study. Am J Geriatr Psychiatry. 2014;22(8):838–44.

[3] Pasco JA, Sanders KM, Hoekstra FM, Henry MJ, Nicholson GC, Kotowicz MA. The human cost of fracture. Osteoporos Int. 2005;16(12):2046–2052.

[4] Bischoff-Ferrari HA, Can U, Staehelin HB, et al. Severe vitamin D deficiency in Swiss hip fracture patients. Bone. 2008;42(3):597–602.

[5] Wong YY, Flicker L. Hypovitaminosis D and frailty: Epiphenomenon or causal? Maturitas. 2015;82(4):328–335. pii:S0378-5122(15)30033-5.

[6] Bischoff-Ferrari HA, Dawson-Hughes B, Staehelin HB, et al. Fall prevention with supplemental and active forms of vitamin D: a meta-analysis of randomised controlled trials. BMJ. 2009;339:b3692.

[7] Trivedi DP, Doll R, Khaw KT. Effect of four monthly oral vitamin D3 (cholecalciferol) supplementation on fractures and mortality in men and women living in the community: randomised double blind controlled trial. BMJ. 2003;326(7387):469.

[8] Holick MF. Vitamin D deficiency. N Engl J Med. 2007;357(3):266–281.

[9] Sanders KM, Stuart AL, Williamson EJ, et al. Annual high-dose oral vitamin D and falls and fractures in older women: a randomised controlled trial. JAMA. 2010;303(18):1815–1822; Erratum: JAMA. 2010;303(23):2357.

[10] Mak JC, Stuart-Harris J, Cameron ID, Mason RS. Oral vitamin D replacement after hip fracture: a comparative review. J Am Geriatr Soc. 2010;58(2) 382–3.

[11] Mak JC, Cameron ID, March LM; National Health and Medical Research Council. Evidence-based guidelines for the management of hip fractures in older persons: an update. Med J Aust. 2010;192(1):37–41.

[12] Ross AC, Taylor CL, Yaktine AL, Valle HBD, Editors. Dietary Reference Intakes for Calcium and Vitamin D. Institute of Medicine. Washington DC: The National Academies Press; 2011. Available from http://www.nap.edu/catalog.php?record_id=13050

[13] Sambrook PN, Cameron ID, Chen JS, Cumming RG, Durvasula S, Herrmann M, Kok C, Lord SR, Macara M, March LM, Mason RS, Seibel MJ, Wilson N, Simpson JM. Does increased sunlight exposure work as a strategy to improve vitamin D status in the elderly: a cluster randomized controlled trial. Osteoporos Int. 2012;23:615–624.

[14] DeLuca HF. Overview of general physiologic features and functions of vitamin D. Am J Clin Nutr. 2004;80(6 Suppl):1689S–1696S.

[15] Durvasula S, Gies P, Mason RS, Chen JS, Henderson S, Seibel MJ, Sambrook PN, March LM, Lord SR, Kok C, Macara M, Parmenter TR, Cameron ID. Vitamin D response of older people in residential aged care to sunlight-derived ultraviolet radiation. Arch Osteoporos. 2014;9:197.

[16] Greinert R, de Vries E, Erdmann F, Espina C, Auvinen A, Kesminiene A, Schüz J. European Code against Cancer 4th Edition: Ultraviolet radiation and cancer. Cancer Epidemiol. 2015;39(Suppl 1):S75–83.

[17] MacLaughlin J, Holick MF. Aging decreases the capacity of human skin to produce vitamin D3. J Clin Invest. 1985;76(4):1536–8.

[18] Holick MF. Sunlight and vitamin D for bone health and prevention of autoimmune diseases, cancers, and cardiovascular disease. Am J Clin Nutr. 2004;80(6 Suppl):1678S–1688S.

http://dx.doi.org/10.5772/66546

[19] Nowson CA, McGrath JJ, Ebeling PR, Haikerwal A, Daly RM, Sanders KM, Seibel MJ, Mason RS; Working Group of Australian and New Zealand Bone and Mineral Society, Endocrine Society of Australia and Osteoporosis Australia. Vitamin D and health in adults in Australia and New Zealand: a position statement. Med J Aust. 2012;196(11):686–7.

[20] Johnson AL, Smith JJ, Smith JM, Sanzone AG. Vitamin D insufficiency in patients with acute hip fractures of all ages and both sexes in a sunny climate. J Orthop Trauma. 2013;27(12):e275–80.

[21] Khadgawat R, Brar KS, Gahlo M, Yadav CS, Malhotra R, Guptat N, Tandon N. High prevalence of vitamin D deficiency in Asian-Indian patients with fragility hip fracture: a pilot study. J Assoc Physicians India. 2010;58:539–42. Erratum: J Assoc Physicians India. 2010;58:630.

[22] de Jong A, Woods K, Van Gestel L, Suresh M, Porteous M. Vitamin D insufficiency in osteoporotic hip fracture patients: rapid substitution therapy with high dose oral cholecalciferol (vitamin D3). Acta Orthop Belg. 2013;79(5):578–86.

[23] Lyles KW, Colón-Emeric CS, Magaziner JS, Adachi JD, Pieper CF, Mautalen C, Hyldstrup L, Recknor C, Nordsletten L, Moore KA, Lavecchia C, Zhang J, Mesenbrink P, Hodgson PK, Abrams K, Orloff JJ, Horowitz Z, Eriksen EF, Boonen S; HORIZON Recurrent Fracture Trial. Zoledronic acid and clinical fractures and mortality after hip fracture. N Engl J Med. 2007;357(18):1799–809.

[24] Mak JC, Mason R, Klein L, Cameron ID. Improving mobility and reducing disability in older people through early high-dose vitamin D replacement following hip fracture: a protocol for a randomized controlled trial and economic evaluation. Geriatr Orthop Surg Rehabil. 2011;2(3):94–9.

[25] Mak JC, Klein LA, Finnegan T, Mason RS, Cameron ID. An initial loading-dose vitamin D versus placebo after hip fracture surgery: baseline characteristics of a randomized controlled trial (REVITAHIP). BMC Geriatr. 2014;14:101.

[26] Mak JC, Klein LA, Mason RS, Cameron ID. Contemporary Pain Management in Elderly Patients After Hip Fracture Surgery: Cross-sectional Analyses at Baseline of a Randomized Controlled Trial. Clin J Pain. 2015;31(9):788–93.

[27] Hollis BW, Wagner CL. Clinical review: The role of the parent compound vitamin D with respect to metabolism and function: Why clinical dose intervals can affect clinical outcomes. J Clin Endocrinol Metab. 2013;98(12):4619–28. doi:10.1210/jc.2013-2653. Epub 2013 Oct 8.

[28] Avenell A, Mak JC, O'Connell D. Vitamin D and vitamin D analogues for preventing fractures in post-menopausal women and older men. Cochrane Database Syst Rev. 2014;4:CD000227.

[29] Huybrechts KF, Ishak KJ, Caro JJ. Assessment of compliance with osteoporosis treatment and its consequences in a managed care population. Bone. 2006;38:922–8.

[30] van Eijken M, Tsang S, Wensing M, de Smet PA, Grol RP. Interventions to improve medication compliance in older patients living in the community: a systemic review of the literature. Drugs Aging. 2003;20:229–40.

[31] Tang BM, Eslick GD, Nowson C, Smith C, Bensoussan A. Use of calcium or calcium in combination with vitamin D supplementation to prevent fractures and bone loss in people aged 50 years and older: a meta-analysis. Lancet. 2007;370(9588):657–66. Erratum: Lancet. 2012;380(9844):806.

Section 4

Vitamin D on Reproductive System

Chapter 12

Vitamin D and Human Reproduction

Fahimeh Ramezani Tehrani and

Samira Behboudi-Gandevani

Additional information is available at the end of the chapter

http://dx.doi.org/10.5772/67394

Abstract

Vitamin D is one of the steroid hormones. The precursor of vitamin D, 7-dehydrocholesterol, which is an intermediary for cholesterol pathway, is available in the skin. Ultraviolet B (UVB) radiation makes the transformation of 7-dehydrocholesterol to provitamin D3, which automatically isomerizes to cholecalciferol (vitamin D3). Vitamin D3 is secreted into blood circulation and carried by the vitamin D–binding protein (VDBP). Around 80–90% of vitamin D is from sunlight-derived production in the skin. A little amount of vitamin D is also extracted from foods and/or additional supplementation. Vitamin D has been well known for its function in maintaining calcium and phosphorus homeostasis and promoting bone mineralization. Accumulating evidence from animal and human studies suggests that vitamin D also modulates reproductive processes in women and men and is involved in many functions of the reproductive system. Vitamin D receptor (VDR) and vitamin D–metabolizing enzymes are found in reproductive tissues of women and men. This chapter presents an up-to-date review for describing the function of vitamin D in female reproduction throughout reproductive ages from menarche to menopause, during pregnancy and lactation, and some disorders affecting women and also the role of vitamin D applied to male fertility.

Keywords: vitamin D, vitamin D receptor, metabolism, female reproduction, male reproduction

1. Introduction

1.1. Overview of vitamin D: production, metabolism and action

Vitamin D (VD) is one of the fat-soluble vitamins from steroid hormones family. While there are various forms of vitamin D, two main forms are necessary for human body: (i) D2 (ergocalciferol) and (ii) D3 (cholecalciferol) [1]. In the presence of ultraviolet radiation, vitamin D2 is derived primarily in plants, yeast, and fungi, and also vitamin D3 is synthesized predominantly by the conversion of 7-dehydrocholesterol, a VD precursor present in the skin, under ultraviolet B radiation with only a small amount of this vitamin with around <10–20% obtained through diet supplements [2]. VD precursor isomerizes into cholecalciferol. Cholecalciferol is bound to serum vitamin D-binding protein (DBP). To become biologically active, two-step enzymatic pathways are necessary; involving 25-hydroxylase of the liver and 1a-hydroxylase (CYP27B1) of the kidney and extra-renal tissues, it is converted to the biologically active hormone calcitriol (1α,25(OH)2D3) [3, 4].

Vitamin D secretion is set out in the renal 1α-hydroxylase phase. Parathyroid hormone (PTH) upregulates the expression of this enzyme and also 1,25(OH)2D3 could suppress itself [5].

Finally, in the kidneys, both 25(OH) D and 1,25(OH) 2D3 convert into an inactive compound of calcitroic acid by 24-hydroxylase, which is water soluble and excreted in bile. Whereas 1α-hydroxylase is predominantly found in the kidneys, it can also be expressed in different extra-renal tissues including bones, colon, breasts, prostate, and placenta. In this respect, it is suggested that macrophages could locally produce 1,25(OH)2D3 [6].

Biological roles of vitamin D are mediated by the VD receptor (VDR), a ligand-dependent transcription factor mainly localized in the target cell nuclei that mediate the genomic action of 1,25(OH)2D3, which influences on the transcription of more than 900 genes [7]. It is widely distributed in over 38 tissues and organs including skeleton, parathyroid glands, and the reproductive tissues indicating its potential role in the regulation of numerous metabolic processes [8].

The regulation of VDR expression is one of the main mechanisms through which target cells respond to vitamin D so that polymorphisms of this receptor can change the usual mode of functioning [9, 10]. After the linking of vitamin D, vitamin D receptors transform to a heterodimer with retinoid x-receptor (RXR) after which this complex binds to specific DNA sequences named vitamin D–response elements (VDREs) in the promoter zone of vitamin D–responsive genes. So, this trimeric complex of VDR-RXRVDRE acts as a molecular switch in nuclear 1,25(OH)2D3 signaling [11].

The genomic response lasts for longer times from a few hours to 1 day for the changes that occur in the transcription of gene. However, the nongenomic response is faster, taking only seconds to a few minutes due to the interaction that occurs with a cell surface receptor and the second messenger [10].

Catabolism of 1,25(OH)2D and 25-OHD to biologically inactive calcitroic acid is mediated by 24-hydroxylase, in the kidney and liver [4]. It is generally accepted that the serum levels of 25(OH)D are considered the best indicator of vitamin D status because of its easy measurement and long half-life in circulation (~2–3 weeks) [12, 13].

The three main steps in vitamin D metabolism, 25-hydroxylation, 1α-hydroxylation, and 24-hydroxylation, are all performed by cytochrome P450 mixed-function oxidases (CYPs) (**Figure 1**).

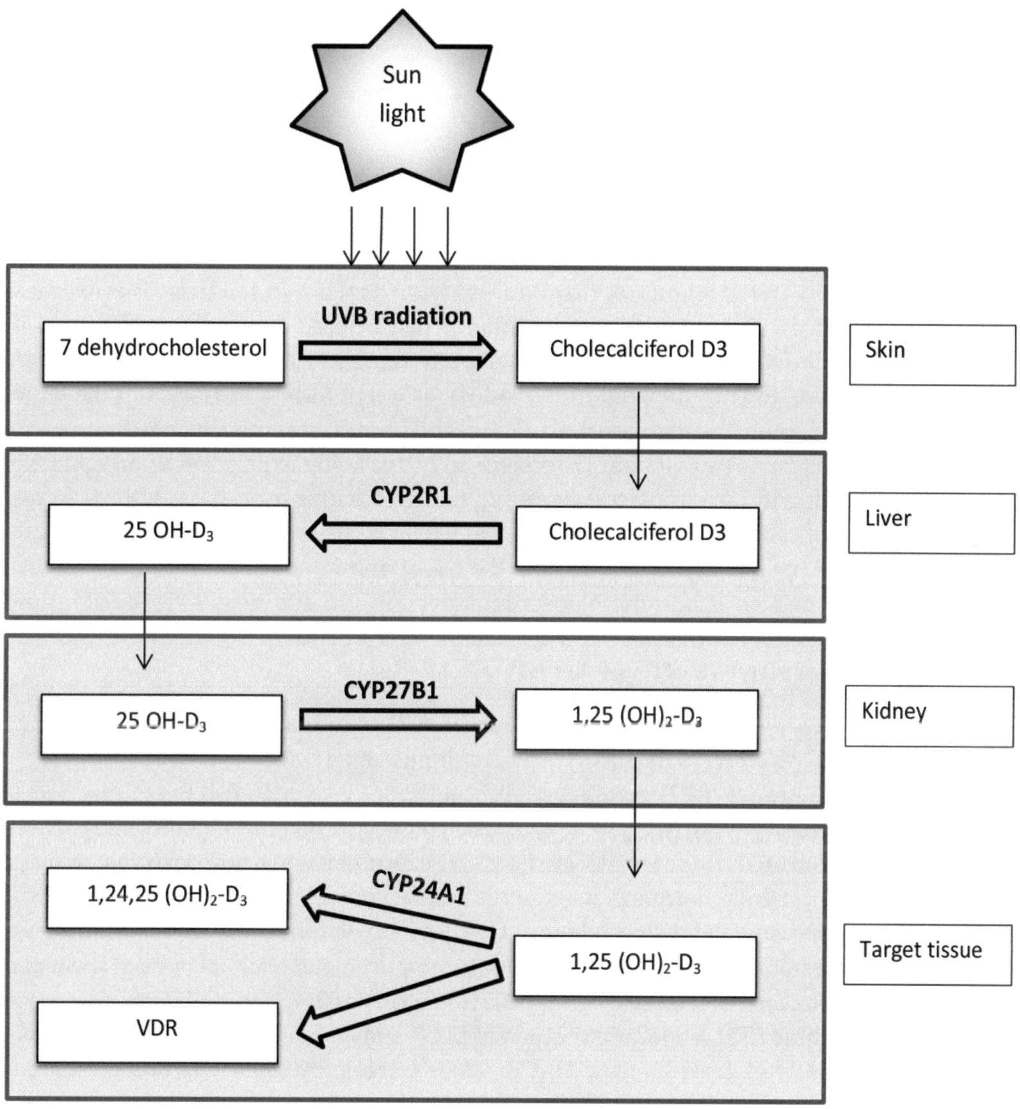

Figure 1. The main steps in vitamin D metabolism.

These enzymes are located either in the endoplasmic reticulum (ER) (e.g., CYP2R1) or in the mitochondria (e.g., CYP27A1, CYP27B1, and CYP24A1). The electron donor for the ER enzymes is the reduced nicotinamide adenine dinucleotide phosphate (NADPH)-dependent P450 reductase. The electron donor chain for the mitochondrial enzymes is composed of ferredoxin and ferredoxin reductase. These are not specific for a given CYP—specificity lies within the CYP. Although the CYPs involved in vitamin D metabolism, only CYP2R1 and CYP24A1, have been crystallized, it is likely that these enzymes contain a number of common structural features. These include 12 helices (A–L) and loops and a common prosthetic group, namely the iron-containing protoporphyrin IX (heme) linked to the thiolate of cysteine. The I helix runs through the center of the enzyme above the heme where a thr(ser) and asp(glu) pair is essential for catalytic activity [14]. CYP2R1, like other microsomal CYPs, contains two extra helices that appear to form a substrate channel in the bilayer of the ER [14]. The B′ helix serves as a gate, closing on substrate binding. Whether a similar substrate channel exists for the mitochondrial CYPs is not clear.

1.2. Physiologic functions of vitamin D

As previously mentioned, once VD is produced in the skin received from the diet, it travels to blood circulation and is bound to vitamin D–binding protein. In the liver, it converts to 25(OH) D as the main circulating form of vitamin D that is measured to determine an individual's vitamin D deficiency. In biological perspectives, 25(OH) D is in an inactive form. However, it converts to [1,25(OH)2D] in the kidney as a free form and travels to the target tissues and participates in the regulation of calcium and phosphorus metabolism. In the intestine, 1,25(OH)2D is bound to vitamin D receptor to increase the expression of an epithelial calcium channel due to increasing the transportation of calcium from the intestinal lumen into the absorptive cell [15]. In this respect, 1,25(OH)2D could enhance the expression of a calcium-binding protein (calbindin9k) to help the transportation of calcium to the intestinal absorptive cell to deposit it into the blood circulation [16]. In addition, 1,25(OH)2D could transport to the skeleton which interacts with VDR in the osteoblast to enhance the expression of RANKL (receptor activator of NFκB ligand).

Monocytic preosteoclasts express the RANK receptor that interacts with RANKL making signal transduction resulting in the formation of multinucleated osteoclasts which are capable of secretion of HCl to save the bone mineral and collagenases to demolish the matrix releasing calcium into the blood circulation.1,25(OH)2D could also directly enhance the calcium tubular reabsorption on the kidneys. As such, 1,25(OH)2D and its receptors reversely regulate the secretion of parathyroid hormone. In addition, among intestine 1,25(OH)2D induces phosphorus absorption. Thus, the main biological functions of vitamin D are to maintain serum calcium and phosphorus in physiological border to support metabolic functions of them and to save the mineralization of the skeleton [17]. The main factors that control the renal secretion of 1,25(OH)2D include PTH, hypocalcemia, and hypophosphatemia that increase its secretion. In this respect, fibroblast growth factor 23 (FGF 23) is secreted by osteocytes and osteoblasts and decreases the renal secretion of 1,25(OH)2D [18].

The dominant physiologic function of vitamin D is the increase of plasma concentration of calcium and phosphate. Both of them are essential for the mineralization of skeleton. Furthermore, the increase in the plasma level of calcium to regulate it in normal levels is also necessary for the functioning of the neuromuscular junction vasodilatation, nerve transmission, as well as hormonal production. Plasma levels of calcium are remained at a very constant level, in supersaturated bone mineral situation. If its plasma concentration becomes lower than saturated level, then mineralization fails, which leads to rickets among children and osteomalacia among adults [17].

The vitamin D could increase the serum level of calcium from three separate pathways:

(i) It could induce the proteins related to active intestinal calcium absorption throughout the entire length of the intestine, although its greatest activity is in the duodenum and jejunum which does not require parathyroid hormone. It is clear that vitamin D directly stimulates intestinal calcium and, independently, phosphate absorption, although it could stimulate active intestinal absorption of phosphate too.

(ii) In no-calcium diet, vitamin D plays an essential role in the mobilization of calcium from bone, a process requiring parathyroid hormone. Vitamin D induces osteoblasts to secretion receptor activator nuclear factor-κ B ligand. RANKL then induces osteoclastogenesis and activates resting osteoclasts for bone resorption [19]. So, vitamin D appears to be implicated in allowing persons to remove calcium from bone when it is not sufficient in the diet [20].

(iii) The distal renal tubule is responsible for reabsorption of the last 1% of the filtered load of calcium, and the two hormones interact to stimulate the reabsorption of this last 1% of the filtered load. Because 7 g of calcium is filtered every day among humans, this represents a major contribution to the calcium pool. Both parathyroid hormone and the vitamin D hormone are required. Calcium physiologic processes are such that a single low concentration of the vitamin D hormone stimulates enterocytes to absorb calcium and phosphate. If the plasma calcium concentration fails to respond, then the parathyroid glands continue to secrete parathyroid hormone, which increases the production of the vitamin D hormone to mobilize bone calcium (acting with parathyroid hormone). Under normal circumstances, environmental calcium is used first; if environmental calcium is absent, then internal stores are used.

Therefore, in general, vitamin D functions on the intestine, bone, and kidney to increase the serum concentration of calcium. If the serum calcium concentrations increase, the parathyroid hormone secretion decreases. In this respect, if serum concentration of calcium elevates too high, the parafollicular cells ("C" cells) of the thyroid produce the calcitonin hormone that could block calcium resorption from bone and lead to maintain serum level of calcium within the normal range. Vitamin D, through its receptor, suppresses parathyroid gene expression and parathyroid cell proliferation, providing important feedback loops that strengthen the direct mechanism of elevated serum concentration of calcium.

Also, it is shown that a deficiency of phosphate stimulates CYP27B1 to produce more vitamin D, which in turn stimulates phosphate absorption in the small intestine, and vitamin D can also induce the secretion of fibroblast-like growth factor-23 (FGF23) by osteocytes in bone, which results in phosphate excretion in the kidney, as well as feedback on vitamin D metabolism [21].

1.3. Vitamin D deficiency

High prevalence of vitamin D deficiency is present in all races, even in temperate areas. It is now recognized that vitamin D deficiency and insufficiency are the most common nutritional deficiency/medical condition in the world, with 20–90% of reproductive-age women being deficient [22]. The recently revised guidelines of the Endocrine Society of North America defined vitamin D deficiency as 25OH-D levels of <20 ng/mL and insufficiency as levels of 20–30 ng/mL [23]. These definitions are based in part on provocative testing in healthy adults. Intestinal calcium absorption is dramatically decreased when vitamin D deficiency occurred. This results in a transient reduction in ionized calcium concentration in the blood that is immediately sensed by the calcium sensor in the parathyroid glands leading to the enhanced secretion of PTH into the blood. As such, PTH enhances the renal tubular reabsorption of calcium and modulates the osteoblasts to increase the expression of RANKL which in turn improves the production of osteoclasts to remove calcium stores from the skeleton. It also increases the kidney secretion of 1,25(OH)2D which in turn transports to the bone and intestine to modulate calcium metabolism. The elevated level of PTH also causes internalization of the sodium-phosphorus cotransporter leading to the loss of phosphorus into the urine [15, 24].

Vitamin D deficiency among children during childhood leads to poor mineralization of the skeletal matrix and contributes to a wide range of skeletal deformities related to rickets such as bowed or knocked knees, rachitic rosary, widened epiphyseal plates at the end of the long bones, and frontal bossing. In older children, it prevents the attainment of the maximum amount of calcium that can be deposited into their skeletons based on their genetic makeup. In addition, vitamin D insufficiency among children leads to osteomalacia. Opposite to osteoporosis, which is an asymptomatic disease, osteomalacia has symptoms including bone pain and muscle weakness that is often misdiagnosed as fibromyalgia, chronic fatigue syndrome, or depression. The mineralization defect is due to the phosphaturic influence of PTH. It decreases the serum phosphorus level to be in the low normal or low range as a result of inadequate calcium-phosphorus secretion for sufficient bone mineralization [25, 26]. In adults, vitamin D deficiency and secondary hyperparathyroidism increase the loss of mineral and matrix which can cause osteopenia, osteoporosis, and an increased risk of fracture. In addition, vitamin D has a wide range of actions that include cell differentiation, apoptosis, antiproliferation, immunosuppression, and anti-inflammation [18, 27].

Mounting evidence suggests that hypovitaminosis D is linked to an increased risk for autoimmune diseases, diabetes, and cardiovascular diseases [28], indicating the importance of sufficient vitamin D levels. In addition, vitamin D deficiency has been linked to an increased risk for several types of cancer including prostate, colon, ovarian, and breast cancer [29]. However, the daily intake of vitamin D for the prevention of those adverse effects is recommended.

However, in this chapter, we focus on the role of vitamin D in three important phases of reproductive women's life span including menarche and adolescence, reproductive period, and menopause. Then, we discuss the role of vitamin D in male fertility.

2. Vitamin D status in adolescent

There is mounting evidence that adolescents are at risk for poor vitamin D status. Since vitamin D is critical for optimal bone mineral accrual in the developing skeleton, poor vitamin D status in adolescence is a matter of concern. In vitamin D deficiency status, calcium homeostasis due to parathyroid and renal regulation at the expense of bone is maintained. In a growing child or adolescent, the lack of calcium accumulation in the skeleton can have negative impact for the attainment of peak bone density [30, 31]. Despite evidence indicating the crucial role of vitamin D in many physiological functions, maintaining adequate vitamin D status in adolescents is challenging in today's food and living environment. It is reported that only 50% of girls (aged 9–13 years) and 32% of girls (aged 14–18 years) are meeting the Dietary Reference Intake (DRI) recommendation for vitamin D (200 IU/d or 5 mg/d) [32]. In adolescents, vitamin D deficiency leads to decreased dietary calcium absorption, altered formation of the growth plate, and defective mineralization of the skeleton, resulting in rickets. Also, it has been determined that subclinical vitamin D deficiency may also result in secondary hyperparathyroidism, lower serum calcium, increased serum alkaline phosphatase, and increased risk of bone abnormalities [33]. A positive correlation between bone mineral density (BMD) and 25(OH)D levels in previous studies supports the important role of vitamin D in protecting adolescent bone [34]. In addition, obesity is a major health problem among children and adolescents. Interestingly, a number of studies have supported the potential role of vitamin D in the modulation of obesity, energy metabolism, and insulin resistance in adolescent and children [35].

Also, vitamin D receptor is expressed in calcium-regulated tissues, including the ovary, and it appears to be necessary for full ovarian functions which indicate that vitamin D plays a key role in estrogen biosynthesis potentially via the maintenance of extracellular calcium concentrations and by direct regulation of aromatase gene expression. This point is discussed later on.

One needs to be aware of the high prevalence of calcium and vitamin D insufficiency in the adolescent age group, and make assessment and management of vitamin D deficiency as the component of routine adolescent health care.

2.1. Vitamin D and puberty (the potential role of vitamin D in hypothalamic hypophyseal ovarian axis)

Puberty is a time of dramatic developmental changes during which a child's body progresses through a sequential set of stages to reach mature adult reproductive function. Although genetic factors play an important role in the timing of puberty, it is well documented that environmental factor may have an effect to the current change in pubertal progression [36]. In this respect, a geographic north-south-gradient effect on age at menarche [37] was proven; young women who live at higher latitudes seem to experience an earlier onset of menses than adolescents who live near to the equator [37]. Although, the time of menarche are influenced by temperature, sun light and socioeconomic situation, but this time also related to the geographic gradient with the especial sun exposure, coincided with vitamin D status [38].

Recent studies reported that vitamin D status was associated with the timing of menarche. Mechanistic explanations of an effect of vitamin D deficiency on early menarche are speculative. However, it is suggested that vitamin D deficiency is associated with the development of adiposity in children [39], and childhood obesity could be a risk factor for early puberty [38]. Thus, vitamin D status could indirectly affect the timing of menarche through its effect on obesity. Biochemical pathways might involve adipose-derived hormones. Some studies indicated that the serum level of leptin increases at early puberty; however, vitamin D is negatively associated with leptin concentrations, but it is unclear whether the leptin or other adipokines derived from adipose tissue could alter in response to vitamin D supplementation [40–42].

Meanwhile, there is possible mechanism involved in the correlation of VD insufficiency with early onset of menarche which is not related to obesity. Insulin-like growth factor-1 (IGF-I) is one of the growth factors which may regulate the timing of puberty and puberty regression by inducing the gonadotropin-releasing hormone (GnRH) pulse, gonadotropin, and sex hormone [43]. It is showed that IGF-I increased the expression of gonadotropin-releasing hormone *in vitro* [43–45].

However, vitamin D receptors have been shown in different parts of the brain including the hypothalamus [46]. Therefore, it is possible that vitamin D plays other unknown roles in the neuroendocrine regulation of the gonadotropic axis [47].

In conclusion, vitamin D status was positively related to age at menarche, and vitamin D insufficiency was associated with earlier menarche. In regard to the serious problems associated with early menarche, the simple inexpensive medication such as vitamin D supplementation may be essential, which is needed to be studied in a randomized trial.

3. Vitamin D and female reproductive system: health implications of vitamin D deficiency in female reproduction

The secosteroid hormone of vitamin D regulates the expression of a large number of genes in reproductive tissues implicating a role for vitamin D in female reproduction. Human and animal data suggest that vitamin D plays an important role in female reproduction. It is demonstrated that VDR is located in several tissues including the immune system, the endocrine system, and the reproductive system [5]. As such, VDR is distributed among nuclei and cytoplasm of granulosa cells of human ovaries which shows that vitamin D is involved in the physiologic functions of ovarian follicles [48]. Also, it is well documented that VDR mRNA is expressed in the ovarian cell and in a purified granulosa cell culture [49]. The expression of VDR in female reproductive organ indicates that vitamin D is involved in female reproductive function.

In this respect, vitamin D deficiency is related to subfertility, endometriosis and polycystic ovary syndrome (PCOS), preeclampsia, preterm delivery, gestational diabetes, and bacterial vaginosis. However, the definition of optimal vitamin D levels in the reproductive period

http://dx.doi.org/10.5772/67394

and the determination of the best dose of vitamin D supplementation need to achieve those levels for several actions of vitamin D through a woman's life are important public health implications.

Current research on the role of vitamin D in earlier age at menarche, fertility impairments, polycystic ovary syndrome, uterine fibroids, endometriosis, maternal, and neonatal adverse outcome even improper semen parameters in the case of *in vitro* treatments suggests that vitamin D deficiency plays an important role in human reproduction processes. Here, in this section, we summarize the recent evidence that vitamin D status influences female reproductive system.

3.1. Vitamin D and ovaries

The physiological role of vitamin D in human reproduction and ovarian steroidogenesis is not well understood. There are several animal studies suggesting the importance of vitamin D in reproduction. It seems that vitamin D induces secretion of progesterone, estrone, and estradiol secretion in ovarian cells independently and, in the case of estradiol, synergistically with insulin. Vitamin D also enhances IGFBP-1 secretion. It is described subsequently.

3.1.1. Vitamin D and follicular development

Recently, it has been shown that vitamin D plays an important role in human follicular development. Vitamin D could downregulate anti-Müllerian hormone (AMH) gene (as the best markers for ovarian reserve) and upregulated FSHR gene expression. An explanation of these findings is as follows: During a women's follicular phase, the follicle that contains the most number of FSH receptors, therefore that is most sensitive to FSH, emerges as dominant at the time of inter-cycle FSH rise during the follicular phase. By inhibiting AMH expression, vitamin D may counteract the repressive effects of AMH on granulosa cell differentiation, thereby allowing follicles to reach terminal maturation and ovulation. A reason for the conflicting results seen between prostate cancer cell line studies (where vitamin D was found to downregulate AMH gene) and granulosa cell studies could be explained by differences in sex and species. In human luteinized granulosa cells, vitamin D decreased AMHR-II and FSH-receptor (FSHR) gene expression. Following follicular selection in a women's late follicular phase, the follicle becomes less dependent on FSH and more dependent on LH, followed by terminal maturation and ovulation. Similar to AMHR-II, FSHR expression in granulosa cells has been found to be the highest in small immature follicles and gradually diminishes during folliculogenesis. FSHR expression decreases along with the progression of maturation of oocytes after human chorionic gonadotropin (hCG) administration. It is not clear if the mechanistic effect of vitamin D on FSHR is happening via AMH signaling. It is well documented that there is an interaction and strong positive correlation between AMHR-II and FSHR gene expression in humans. It could be that vitamin D alters common intracellular mechanistic pathways involved in the regulation of both AMHR-II and FSHR. Clearly, a complicated interrelationship exists between these parameters. These findings suggest that vitamin D might promote the differentiation and development of human granulosa cells [50–54].

Of importance is the fact that there is a seasonal variation in serum AMH (being 18% lower in the winter than in the summer) that correlated with changes in seasonal serum 25OH-D. Also, 25-dihydroxyvitamin D3 supplements were sufficient to block the seasonal changes in both 25OH-D and AMH levels [55].

3.1.2. Vitamin D and steroidogenesis

All sex hormones are derived from cholesterol as the common precursor, which can be obtained through dietary sources or synthesized *de novo* from acetyl CoA. Sex hormone production is controlled by multiple enzymes. There are growing literature body suggesting that vitamin D affects the expression and activity of some of these enzymes. For example, it is reported that the treatment of human granulosa cells with 1,25-dihydroxy vitamin D3 *in vitro* increased progesterone production in the presence of the precursor substrate pregnenolone [54]. Also, it has been shown that vitamin D increased progesterone, estrogen, estrone, and insulin-like growth factor-binding protein 1 production in human ovarian cells. Moreover, 1,25-dihydroxyvitamin D3 stimulated estrogen and progesterone production in human placenta [56].

However, two of these steroids are explained particularly as follows:

- *17β-hydroxy steroid dehydrogenase (17β-HSD):* the biological active form of androgens and estrogens are biologically active in their 17β-hydroxy configuration. As well, 17-oxo derivatives are not capable to bind to their receptors. The ribozymes is the convertor enzyme which is one of the 17β-hydroxy steroid dehydrogenase (17β-HSD) families. These isozymes modulate intracellular level of steroid hormones in target tissues [57]. However, vitamin D application *in vitro* increased 3β-HSD mRNA levels and progesterone production. These suggest that vitamin D may play a role in enhancing certain key steroidogenic enzymes such as 3b-HSD. During the normal menstrual cycle, luteinized human granulosa cells usually form the corpus luteum which produces large amounts of progesterone (and some estrogens) and induces endometrial changes such as decidualization to support a pregnancy. Literature suggests that 1,25-dihydroxyvitamin D3 may potentiate granulosa cell luteinization as reflected by increased progesterone production, thus providing a better endometrial environment. Whether this is clinically relevant still needs to be determined *in vivo* [50].
- *Aromatase:* Aromatase is an estrogen synthetize, which catalyzes estrogen biosynthesis from androgen precursors. Aromatase is found in several tissues including the ovaries, liver, breasts, brain, and adipose tissue [58–60].

3.1.3. Vitamin D and ovarian reserve

According to recent evidences, in the serum the positive correlation exists between circulating vitamin D and ovarian reserve markers, particularly anti-Müllerian hormone as the best predictor of ovarian reserve.

Gonadal-specific glycoprotein of AMH is a kind of transforming growth factor (TGF) superfamily. In men, sertoli cells produce the AMH during male fetal sex differentiation which stimulates the regression of the Müllerian ducts [61]. In women, AMH is secreted by granulosa cells in growing primary and prenatal follicles but does not produce until near birth. Slight variations in AMH concentration during the menstrual cycle and its unique secretion by growing ovarian follicles make it a suitable predictive indicator for assisted reproductive technology (ART). However, some studies have been reported that environmental factors including vitamin D deficiency may change its expression and serum concentration [54, 62, 63]. The fact that vitamin D supplementation prevented the seasonal changes in serum AMH strongly indicates that AMH production in adults may be regulated by vitamin D. Thus, the assessment of vitamin D status, theoretically, might be considered as part of the routine workup in infertile women. Additionally, appropriate supplementation of patients with vitamin D deficiency might translate to better ovarian reserve markers and better ovarian follicular dynamics. However, most of the studies to date used markers of ovarian reserve/function rather than pregnancy as an outcome, which limits the translational significance of the findings [50].

3.2. Vitamin D and fertility

A seasonal distribution in human natural conception and birth rates has been consistently demonstrated, showing a peak conception rate during summer in northern countries with strong seasonal contrast in luminosity [64].

Some studies have demonstrated that the low level of vitamin D leads to a 75% decrease in fertility of female rats which is associated with 50% reduction in fecundation and enhancement of the probability of complications during pregnancy [65]. Also, vitamin D deficiency may lead to uterine hypoplasia and impaired folliculogenesis [66]. In this respect, vitamin D modulates estrogen biosynthesis through the maintenance of calcium homeostasis [65, 67]. It is shown that infertility was a secondary consequence of the low level of calcium rather than a direct result of the non-functional VDR [67, 68]. The indirect consequence of vitamin D deficiency on fertility through the regulation of calcium level in reproductive organ was also shown by literature in diet-stimulated vitamin D–deficient animals, in which both vitamin D and a diet supplemented with high levels of calcium repaired fertility. As such, some studies have been reported that the low level of vitamin D itself and not hypocalcemia is responsible for subfertility in vitamin D–deficient rats, exposed to different levels of serum calcium and phosphorus [68, 69]. Different studies have examined the role of vitamin D in a spectrum of female reproductive system disorders, such as adverse effect on pregnancy, endometriosis, and subfertility treated by IVF and PCOS.

3.3. Vitamin D and pregnancy: adaptations and metabolism during gestation

During pregnancy and lactation, there is an increase in the rate of synthesis and plasma levels of active form of vitamin D, which presumably functions to increase the intestinal absorption of calcium and the mobilization of maternal bone. The human embryos consume 30 g of calcium. More than 99% of this calcium is contained within the skeleton. Nearly 150 mg/kg/day of this calcium is transferred by placenta during the last trimester [70]. Serum protein-bound and complexed fraction calcium levels decrease during pregnancy due to the decrease in serum albumin. This physiological decrease is not an evidence of real hypocalcemia. The physiologically active form of calcium, ionized calcium, does not change during pregnancy. Parathyroid hormone decreases to the lower limit of the normal range and can become undetectable. Serum levels of other hormones potentially regulate the calcium including estradiol, prolactin, human placental lactogen (hPL), and parathyroid hormone-related protein (PTHrP), and they increase during pregnancy [70]. Doubling the amount of intestinal calcium absorption starting early in pregnancy seems to meet the fetal requirement for calcium. Skeletal resorption can also provide mineral to the blood, but evidence is controversial on whether the maternal skeleton contributes considerable amounts of calcium to the fetus. In this respect, bone resorption indicators are enhanced during pregnancy. The maternal kidneys do not reclaim calcium avidly during pregnancy; instead, urinary calcium excretion increases in parallel with the increase in intestinal calcium absorption [70]. Ionized calcium levels are stable until third trimester of pregnancy. PTH is decreased early in pregnancy but can increase in the third trimester of pregnancy. Skeletal mineral enhances in early pregnancy. It is well known that vitamin D deficiency is prevalent among pregnant women. Decrease of plasma vitamin D could contribute to the reduction in plasma calcium level during pregnancy and may result from increased maternal metabolism or enhanced utilization of vitamin D by the fetus [71]. Moreover, maternal low level of vitamin D might be independently correlated with an increased risk for gestational diabetes mellitus (GDM), preeclampsia, and small-for-gestational age (SGS) births [72–74] as well as with offspring rickets [75]. 25-hydroxyvitamin D [25(OH)D], the storage form of vitamin D, easily could pass from placentas in rats [76] and probably crosses the hemochorial human placenta easily. As well, the cord blood 25(OH)D levels are similar to or up to 20% lower than maternal level [77]. Thus, neonates with an adult level of normal 25(OH)D, their mothers have the sufficient level of vitamin D. Maternal transient of 25(OH)D to fetus could decrease maternal levels, especially if the mother has normal and adequate vitamin D concentration, whereas some studies have been demonstrated that either no change or a modest decrease in maternal 25(OH)D levels during pregnancy. The low fetal level of 1,25(OH)2D shows the low fetal PTH and high phosphorus level, which suppress renal 1α-hydroxylase. Although PTHrP is increased in the fetal blood circulation, it seems to be less able to induce the renal 1α-hydroxylase than PTH [78, 79].

The total serum level of 1,25(OH)2D doubled or tripled in the maternal circulation starting in the first trimester of pregnancy, but studies have only shown increased free concentrations during the last trimester. This elevation is due to maternal synthesis by the renal 1α-hydroxylase.

In addition, intestinal calcium absorption doubles in humans and rodents early in pregnancy, well before free 1,25(OH)2D concentrations increase late in pregnancy [80].

3.4. Vitamin D and maternal outcomes in pregnancy

It is shown that during pregnancy, vitamin D deficiency has been related to increased risks of adverse pregnancy outcome including gestational diabetes, recurrent pregnancy loss (RPL), preeclampsia, and small-for-gestational-age babies (**Figure 2**).

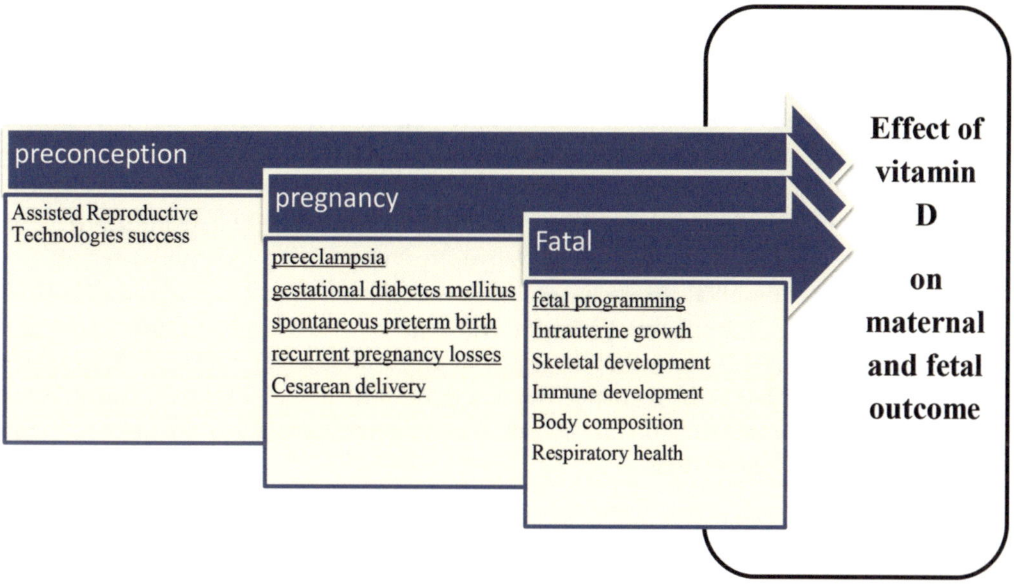

Figure 2. Serum level of vitamin D and pregnancy outcome.

3.4.1. Maternal plasma vitamin D levels and preeclampsia

Preeclampsia is a pregnancy-induced multi-systemic problem characterized by *de novo*-onset hypertension and proteinuria after 20 weeks of gestation which is prevalent in 2–8% of all pregnancies. It is one of the major acute and long-term health risks for maternal and perinatal mortality and morbidities [81, 82]. The underlying etiologies of preeclampsia are not completely understood. It has been hypothesized that abnormal trophoblastic invasion, oxidative stress, inflammatory responses, and endothelial dysfunction are possible contributing factors [83]. Maternal low level of vitamin D is so prevalent during pregnancy and is a kind of worldwide public health problem [84, 85]. As we state before, vitamin D effects on placental function and inflammatory response [86]. Recently, epidemiological studies have demonstrated an association between low maternal vitamin D status during pregnancy and the incidence of preeclampsia and suggest that vitamin deficiency may be an independent risk factor for preeclampsia [74, 87].

However, the underlying mechanisms remain unknown. Vitamin D deficiency is associated with inflammation-linked vascular endothelial dysfunction. Proinflammatory cytokines such as tumor necrosis factor-a, interleukin-6, and interferon-c have been reported to be increased in pregnancies with vitamin D deficiency. The molecular mechanisms involving hypovitaminosis D in endothelial dysfunction might be regulated, in part, by proinflammatory transcription factor nuclear factor-κB (NF-κB), as a major proinflammatory nuclear transcription factor, and interleukin-6, low VDR, 1α-hydroxylase, and hypocalcemia [88]. The endothelial cell expression of NF-κB and interleukin-6 and downregulation of NF-κB were more in the low level of vitamin D. Interleukin-6 expression in endothelial cells was strong negatively associated with 25(OH)D [89].

The levels of other circulating proinflammatory cytokines, including tumor necrosis factor-a33and C-reactive protein (CRP), are inversely associated with serum level of 25(OH)D [90]. Oxidative stress is elevated in vitamin D deficiency. Vitamin D supplementation could reverse this effect. High levels of thiobarbituric-acid-reactive substances, which indicate lipid peroxidation, have been shown in women with low level of vitamin D [91]. Endothelial cell damage or dysfunction appears to be a basic pathophysiological event of the maternal vascular system in women with preeclampsia [89, 92]. Maternal low level of vitamin D may influence a pro-inflammatory response, enhancement of oxidative stress, and lead to endothelial dysfunction and finally preeclampsia [93]. As such, some evidence demonstrates that vitamin D affects the genes responsible for trophoblast invasion and angiogenesis critical for implantation that appears to be implicated in the pathophysiology of preeclampsia [87]. Although calcium and vitamin D supplementation in pregnancy was associated with a significant reduction in blood pressure, the effect of intervention on the incidence of preeclampsia is controversial. In this respect, more research is needed.

3.4.2. Vitamin D and small-for-gestational age

Fetal growth restriction, most often estimated by the incidence of a birth weight that is small-for-gestational age, is a major public health issue across the globe. Infants suffering from SGA have a higher risk for serious neonatal morbidities and mortalities among infancy into adulthood. Growth restriction is related to a wide range of maternal factors including nutritional status, obesity, age, smoking, and infection, although there are insufficient effective interventions for prevention [94]. Several observational studies have linked maternal 25(OH)D concentrations and the risk of SGA in general obstetric populations [95]. It has been shown that maternal second trimester vitamin D status was inversely associated with the risk of SGA in singleton pregnancies.

The biologic mechanism that may connect maternal vitamin D status to fetal growth remains elusive. A plausible mechanism for the impact of maternal vitamin D on fetal growth is placental vascularization, which has received considerable attention in its association with fetal growth [96, 97]. Vitamin D has several biologically possible roles in fetal growth. In this respect, the vitamin D–activating enzyme CYP27B1 and VDR are expressed in human placenta [98]. The active form of vitamin D, 1,25-dihydroxyvitamin D, which acts through the VDR and the cAMP/protein kinase A (PKA)-signaling pathway, modulates human chorionic gonadotropin production in human syncytiotrophoblast and enhances placental sex steroid secretion. Vitamin D is also essential in glucose and insulin metabolism in glucose availability for transplacental transport and fetal use. As a regulator of calcium homeostasis and transport, calcitriol also can effect on fetal growth directly due to impacts on skeletal muscle and bone growth and development [56, 99].

In addition, several observational studies have connected poor vitamin D status with a higher risk of preeclampsia, which, like fetal growth restriction, has placental origins related to angiogenesis and uterine blood flow [97]. Vitamin D deficiency makes labyrinth of placental vessels narrower, indicating dysregulated vascularization [100].

Also, it is suggested that fetal VDR gene may play a role in the regulation of fetal growth. It is observed that sequence variation in the VDR gene modified the effect of maternal vitamin D deficiency on infant size at birth. Low 25(OH)D concentrations were associated with lower-birth-weight infants only among infants that were either homozygous for the *Fok*I major allele or heterozygotes [95].

More basic science research is needed in this area as well as studies with multiple measurements of fetal growth and placental vascularization.

3.4.3. Vitamin D and gestational diabetes mellitus

Gestational diabetes mellitus is one of the most prevalent disorders which have long-term consequence for the health of mothers and their children. It can increase the risk of developing type 2 diabetes, while their children may be at risk of obesity and diabetes later in life [101].

Polymorphisms of vitamin D have been related to insulin release and glucose tolerance [102]. A genetic influence of CYP27B1 polymorphisms may modulate 25(OH)D3 concentration in gestational diabetes patients [103]. Recent evidence from meta-analysis indicated a significant inverse relation of serum 25OHD and the incidence of GDM [104]. It is reported that 25(OH) D3 levels of <50 nmol/l at 16 weeks of gestation before the onset of GDM were associated with a 2.7-fold increased risk for the development of GDM later in pregnancy independent of measured confounders [72]. There are several mechanisms that explain the association between vitamin D deficiency and gestational diabetes risk: (i) vitamin D could directly or indirectly regulate pancreatic β-cell function and production by binding its circulating active form, 1,25-(OH)D, to β-cell vitamin D receptor and controlling the balance between the extracellular and intracellular β-cell calcium pools [105, 106]. (ii) Vitamin D can stimulate insulin sensitivity by inducing the expression of insulin receptors and increasing insulin responsiveness for glucose transportation. It also controls extracellular calcium to ensure normal calcium entry through cell membranes and a sufficient intracellular cytosolic calcium pool, which is crucial for insulin-mediated intracellular processes in insulin-responsive organs [72]. Finally, (iii) it is probable that the negative association of 25-[OH] D concentration with GDM risk shows the impact of other components of major endogenous and exogenous sources of vitamin D on glucose homeostasis due to other mechanisms. For instance, endogenous secretion of vitamin D in the skin with the sun exposure is a main source of plasma vitamin D. Sun exposure could be positively associated with outside home physical activity, a protective factor for insulin resistance, impaired glucose tolerance, and GDM [107].

However, randomized controlled trials (RCTs) of vitamin D supplementation, initiated early in pregnancy, are now required to demonstrate whether vitamin D supplementation might reduce the incidence or severity of GDM.

3.4.4. Vitamin D and spontaneous preterm birth

Spontaneous preterm birth (SPB) happens before 37 weeks of gestation. Intrauterine infection and inflammation is one of the main factors underlying this disorder. One important factor is bacterial vaginosis, which could disturb the normal balance of vaginal flora with enhanced growth of anaerobic bacteria responsible for the secretion of inflammatory cytokines, prostaglandins, and phospholipase A2 [108]. In this respect, studies have shown a linear inverse association between maternal vitamin D status and the prevalence of bacterial vaginosis among pregnant women [109–111]. Vitamin D has immunomodulatory and anti-inflammatory effects, including the control of the secretion and function of cytokines and neutrophil degranulation products that is important and relevant to prevent microbial invasion which may have a protective effect on SPB risk [27, 112]. Several cells of the immune system express VDRs and are regulated by vitamin D [113]. Although vitamin D function adjusts the activation of the acquired immune system in response to autoimmunity, it has key role to increase the innate immune system. It is involved in cell-mediated immunity by decreasing the secretion of inflammatory cytokines including IL-1, 6 and TNF-α that are involved in SPB [114, 115].

Human decidual cells are capable to synthesize the active form of vitamin D. Therefore, some studies demonstrated that vitamin D is involved in the modulation of acquired and innate immune responses at the fetal-maternal interface across gestation [116]. Vitamin D might decrease the risk of SPB also by helping to maintain myometrial quiescence. Myometrial contractility is related to calcium within the muscle cell and this process is manipulated by vitamin D [117]. The prevalence of SPB was lowest among women who conceived in summer and fall and was highest among winter and spring conceptions [118] and vitamin D supplementation in early pregnancy may protect against preterm birth [119]. More large studies are awaited to validate these important findings that might represent vitamin D supplementation as a simple and inexpensive method to reduce the risk of this adverse pregnancy outcome.

3.4.5. Vitamin D and recurrent pregnancy losses

Recurrent pregnancy loss is a devastating reproductive problem affecting approximately 5% of couples trying to conceive [120, 121]. RPL is typically defined as two or three or more consecutive pregnancy losses. Genetic, hormonal, metabolic, uterine anatomical, infectious, environmental, occupational and personal habits, thrombophilia, or immune disorders were reported as possible etiologies [121]. Despite the many etiologies, a majority of women with recurrent miscarriage have no discernible cause. It has been postulated that immunologic aberrations may be the cause in many of such cases.

Tissue responses to vitamin D include the regulation of hormone secretion, the modulation of immune responses, and a control of cellular proliferation and differentiation [122]. Vitamin D could inhibit the proliferation of T helper 1 (Th1) cells and limit the secretion of cytokines, such as interferon gamma (IFN-g), interleukin-2 (IL-2), and tumor necrosis factor-alpha (TNF-a). Also, vitamin D stimulates T helper 2 (Th2) cytokines, such as IL-4, IL-5, IL-6, IL-9, IL-10, and IL-13 [123]. Furthermore, in many studies vitamin D has been presented as a modifiable environmental factor for Th1-mediated autoimmune disease and appears to be important for susceptibility to and severity of the disease. Vitamin D also regulates B-cell immunity. It downregulates the proliferation and differentiation of B lymphocytes and inhibits IgG production [123].

With these immune-regulatory effects of vitamin D, it has been hypothesized that vitamin D could regulate immune response during implantation. In early pregnancy, trophoblasts secrete and respond to vitamin D, which influences local anti-inflammatory responses and stimulates decidualization for successful pregnancy [124]. A dominant Th2 immune response is important to maintain maternal-fetal relationship for successful pregnancy. By contrast, autoimmunity and dysregulated cellular immune reactions may be responsible for immunological alterations leading to recurrent pregnancy losses (RPL). High proportion of women with RPL has vitamin D deficiency, which is associated with increased cellular and autoimmunity. Women with RPL have increased prevalence of various autoantibodies, such as APA, ANA, and TPO antibody [124]. Vitamin D was shown to prevent autoimmune thyroiditis by inhibiting lymphocyte proliferation and secretion of inflammatory cytokines [125].

Low vitamin D appears to be important for autoimmune disease susceptibility and severity and vitamin D deficiency was associated with an increased presence of autoantibodies [126] via B-cell hyperactivation and autoantibody production [127]. It can be inferred that vitamin D plays a role in regulating B-cell proliferation and function during successful pregnancy.

Several studies have reported a link between RPL and altered cytotoxicity and level of peripheral natural killer (NK) cells [128]. Preconception evaluation of NK cell activity in women with RPL has been reported to predict pregnancy outcome of the subsequent pregnancy [128]. Furthermore, elevated peripheral NK cells in pregnant women predict spontaneous abortions with normal karyotype in index pregnancy. Recently, increasing evidence supports a novel immune-regulatory role of vitamin D [129]. Vitamin D and NK cytotoxicity seem to have a direct inverse relationship, which might associate with RPL. Vitamin D may be regulating the NK cell population and cytotoxicity. Elevation of NK cell and/or cytotoxicity is risk factor for abortion.

Briefly, the high prevalence of hypovitaminosis D was detected among women with RPL. Vitamin D deficiency has immunological function in RPL. Vitamin D is related to B- and NK-cell immunity and Th1/Th2 balance, and vitamin D deficiency leads to a tendency to develop APA and other autoantibodies, which are associated with autoimmune disease and adverse reproductive outcome. Therefore, the assessment of vitamin D status is very important in women with RPL and autoimmune or cellular immune abnormalities. As such, vitamin D–deficient women have significantly increased risk for autoimmune abnormalities that is a risk factor for RPL and infertility. NK-cell cytotoxicity and Th1 polarization were significantly decreased *in vitro* by vitamin D, and vitamin D decreased perforin production and polarization in NK cells. Also, vitamin D suppressed type 1 cytokine secretion and increased type 2 and growth factors from NK cells. So, these results raise the likelihood that vitamin D could be available as a new therapeutic choice for RPL and infertility [124]. Further study is required to elucidate immune-regulatory function of vitamin D.

3.4.6. *Vitamin D and mode of delivery*

Serum calcium status, which is regulated by vitamin D, plays a role in smooth muscle function in early labor [130]. It was speculated that the higher serum calcium levels played a role in the mechanism of the initiation of labor [130].

An inverse association with having a primary cesarean section and vitamin D deficiency is shown. Severe vitamin D–deficient women with levels of 25(OH)D3 of <37.5 nmol/l delivered nearly four times as often by cesarean section than those with 37.5 nmol/l or greater (odds ratio (OR): 3.84) [131]. Because vitamin D is essential for the maintenance of calcium homeostasis, it is possible that the low level of vitamin D, which leads to modest lowering of the serum calcium, is associated with both skeletal muscle and smooth muscle strength and may have a role in the initiation of early labor. Poor maternal vitamin D status might reduce the strength of the pelvic musculature and the mother's ability to push and deliver vaginally [132]. Also, vitamin D status is associated with preeclampsia [123] and gestational diabetes [133], which may increase the odds of cesarean [134]. It is also suggested that the low level of vitamin D is associated with cesareans due to cephalopelvic disproportion or failure to progress, although there is some controversy.

However, it should be noted that intravenous hydration would have diluted the blood and artificially imitate the lower level of 25(OH) D. However, the volume of intravenous fluids associated with blood loss is essentially the same.

However, vitamin D status and the mode of delivery should be examined and therefore the issue needs further investigation.

3.4.7. *Vitamin D and fetal programming*

Vitamin D stimulates more than 3000 genes. Several of them play a role in fetal growth and development [135]. It might be possible that vitamin D may be specifically relevant to the fetal programming indicating that vitamin D as an environmental factor may influence the genomic programming of fetal and neonatal development and following disease risk in childhood and adult life [136]. In this respect, in later life, mother and child who suffered from vitamin D deficiency during pregnancy suffer more often from chronic diseases such as wheezing and asthma [137, 138], schizophrenia [139], multiple sclerosis [140], type 1 diabetes mellitus, and insulin resistance [141, 142]. Mechanisms underlying this long-term effect of the intrauterine environment are not known [143–145] yet, but epigenetic mechanisms that lead to persistent changes in structure and function in endocrine systems are hypothesized [101].

3.4.8. *Vitamin D and infertility*

Infertility is a complex disorder with significant medical, psychosocial, and economic aspects, which affects about 15% of couples [146]. In accordance with preview findings, vitamin D deficiency has emerged as a factor that influences female infertility. The role of vitamin D in reproduction processes and its significance in infertility therapy covering topics of polycystic ovary syndrome, endometriosis infertility, myoma infertility, male infertility, premature ovary failure, and *in vitro* fertilization (IVF) techniques will be discussed.

3.4.9. *Vitamin D and polycystic ovary syndrome*

PCOS is one of the most common female endocrinopathies in reproductive-aged women [147, 148] which is characterized by elevated ovarian and adrenal androgen production,

http://dx.doi.org/10.5772/67394

hyperandrogenic symptoms such as hirsutism, acne and/or alopecia, irregular menstruation, and polycystic ovaries morphology [149]. Women with PCOS typically produce an increased number of oocytes, often of poor quality, resulting in lower fertilization and implantation and higher miscarriage rate. In addition, insulin resistance is common in PCOS women who are therefore at an increased risk of type 2 diabetes [150, 151]. PCOS is the most common cause of anovulatory infertility in women [152].

There might be a relationship between vitamin D deficiency and PCOS phenotype. In this respect, several studies have demonstrated that vitamin D deficiency is more common in women with PCOS compared with control women [153, 154]. Also, vitamin D deficiency might be a contributing factor to insulin resistance, obesity, and metabolic syndrome, all of which are commonly observed in PCOS and associated with ovulatory dysfunction [151]. Interestingly, vitamin D supplementation might improve menstrual irregularity, follicular development, and pregnancy rate in women with PCOS [155, 156]. The mechanisms underlying the association of low 25(OH)D levels with PCOS are not fully understood. It is briefly discussed in the subsequent text:

- *Vitamin D and insulin resistance and obesity*: There is some evidence suggesting that vitamin D deficiency might be involved in the pathogenesis of insulin resistance and the metabolic syndrome in PCOS [157]. Several studies have shown that there is an inverse correlation between 25OH-D and IR, obesity, and free androgen index [154, 158]. Additionally, several studies have shown that vitamin D supplementation might improve IR and reduce serum androgens [159]. The mechanisms underlying the association of low 25(OH)D levels and insulin resistance are unclear. As obesity is related to insulin resistance in PCOS [160], it may contribute to low circulating vitamin D levels by trapping vitamin D in fat tissues. There are some mechanisms explaining the correlation between low level of vitamin D and insulin resistance. Vitamin D may have a positive influence on insulin action by inducing the expression of insulin receptors and so increasing insulin responsiveness for glucose transport [161]. In addition, vitamin D modulate extracellular and intracellular calcium which is necessary for insulin-mediated intracellular processes in insulin-responsive organ including skeletal muscle and adipose tissue. Moreover, changes in calcium flux can have negative effects on insulin production, which is a calcium-dependent process.
- *Vitamin D and sRAGE*: Advanced glycation end products (AGEs) have been shown to be involved in the pathogenesis of PCOS, and their serum levels are elevated in women with PCOS. AGEs accumulate in ovarian theca and granulosa layers of women with PCOS. This accumulation may be implicated in worsening ovarian follicular growth [162, 163]. However, significant increase in serum 25OH-D following replacement was associated with a significant increase in receptor for advanced glycation end product (sRAGE) levels, and a significant decrease in the abnormally elevated serum AMH levels that are usually observed in PCOS. The increase in sRAGE is usually beneficial because it binds circulating AGEs and inhibits their inflammatory deleterious effects. Lower serum AMH level in PCOS might potentially improve the ovulatory process because it decreases intrafollicular androgens and increases follicular sensitivity to FSH [164, 165].

- *Vitamin D receptor polymorphism and PCOS*: Vitamin D receptors modulate more than 3% of the human genome, including genes that are fundamental for glucose metabolism. In this atmosphere, it has been reported that VDR-related polymorphisms (Cdx2, Bsm-I, Fok-I, Apa-I, and Taq-I) are associated with vitamin D metabolism and might participate to PCOS susceptibility [166, 167]. It seems possible that variants in the VDR through their effect on luteinizing hormone, sex hormone-binding globulin (SHBG) levels, and testosterone are involved in the pathogenesis of PCOS.
- *Vitamin D and gene product of PCOS*: Phosphoprotein enriched in diabetes gene product (PED/PEA-15), an antiapoptotic protein, has been shown to be overexpressed in insulin resistance, DM type 2, and PCOS. Recent data suggested that the low level of vitamin D may elevate the serum levels of this antiapoptotic protein, contributing to the impairment of the ovarian apoptotic mechanism. In addition, the low level of adiponectin that is present in PCOS has been related to vitamin D concentrations, due to body mass index (BMI)-dependent mechanisms. Further genes involved in vitamin D synthesis, hydroxylation, and transport, and their role in PCOS are currently under investigation [168–170].

3.5. Vitamin D and uterine leiomyoma

Leiomyoma (fibroids) are benign tumors that develop in the uterine muscle of premenopausal women. The most common symptoms are pain and bleeding with associated anemia [171]. Although fibroids are hormonally dependent, factors that stimulate development are largely unknown. Vitamin D status has recently been related to the development of uterine leiomyomas, with observations showing that lower 25(OH)D levels correlate with a higher risk and a greater volume of uterine [171, 172]. Recent studies showed that both myometrial and leiomyoma cells are highly sensitive to the regulatory effect of 1,25-dihydroxyvitamin D3 [173]. The signaling of 1,25(OH)2D3 is mediated via its ubiquitously expressed nuclear receptor, the vitamin D receptor, which is expressed in both the myometrium and endometrium of the human uterus throughout the menstrual cycle [174].

The pathogenesis of fibroids has been hypothesized to involve a positive feedback loop between extracellular matrix production and cell proliferation, and vitamin D might act to block the positive feedback [175]. Vitamin D deficiency may stimulate cell proliferation [176]. The vitamin D [1,25(OH)2D3]-induced antiproliferative action is mediated predominantly through a G1/S phase block of the cell cycle. Because 1,25(OH)2D3 regulates many of the cell cycle–regulatory genes and reduces or increases the kinase activities of cyclin-dependent kinases (CDKs), this results in a decreased number of cells in the S phase and an accumulation of cells in the G0–G1 phase [177]. The cyclin-dependent kinase inhibitors p21 and/or p27 are genomic targets of the 1,25(OH)2D3-VDR complex in many cell types. Also, 1,25(OH)2D3 blocks mitogenic signaling, including that of estrogen, epidermal growth factor (EGF), and insulin-like growth factor 1, and upregulates growth inhibitors such as transforming growth factor β (TGF-β) [178]. In addition, 1,25(OH)2D3 activates VDR-mediated apoptosis [179].

Myometrial and leiomyoma cells are clearly target cells of 1,25(OH)2D3. The data are consistent with the observed expression of VDR protein in the myometrial and leiomyoma tissues and cultivated cells and with our previous description of VDR mRNA expression in myometrial biopsies [180]. The punctuate pattern of expression within the nuclei of cultured cells has also been observed in other cell types and may display specific binding sites of VDR to target genes [181].

More research is needed to find out whether women with hypovitaminosis D also have more uterine leiomyomas than women with efficient vitamin D supplies.

3.6. Vitamin D and endometriosis

Endometriosis is one of the estrogen-dependent inflammatory problems characterized by the expression of endometrial tissue outside the uterine related to chronic pelvic pain and subfertility. The prevalence of this disorder is 10% of all women and 40% of infertile women. Although endometriosis is not a malignant disorder, disturbances in cellular proliferation, cellular migration, cellular invasion, and neoangiogenesis are common [182]. Endometriosis is dependent on a following complex interaction of immunologic, hormonal, genetic, and environmental factors; however, the etiology of endometriosis is not completely understood [182].

It is documented that the regulatory network of vitamin is involved in the pathogenesis of endometriosis [69]. The higher 25(OH)D levels in women with endometriosis are detected. The proposed associations of vitamin D status and endometriosis are as follows:

I: it has been shown that the VDR and 1a-hydroxylase are expressed in the endometrium [24], suggesting that endometrium is an extra renal site of vitamin D synthesis and vitamin D action which leads to overexpressing of them [183]. It has been shown that VDR and 1α-hydroxylase are expressed in both the orthotopic and ectopic endometria [183].

II: Genetic variation in the VDR could be involved as a potential link between the vitamin D–regulatory network and endometriosis pathogenesis. VDR polymorphism has been investigated as a potential link between vitamin D–regulatory network and endometriosis pathogenesis. It was indicated that VDR dysregulation compromises innate immune response, involving VDR in the pathogenesis of endometriosis. DNA methylation and transcriptional repression signaling have been suggested as the most affected pathways involving VDR dysregulation in women with endometriosis [69]. In this respect, the expressions were the highest in the endometria of women with endometriosis involving the epigenome of steroid hormone response in the pathogenesis of the disease, and also VDRmRNA expression has the upregulation of VDRmRNA expression in the ovarian tissue of patients with endometriosis [184]. DNA methylation and transcriptional repression signaling have been suggested as the most affected pathways involving VDR dysregulation in women with endometriosis, involving the epigenome of steroid hormone response in the pathogenesis of the disease.

III: Vitamin D is involved in the regulation of the immune system, which may be speculated about an influence of vitamin D in the local immune suppression and development of endometriosis. This finding may be explained by an influence of vitamin D on the local activity of immune cells and cytokines maintaining endometriosis and an insufficiency to activate macrophage's phagocytotic function in those carrying the GC*2 polymorphism.

However, the hypothesis of a beneficial effect of vitamin D supplementation in the treatment of patients with endometriosis has not yet been clinically tested.

3.7. Vitamin D and assisted reproductive technologies

Vitamin D has also been shown to be involved in the pathophysiology of some disorders of women of childbearing age that are most commonly encountered among women undergoing *in vitro* fertilization procedures [185].

The issue of whether vitamin D levels are reliable predictors of ART outcomes is still controversial. In some studies, among infertile women undergoing IVF, women with higher serum concentration of vitamin D and follicular fluid were significantly more likely to achieve clinical pregnancy following IVF, and also high serum vitamin D concentration was significantly related to improved parameters of ovarian hyperstimulation [186].

The mechanism by which vitamin D affects fertility is unclear. However, vitamin D acts through the endometrium to influence IVF success and is supported by biological evidence. Postulated mechanisms include its effect on ovarian steroidogenesis and implantation [186, 187]. In addition, vitamin D signaling is involved in the cross-talk between the embryo and endometrium; in response to interleukin (IL)-1B secreted by the blastocyst, endometrial dendritic cells and macrophages produce 1-alpha hydroxylase and calcitriol (the active form of vitamin D) [188]. Calcitriol binds to the vitamin D receptor in the endometrium to regulate target genes such as calbindin, osteopontin, and HOX10A, genes critical for embryo implantation and placentation [189]. Endometrial HOX10A expression parallels that of the vitamin D–signaling pathway; both increase mid-cycle shortly before expected implantation, at the time of maximal endometrial differentiation [189]. Vitamin D also has immunomodulatory effects that may contribute to implantation [190]. Calcitriol attenuates decidual T-cell function. Decidual natural killer cells treated with calcitriol show decreased synthesis of cytokines CSF2, IL1, IL6, and TNF. Calcitriol has also been shown to interfere with the production of cytokines in whole endometrial cells isolated from women with a history of recurrent miscarriage, leading some investigators to hypothesize that vitamin D could play a role in the treatment for recurrent miscarriage [191]. VDR and 1-alpha hydroxylase expression continue to increase in the first and second trimesters. In cultured syncytiotrophoblasts, calcitriol regulates hCG expression and secretion, and it stimulates E2 and P secretion from trophoblasts in a dose-dependent manner [99]. Abnormal expression of 1-alpha hydroxylase has been observed in pregnancies complicated by preeclampsia, suggesting that calcitriol may regulate placental development. Thus, the impact of vitamin D deficiency may extend to the entire placental-decidual unit.

Further research is needed to elucidate the mechanism by which vitamin D acts to influence IVF success and to determine whether repleting one's vitamin D stores will improve pregnancy rates.

http://dx.doi.org/10.5772/67394

3.8. Vitamin D and fetal outcomes

Although research into fetal origins of disease in later life remains in its infancy, there is increasing suspicion that gestational nutritional sufficiency may be a determinant of health in later life. Vitamin D deficiency has been related to various adverse maternal, fetal, and postnatal outcomes.

Recent research has focused on the role of gestational vitamin D status in modulating intrauterine growth, body composition, skeletal development, immune development, and respiratory health of the offspring:

- *Intrauterine growth*

It is shown that low vitamin D levels during pregnancy may account for reduced fetal growth and for altered neonatal development [192]. Variation in the maternal VDR gene polymorphisms contributes to vitamin D–related disparities in fetal growth [193]. Maternal VDR genotype was significantly and independently associated with the risk of SGA, with implicated SNPs differing in white and black women. In this respect, single VDR SNP (rs7975232) has association with birth weight. rs7975232, an anonymous polymorphism, is part of a VDR gene haplotype associated with variation in mRNA stability. mRNA stability can directly affect the amount of protein produced, thus directly affecting vitamin D levels and calcium homeostasis [194]. In early pregnancy, more than 300 genes were differentially expressed in women indicating a role of vitamin D in the genetic regulation of processes contributed in fetal development [195]. Further research identifying the functionality of VDR gene polymorphisms in pregnant women will improve our understanding of the underlying mechanisms influencing birth weight.

- *Body composition*

Increasing evidence that vitamin D affects cell development and differentiation in tissues including bone, muscle, and fat suggests that the in utero vitamin D environment may influence body composition and cardiovascular disease risk factors in the offspring [196]. In this respect, greater adiposity was found for men and women born in winter-spring, possibly reflecting fetal exposure to low vitamin D during the second or third trimester of pregnancy [196]. Vitamin D deficiency is also emerging as a risk factor for the metabolic syndrome in adults. The evidence supports an inverse relationship between serum 25OHD and components of the metabolic syndrome, including blood glucose concentration, insulin resistance, dyslipidemia, raised blood pressure, and abdominal obesity [197]. The highly active form of vitamin D, 1,25(OH)2D3, exerts a coordinated control over lipogenesis and lypolysis [198]. Given current concerns about childhood obesity and the increasing prevalence of vitamin D deficiency with urbanization, it is important to explore the hypothesis that maternal vitamin D level may affect later body size and composition.

- *Skeletal development*

Advances in bone assessment technology have prompted research on gestational vitamin D status and offspring skeletal development. It is reported that maternal serum 25(OH)D was inversely correlated with fetal femoral distal metaphyseal cross-sectional area and splaying index. Fetal femoral splaying is analogous to that seen in childhood rickets, suggesting that effects of vitamin D deficiency on bone development may initiate early in gestation [199]. Because bone size is related to bone strength, it is hypothesized that lasting differences in femoral distal metaphyseal cross-sectional area may have implications on future fracture risk [35].

- *Immune development*

The role of prenatal vitamin D status in fetal and neonatal immune development comprises a growing area of research. Cord blood gene expression of tolerogenic immunoglobulin-like transcripts 3 and 4 (ILT3 and ILT4) was significantly higher when mothers were supplemented with vitamin D during pregnancy [200]. Also, a researcher showed that there is a weak but significant positive correlation between cord plasma 25(OH)D and cord blood mononuclear cell release of IFN-ɣ, a cytokine that plays a key role in Th1 cell development, upon stimulation with lipopolysaccharide [200]. It suggests that prenatal vitamin D status could influence immune development and predisposition for allergy [201]. A recent body of work has begun to suggest that lower gestational vitamin D levels may also be associated with higher rates of pediatric atopic disease [202], food sensitivities [203], atopic dermatitis, eczema, asthma, impaired lung function, allergic disease, and other conditions frequently characterized by a hypersensitive immune state [204–206]. It appears that fetal vitamin D levels may play a modulating role in immune functions involved in atopic disorders. As hypersensitivity outcomes may also be seen in those children born to mothers contaminated with assorted xenobiotics in pregnancy [207, 208], however, it is not known whether the immune dysregulation and hypersensitivity may be the consequence of a primary gestational insufficiency of vitamin D, or whether various chemical toxicants might play a role by impairing vitamin D uptake, renal synthesis, and assimilation [209] while at the same time inducing immune compromise and hypersensitivity through other mechanisms [210].

- *Respiratory health*

Earlier findings of an inverse correlation between maternal intake of vitamin D during pregnancy and incidence of wheeze or asthma in the offspring [137] have raised interest in the role of vitamin D in childhood respiratory health. It is reported that maternal vitamin D deficiency increases the risk of both respiratory and general infections in the first 3 months of life, and also cord blood vitamin D status was inversely associated with wheeze during the first 5 years of life [211]. Similarly, infants born to mothers with a vitamin intake during pregnancy had significantly lower odds of developing wheeze or eczema [212]. It seems that vitamin D plays an essential role in myriad genes that encode for health and well-being in the offspring, it behooves the medical and public health community to endeavor to secure vitamin D adequacy in the gestational period.

3.9. Vitamin D and lactation

Near-exclusive breastfeeding for 6 months leads, on average, to maternal calcium loss four times higher than in pregnancy because lactation can require 150–300 mgCa/kg/day. Vitamin D goes across easily into breast milk, but 25(OH)D passes very poorly, and 1,25(OH)2D does not seem to pass at all [213]. 1,25(OH)2D level decreases quickly after pregnancy and are normal during lactation [213]. 25(OH)D concentrations were decreased during lactation [214]. In lactating rats and mice, 1,25(OH)2D levels remain elevated until weaning [213].

Studies have generally shown that providing vitamin D to lactating mothers increases their 25(OH)D concentrations but has no significant effect on any other maternal outcome [215, 216]. Animal studies showed that skeletal resorption prepared most of the calcium required during lactation, irrespective of dietary calcium intake. The obligatory increase in PTHrP and decrease in estradiol program the lactational loss of skeletal calcium content, and vitamin D status does not affect this loss. Increasing calcium and vitamin D intake during lactation might simply increase urinary calcium and, thereby, increase the kidney stone risk [70].

4. Vitamin D and menopause

During menopause, the estrogen deprivation results in elevated bone turnover, reduction in bone mineral density, and increase in the risk of fracture. Musculoskeletal discomfort may impair health-related quality of life. Moreover, body composition changes including increased fat mass and decreased lean mass, which may be related to elevated risk of VD deficiency. However, we discuss the adverse health outcomes related to both menopause and VD deficiency and the possible interaction of both risk factors in these conditions.

4.1. Vitamin D and vasomotor climacteric symptoms

Hot flashes are the most common menopausal symptom. Although their exact pathophysiological mechanism is unclear, estrogen deprivation is suggested to cause stimulation in noradrenergic hyperactivity, which leads to a heat loss response and the sensation of warmth throughout the body followed by sweats [217]. There are several lines of evidence indicating shared complications of women affected by vasomotor climacteric symptoms and vitamin D deficiency such as accelerated bone turnover, increased loss of bone mass, hypertension, and depression. Also, it is suggested that decline in serotonin, as a neurotransmitter with known effects on thermoregulation, is an alternative underlying mechanism in vasomotor climacteric symptoms. In this respect, vitamin D can protect against experimental serotonin depletion; one proposed mechanism for symptom alleviation is the prevention of serotonin decline in menopause [218]. RCTs investigating the effect of VD supplementation using adequate doses in peri- or early postmenopausal women are warranted.

4.2. Vitamin D, obesity, and menopause

Some studies demonstrated that menopause is related to obesity and changed body fat distribution. Obesity occurs because fat-free mass was lost after menopause, due to lesser exercise and greater increases in fat mass [219]. This could increase the risk of cardiometabolic disease, cancer, and consecutive mortality [220]. It has been related to estrogen effects on lipolysis and lipogenesis in visceral adipocytes [221] and the SHBG-lowering effect of estrogen deficiency resulting in elevated free testosterone levels. Hyperandrogenemia is related to visceral fat accumulation [221]. Although the serum level of estrogens and androgen levels decrease during menopause, but the more pronounced decrease in estrogen levels might result in increased visceral fat accumulation [219]. In addition, estrogen deprivation may also effect on energy balance, metabolic rate, fat oxidation, and total body weight [219]. It is well documented that obesity is associated with vitamin D deficiency. It is demonstrated that a higher degree of obesity leads to lower 25(OH)D, whereas any effects of lower 25(OH)D increasing BMI are likely to be small. Interestingly, vitamin D supplementation decreases body fat mass without any change in body weight or waist circumference [222]. Further, physical activity and thus sun exposure, which is essential for vitamin D production in the skin [223], decline during menopause [224].

4.3. Vitamin D, cardiovascular disease, and menopause

Cardiovascular diseases are not common among premenopausal women. Sex difference between cardiovascular outcomes in female and male may be associated with protective effects of endogenous estrogens. The estrogen deprivation during menopause may be related to the unfavorable changes of lipid and carbohydrate metabolism during menopause leading to the increased incidence of cardiovascular events [220]. There is large evidence from observational studies linking low vitamin D levels with cardiovascular risk factors as well as with cardiovascular events [225]. Vitamin D has been suggested to be involved in insulin resistance, type 2 diabetes, and the MetS in premenopausal as well as in postmenopausal women [225]. This association might in part be caused by the relation of hypovitaminosis D with obesity (**Figure 3**).

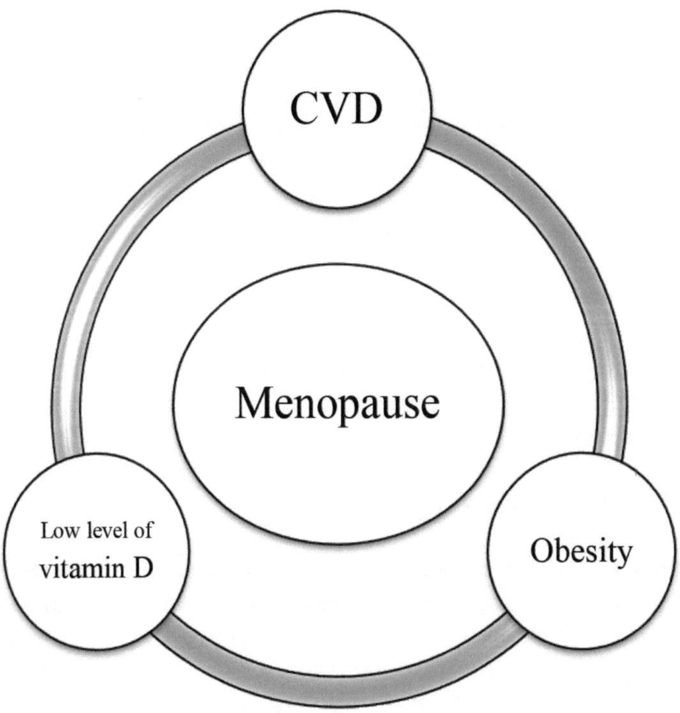

Figure 3. Menopause, low serum vitamin D, and cardiovascular diseases.

There are, however, mechanisms beyond obesity such as a beneficial VD effect on insulin action and VD-related genetic variants are associated with insulin resistance and insulin sensitivity [153]. Hypovitaminosis D has also been associated with hypercholesterolemia in several studies [226]. As low VD levels are associated with an unhealthy lifestyle such as few physical outdoor activities, a sedentary lifestyle, and obesity, it is difficult to interpret these findings.

4.4. Vitamin D, musculoskeletal disease, and menopause

Estrogen deprivation in menopause has been suggested as risk factor developing musculoskeletal symptoms such as aches and pain, joint pain, muscle stiffness, and skull and neck aching [227, 228]. Accelerated loss of bone mass occurs during the menopause as a result of naturally decreasing estrogen levels, putting women at risk of osteoporosis and fracture. Lifestyle advice including adequate calcium and VD intake, physical activity, encouraging nonsmoking, and only moderate alcohol consumption has been recommended for bone health in postmenopausal women [224]. It has been demonstrated that vitamin D has a dual effect on the musculoskeletal system: (i) on the bone mass, bone density, and bone quality and also (ii) on the muscle mass, muscle strength, and muscle function. In addition, adequate vitamin D status decreases the risk of falling in older individuals, due to improved neuromuscular function [225]. Hypovitaminosis D myopathy is a prominent symptom of vitamin D deficiency, and severely impaired muscle function may be present even before biochemical signs of bone disease develop [229]. There is evidence showing that hormones including VD as well as sex hormones modulate the functional relation between bone and muscle tissues. Also, there is evidence suggesting a relationship between osteoporosis, cardiovascular disease, and mortality, and low VD levels may be an underlying mechanism that are associated with increased bone turnover and increased risk of mortality and mortality [230]. Meanwhile, as both menopause and low level of vitamin D are related to musculoskeletal symptoms, it is suggested that vitamin D supplementation may benefit for joint pain, muscle mass, and function in peri- and postmenopausal women. Moreover, vitamin D supplementation might recover muscle function which leads to better bone mineral density and a lower risk of falling. Also, it has a positive effect on impaired cognitive function and depression, which may *per se* decrease the risk of falling [225]. Thus, randomized trials investigating vitamin D effects in peri- and early postmenopausal women on musculoskeletal symptoms and diseases are highly needed.

4.5. Vitamin D, cancer, and menopause

The incidence of cancer rises in women with increasing age. Besides other factors, this is aggravated by several lifestyle aspects such as reduced physical activity, a sedentary lifestyle, increased caloric intake, as well as obesity [224]. There is accumulating evidence suggesting that the low level of vitamin D is one of important risk factors for cancer and cancer-related mortality [231]. High 25(OH)D levels are related to reduction in mortality in patients suffering from colon, lung, and breast cancer [232, 233]. The underlying pathophysiology is not completely understood, but it is suggested that antiproliferative and apoptotic effects of vitamin D on cancer cells, inhibition of metastatic distribution and tumor invasion, and promotion of sensitivity to radiation and chemotherapy influence on decreased mortality in cancers [233]. Since the increased risk of hypovitaminosis D in peri- and early postmenopausal women as well as the fact that underlying risk factors are also associated with an increased risk of cancer, one might speculate that an adequate vitamin D levels in those women might also be helpful regarding the risk of cancer.

4.6. Vitamin D, menopause, and mood disorders

It is well documented that women are more at a higher risk compared to men who develop psychological problems such as mood disorders and depression, which has been related to the effect of estrogen fluctuation during reproductive cycle. As well, some studies reported an increased prevalence of depression and anxiety in women across the menopausal transition [234]. It has been suggested that vitamin D may affect mood and cerebral function and the low level of vitamin D is related to vascular neuropathology [225]. Increased phagocytosis of amyloid plaques, antioxidative effects, modulation of neurotrophins, neuronal calcium regulation, immunomodulation and vascular protection [235], and changes in calcium homeostasis [236] are observed. Vitamin D deficiency and the decline in estrogens during menopausal transition are conditions associated with an increased risk of mood disorders such as depression. In this respect, more research is needed.

5. Vitamin D and male fertility

There is extensive evidence demonstrating that calcium is essential in the male reproductive system, where it is crucial for spermatogenesis, sperm motility, and acrosome reaction [237]. In this respect, the role of vitamin D, as an important modulator of calcium metabolism, in semen quality and spermatogenesis is not completely understood.

The basis of the interplay between vitamin D and reproduction lays on the presence of both VDR and 1α-hydroxylase (CYP27B1) in various tissues of the reproductive system in both sexes. VDR expression has been shown in the testis of rat [238]. Human studies demonstrated that vitamin D receptors are found in testis, epididymis, prostate, seminal vesicles, and Leydig cells. The level of expression differs, which is higher in epididymis and seminal vesicles compared to others [239]. Vitamin D receptors were also expressed in normal and abnormal sperm [239]. Acrosomal region, head, especially the nucleus, and the neck of the sperm are the sites with the most numerous mRNA VDR expression [240].

The precise role of vitamin D receptors in the sperm nucleus is unclear. It has been shown that it has a protective genomic factor, which is essential for the proper control of sperm DNA integrity and maintenance of genome stability [240]. The 1,25 (OH) 2D3 molecules seem to regulate cholesterol efflux in human sperm, affect tyrosine and threonine phosphorylation of sperm proteins, and enhance sperm bioavailability. Further, it increases intracellular calcium levels, sperm mobility, and acrosin activity, and decreases triglyceride in sperm to contribute in fertilizing capacity within the female reproductive system [241]. Vitamin D receptors are also found in the cytoplasm of epithelial cells of the epididymis and ductal prostate epithelium [242].

Meanwhile, it is shown that the mRNA encoding CYP2R1 [243] and CYPB1 is presented in all tissues of the reproductive tract [243]. The exact role of CYPR1 is unclear, but it has been suggested that it is related to vitamin D functions, as its expression progressively reduces in testicular damage [243]. Vitamin D appears to be implicated in amino acid accumulation, which is achieved either through its genomic effect, triggered by protein kinase A and C (PKC), or by a rapid, nongenomic effect, contributing in calcium/potassium channels in the plasma membrane [244]. The cyclic AMP/PKA complex is a mediator of 1,25(OH)2D3 in both genomic and nongenomic actions. As such, 1,25(OH)2D3, membrane depolarization occurs, inducing the opening of L-calcium channels and entry of calcium [241]. However, in sertoli cells, vitamin D could induce calcium uptake through an unknown receptor activity [245]. Further, vitamin D acts on sertoli cells through chloride channel activation, which is mediated through a PKA/PKC-dependent, nongenomic pathway [246]. Vitamin D enhances gamma-glutamyl transpeptidase activity, an enzyme contributed in the synthesis of proteins produced by sertoli cells. Literature supported a protective effect of vitamin D from oxidative stress and cellular toxicity, as well as maintenance of the number and motility of sperm [247]. Lastly, it has been suggested that vitamin D induces the expression of calcium-binding protein CaBP28k in testis, which is contributed in the process of spermatogenesis and steroidogenesis [248].

Recent literature suggested that men suffering from severe hypospermatogenesis or idiopathic sertoli cell-only syndrome (SCOS), despite normal levels of total testosterone and estradiol, had lower plasma 25(OH)D concentrations, higher concentrations of bone resorption markers, and lower T-scores both in femoral neck and in lumbar spine compared to healthy controls [240, 249]. Researchers showed that there are positive correlation of 25(OH) D serum levels with sperm motility and progressive motility. Moreover, men with vitamin D deficiency (<10 ng/ml) had a lower proportion of motile, progressive motile, and morphologically normal spermatozoa [250]. Further investigations are needed to evaluate the positive role of vitamin D supplementation in men's infertility.

6. Vitamin D supplementation

There are no specific guidelines regarding vitamin D supplementation for women or men affected by endocrine disturbances. Thus, according to positive vitamin D effects on bone health, the Institute of Medicine [251] and the Endocrine Society [22] suggest a vitamin D level of at least 50 nmol/l (20 ng/ml). Based on the Recommended Daily Allowance (RDA, covering requirements of R97.5% of the population), the daily intake of vitamin D should be 600 IU/day for each person up >70 years and 800 IU/day for older adults. The Endocrine Practice Guidelines Committee [22] recommend a daily intake of 1500–2000 IU vitamin D3 daily for adults older than 18 years up to 70 years in order to raise the blood level of 25(OH) D to more than 30 ng/ml. It is documented that vitamin D supplementation with 1000 IU/day increases 25(OH)D levels/10 ng/ml [133]. However, in severe vitamin D deficiency, higher doses of vitamin D 50,000 IU weekly for up to 8 weeks are recommended. Notably, vitamin D intoxication, which leads to hypercalcemia, renal damage, and vascular calcification, occurred in 25(OH)D levels to more than 150 ng/ml [18]. Regarding several adverse effects of the low level of vitamin D on different health aspects, vitamin D supplementation in order to reach an adequate vitamin D level is highly recommended.

Author details

Fahimeh Ramezani Tehrani* and Samira Behboudi-Gandevani

*Address all correspondence to: ramezani@endocrine.ac.ir

Reproductive Endocrinology Research Center, Research Institute for Endocrine Sciences, Shahid Beheshti University of Medical Sciences, Tehran, Iran

References

[1] Vitamin D. Native, aktive und selektive Vitamin D-Präparate und-Rezeptor-Aktivatoren für Patienten mit Niereninsuffizienz. N Engl J Med. 2007;357:266–81.

[2] Wagner CL, Taylor SN, Dawodu A, Johnson DD, Hollis BW. Vitamin D and its role during pregnancy in attaining optimal health of mother and fetus. Nutrients. 2012;4(3):208–30.

[3] Harris HR, Chavarro JE, Malspeis S, Willett WC, Missmer SA. Dairy-food, calcium, magnesium, and vitamin D intake and endometriosis: a prospective cohort study. American Journal of Epidemiology. 2013;177(5):420–30.

[4] Jones G, Strugnell SA, DeLuca HF. Current understanding of the molecular actions of vitamin D. Physiological Reviews. 1998;78(4):1193–231.

[5] Henry HL. Regulation of vitamin D metabolism. Best Practice & Research Clinical Endocrinology & Metabolism. 2011;25(4):531–41.

[6] Chen TC, Wang L, Whitlatch LW, Flanagan JN, Holick MF. Prostatic 25-hydroxyvitamin D-1α-hydroxylase and its implication in prostate cancer. Journal of Cellular Biochemistry. 2003;88(2):315–22.

[7] Minghetti PP, Norman AW. 1,25 (OH) 2-vitamin D3 receptors: gene regulation and genetic circuitry. The FASEB Journal. 1988;2(15):3043–53.

[8] Kinuta K, Tanaka H, Moriwake T, Aya K, Kato S, Seino Y. Vitamin D is an important factor in estrogen biosynthesis of both female and male gonads 1. Endocrinology. 2000;141(4):1317–24.

[9] Heaney RP. Vitamin D in health and disease. Clinical Journal of the American Society of Nephrology. 2008;3(5):1535–41.

[10] Bouillon R, Carmeliet G, Verlinden L, van Etten E, Verstuyf A, Luderer HF, et al. Vitamin D and human health: lessons from vitamin D receptor null mice. Endocrine Reviews. 2008;29(6):726–76.

[11] Carlberg C. Molecular basis of the selective activity of vitamin D analogues. Journal of Cellular Biochemistry. 2003;88(2):274–81.

[12] Vieth R. Vitamin D supplementation, 25-hydroxyvitamin D concentrations, and safety. The American Journal of Clinical Nutrition. 1999;69(5):842–56.

[13] Haussler MR, Jurutka PW, Mizwicki M, Norman AW. Vitamin D receptor (VDR)-mediated actions of 1,25 (OH) 2 vitamin D 3: genomic and non-genomic mechanisms. Best Practice & Research Clinical Endocrinology & Metabolism. 25(4):543–59.

[14] Sugimoto H, Shiro Y. Diversity and substrate specificity in the structures of steroidogenic cytochrome P450 enzymes. Biological and Pharmaceutical Bulletin. 2012;35(6):818–23.

[15] F Holick M. Vitamin D: evolutionary, physiological and health perspectives. Current Drug Targets. 2011;12(1):4–18.

[16] Christakos S, Dhawan P, Liu Y, Peng X, Porta A. New insights into the mechanisms of vitamin D action. Journal of Cellular Biochemistry. 2003;88(4):695–705.

[17] DeLuca HF. Overview of general physiologic features and functions of vitamin D. The American Journal of Clinical Nutrition. 2004;80(6):1689S–96S.

[18] Holick MF. Vitamin D deficiency. New England Journal of Medicine. 2007;357(3):266–81.

[19] Suda T, Ueno Y, Fujii K, Shinki T. Vitamin D and bone. Journal of Cellular Biochemistry. 2003;88(2):259–66.

[20] Christakos S, Ajibade DV, Dhawan P, Fechner AJ, Mady LJ. Vitamin D: metabolism. Endocrinology Metabolism Clinics of North America. 2010; 39(2):243–53

[21] Liu N, Nguyen L, Chun RF, Lagishetty V, Ren S, Wu S, et al. Altered endocrine and autocrine metabolism of vitamin D in a mouse model of gastrointestinal inflammation. Endocrinology. 2008;149(10):4799–808.

[22] Holick MF, Binkley NC, Bischoff-Ferrari HA, Gordon CM, Hanley DA, Heaney RP, et al. Evaluation, treatment, and prevention of vitamin D deficiency: an Endocrine Society clinical practice guideline. The Journal of Clinical Endocrinology & Metabolism. 2011;96(7):1911–30.

[23] Lips P, Bouillon R, Van Schoor NM, Vanderschueren D, Verschueren S, Kuchuk N, et al. Review article: Reducing fracture risk with calcium and vitamin D. Clinical Endocrinology. 2010;73(3):277–85.

[24] Holick MF. Resurrection of vitamin D deficiency and rickets. The Journal of Clinical Investigation. 2006;116(8):2062.

[25] Plotnikoff GA, Quigley JM, editors. Prevalence of severe hypovitaminosis D in patients with persistent, nonspecific musculoskeletal pain. Mayo Clinic Proceedings; 2003. Elsevier.

[26] Lips P. Vitamin D physiology. Progress in Biophysics & Molecular Biology. 2006;92(1):4–8.

[27] Liu PT, Stenger S, Li H, Wenzel L, Tan BH, Krutzik SR, et al. Toll-like receptor triggering of a vitamin D-mediated human antimicrobial response. Science. 2006;311(5768):1770–3.

[28] Pilz S, Dobnig H, Winklhofer-Roob B, Riedmüller G, Fischer JE, Seelhorst U, et al. Low serum levels of 25-hydroxyvitamin D predict fatal cancer in patients referred to coronary angiography. Cancer Epidemiology Biomarkers & Prevention. 2008;17(5):1228–33.

[29] Knight JA, Lesosky M, Barnett H, Raboud JM, Vieth R. Vitamin D and reduced risk of breast cancer: a population-based case-control study. Cancer Epidemiology Biomarkers & Prevention. 2007;16(3):422–9.

[30] Harkness LS, Bonny AE. Calcium and vitamin D status in the adolescent: key roles for bone, body weight, glucose tolerance, and estrogen biosynthesis. Journal of Pediatric and Adolescent Gynecology. 2005;18(5):305–11.

[31] Bachrach LK. Acquisition of optimal bone mass in childhood and adolescence. Trends in Endocrinology & Metabolism. 2001;12(1):22–8.

[32] Moore C, Murphy MM, Keast DR, Holick MF. Vitamin D intake in the United States. Journal of the American Dietetic Association. 2004;104(6):980–3.

[33] Holick MF. Sunlight and vitamin D for bone health and prevention of autoimmune diseases, cancers, and cardiovascular disease. The American Journal of Clinical Nutrition. 2004;80(6):1678S–88S.

[34] Outila TA, Kärkkäinen MUM, Lamberg-Allardt CJE. Vitamin D status affects serum parathyroid hormone concentrations during winter in female adolescents: associations with forearm bone mineral density. The American Journal of Clinical Nutrition. 2001;74(2):206–10.

[35] Skinner JD, Bounds W, Carruth BR, Ziegler P. Longitudinal calcium intake is negatively related to children's body fat indexes. Journal of the American Dietetic Association. 2003;103(12):1626–31.

[36] Lee HS, Kim YJ, Shim YS, Jeong HR, Kwon E, Hwang JS. Associations between serum vitamin D levels and precocious puberty in girls. Annals of Pediatric Endocrinology & Metabolism. 2014;19(2):91–5.

[37] Grivas TB, Vasiliadis E, Mouzakis V, Mihas C, Koufopoulos G. Association between adolescent idiopathic scoliosis prevalence and age at menarche in different geographic latitudes. Scoliosis. 2006;1(1):1.

[38] Lee JM, Appugliese D, Kaciroti N, Corwyn RF, Bradley RH, Lumeng JC. Weight status in young girls and the onset of puberty. Pediatrics. 2007;119(3):e624–e30.

[39] Gilbert-Diamond D, Baylin A, Mora-Plazas M, Marin C, Arsenault JE, Hughes MD, et al. Vitamin D deficiency and anthropometric indicators of adiposity in school-age children: a prospective study. The American Journal of Clinical Nutrition. 2010;92(6):1446–51.

[40] Yura S, Ogawa Y, Sagawa N, Masuzaki H, Itoh H, Ebihara K, et al. Accelerated puberty and late-onset hypothalamic hypogonadism in female transgenic skinny mice overexpressing leptin. The Journal of Clinical Investigation. 2000;105(6):749–55.

[41] Donoso M, Muñoz-Calvo M, Barrios V, Garrido G, Hawkins F, Argente JS. Increased circulating adiponectin levels and decreased leptin/soluble leptin receptor ratio throughout puberty in female ballet dancers: association with body composition and the delay in puberty. European Journal of Endocrinology. 2010;162(5):905–11.

[42] Maetani M, Maskarinec G, Franke AA, Cooney RV. Association of leptin, 25-hydroxyvitamin D, and parathyroid hormone in women. Nutrition and Cancer. 2009;61(2):225–31.

[43] DiVall SA, Radovick S. Pubertal development and menarche. Annals of the New York Academy of Sciences. 2008;1135(1):19–28.

[44] Zhen S, Zakaria M, Wolfe A, Radovick S. Regulation of gonadotropin-releasing hormone (GnRH) gene expression by insulin-like growth factor I in a cultured GnRH-expressing neuronal cell line. Molecular Endocrinology. 1997;11(8):1145–55.

[45] Breen ME, Laing EM, Hall DB, Hausman DB, Taylor RG, Isales CM, et al. 25-Hydroxyvitamin D, insulin-like growth factor-I, and bone mineral accrual during growth. The Journal of Clinical Endocrinology & Metabolism. 2011;96(1):E89–E98.

[46] Eyles DW, Smith S, Kinobe R, Hewison M, McGrath JJ. Distribution of the vitamin D receptor and 1α-hydroxylase in human brain. Journal of Chemical Neuroanatomy. 2005;29(1):21–30.

[47] Villamor E, Marin C, Mora-Plazas M, Baylin A. Vitamin D deficiency and age at menarche: a prospective study. The American Journal of Clinical Nutrition. 2011;94(4):1020–5.

[48] Thill M, Becker S, Fischer D, Cordes T, Hornemann A, Diedrich K, et al. Expression of prostaglandin metabolising enzymes COX-2 and 15-PGDH and VDR in human granulosa cells. Anticancer Research. 2009;29(9):3611–8.

[49] Parikh G, Varadinova M, Suwandhi P, Araki T, Rosenwaks Z, Poretsky L, et al. Vitamin D regulates steroidogenesis and insulin-like growth factor binding protein-1 (IGFBP-1) production in human ovarian cells. Hormone and Metabolic Research. 2010;42(10):754.

[50] Irani M, Merhi Z. Role of vitamin D in ovarian physiology and its implication in reproduction: a systematic review. Fertility and Sterility. 2014;102(2):460–8.e3.

[51] Merhi ZO, Minkoff H, Feldman J, Macura J, Rodriguez C, Seifer DB. Relationship of bariatric surgery to Müllerian-inhibiting substance levels. Fertility and Sterility. 2008;90(1):221–4.

[52] Merhi ZO. Impact of bariatric surgery on female reproduction. Fertility and Sterility. 2009;92(5):1501–8.

[53] Merhi ZO, Seifer DB, Weedon J, Adeyemi O, Holman S, Anastos K, et al. Circulating vitamin D correlates with serum antimüllerian hormone levels in late-reproductive-aged women: Women's Interagency HIV Study. Fertility and Sterility. 2012;98(1):228–34.

[54] Merhi Z, Doswell A, Krebs K, Cipolla M. Vitamin D alters genes involved in follicular development and steroidogenesis in human cumulus granulosa cells. The Journal of Clinical Endocrinology & Metabolism. 2014;99(6):E1137–E45.

[55] Dennis NA, Houghton LA, Jones GT, van Rij AM, Morgan K, McLennan IS. The level of serum anti-Müllerian hormone correlates with vitamin D status in men and women but not in boys. The Journal of Clinical Endocrinology & Metabolism. 2012;97(7):2450–5.

[56] Barrera D, Avila E, HernJández G, Halhali A, Biruete Bn, Larrea F, et al. Estradiol and progesterone synthesis in human placenta is stimulated by calcitriol. The Journal of Steroid Biochemistry and Molecular Biology. 2007;103(3):529–32.

[57] Nobel S, Abrahmsen L, Oppermann U. Metabolic conversion as a pre-receptor control mechanism for lipophilic hormones. European Journal of Biochemistry. 2001;268(15):4113–25.

[58] Gruber CJ, Tschugguel W, Schneeberger C, Huber JC. Production and actions of estrogens. New England Journal of Medicine. 2002;346(5):340–52.

[59] Usluogullari B, Duvan CZ, Usluogullari CA. Use of aromatase inhibitors in practice of gynecology. Journal of Ovarian Research. 2015;8(1):1.

[60] Krishnan AV, Swami S, Peng L, Wang J, Moreno J, Feldman D. Tissue-selective regulation of aromatase expression by calcitriol: implications for breast cancer therapy. Endocrinology. 2010;151(1):32–42.

[61] Josso N, Picard JY, Rey R, Di Clemente N. Testicular anti-Mullerian hormone: history, genetics, regulation and clinical applications. Pediatric Endocrinology Reviews: PER. 2006;3(4):347–58.

[62] Zec I, Tislaric-Medenjak D, Bukovec Megla Z, Kucak I. Anti-Mullerian hormone: a unique biochemical marker of gonadal development and fertility in humans. Biochemia Medica. 2011;21(3):219–30.

[63] Krishnan AV, Moreno J, Nonn L, Malloy P, Swami S, Peng L, et al. Novel pathways that contribute to the anti-proliferative and chemopreventive activities of calcitriol in prostate cancer. The Journal of Steroid Biochemistry and Molecular Biology. 2007;103(3):694–702.

[64] Rojansky N, Brzezinski A, Schenker JG. Seasonality in human reproduction: an update. Human Reproduction. 1992;7(6):735–45.

[65] Panda DK, Miao D, Tremblay ML, Sirois J, Farookhi R, Hendy GN, et al. Targeted ablation of the 25-hydroxyvitamin D 1α-hydroxylase enzyme: evidence for skeletal, reproductive, and immune dysfunction. Proceedings of the National Academy of Sciences. 2001;98(13):7498–503.

[66] Wojtusik J, Johnson PA. Vitamin D regulates anti-Mullerian hormone expression in granulosa cells of the hen. Biology of Reproduction. 2012;86(3):91.

[67] Sun W, Xie H, Ji J, Zhou X, Goltzman D, Miao D. Defective female reproductive function in 1,25 (OH) 2D-deficient mice results from indirect effect mediated by extracellular calcium and/or phosphorus. American Journal of Physiology-Endocrinology and Metabolism. 2010;299(6):E928–E35.

[68] Johnson LE, DeLuca HF. Vitamin D receptor null mutant mice fed high levels of calcium are fertile. The Journal of Nutrition. 2001;131(6):1787–91.

[69] Anagnostis P, Karras S, Goulis DG. Vitamin D in human reproduction: a narrative review. International Journal of Clinical Practice. 2013;67(3):225–35.

[70] Kovacs CS. Vitamin D in pregnancy and lactation: maternal, fetal, and neonatal outcomes from human and animal studies. The American Journal of Clinical Nutrition. 2008;88(2):520S–8S.

[71] Lerchbaum E, Obermayer-Pietsch B. Mechanisms in endocrinology: Vitamin D and fertility: a systematic review. European Journal of Endocrinology. 2012;166(5):765–78.

[72] Zhang C, Qiu C, Hu FB, David RM, Van Dam RM, Bralley A, et al. Maternal plasma 25-hydroxyvitamin D concentrations and the risk for gestational diabetes mellitus. PloS One. 2008;3(11):e3753.

[73] Maghbooli Z, Hosseinâ€ Nezhad A, Karimi F, Shafaei AR, Larijani B. Correlation between vitamin D3 deficiency and insulin resistance in pregnancy. Diabetes/Metabolism Research and Reviews. 2008;24(1):27–32.

[74] Bodnar LM, Catov JM, Simhan HN, Holick MF, Powers RW, Roberts JM. Maternal vitamin D deficiency increases the risk of preeclampsia. The Journal of Clinical Endocrinology & Metabolism. 2007;92(9):3517–22.

[75] Wagner CL, Greer FR. Prevention of rickets and vitamin D deficiency in infants, children, and adolescents. Pediatrics. 2008;122(5):1142–52.

[76] Haddad Jr JG, Boisseau V, Avioli LV. Placental transfer of vitamin D3 and 25-hydroxycholecalciferol in the rat. Journal of Laboratory and Clinical Medicine. 1971;77(6):908–15.

[77] Fleischman AR, Rosen JF, Cole J, Smith CM, Deluca HF. Maternal and fetal serum 1, 25-dihydroxyvitamin D levels at term. The Journal of Pediatrics. 1980;97(4):640–2.

[78] Horwitz MJ, Tedesco MB, Sereika SM, Hollis BW, Garcia-Ocaña A, Stewart AF. Direct comparison of sustained infusion of human parathyroid hormone-related protein-(1-36) [hPTHrP-(1-36)] versus hPTH-(1-34) on serum calcium, plasma 1, 25-dihydroxyvitamin D concentrations, and fractional calcium excretion in healthy human volunteers. The Journal of Clinical Endocrinology & Metabolism. 2003;88(4):1603–9.

[79] Horwitz MJ, Tedesco MB, Sereika SM, Syed MA, Garcia-Ocaña A, Bisello A, et al. Continuous PTH and PTHrP infusion causes suppression of bone formation and discordant effects on 1,25 (OH) 2Vitamin D. Journal of Bone and Mineral Research. 2005;20(10):1792–803.

[80] Kovacs CS. Calcium and bone metabolism in pregnancy and lactation. The Journal of Clinical Endocrinology & Metabolism. 2001;86(6):2344–8.

[81] Steegers EAP, von Dadelszen P, Duvekot JJ, Pijnenborg R. Pre-eclampsia. The Lancet. 2010;376(9741):631–44.

[82] James JL, Whitley GS, Cartwright JE. Pre-eclampsia: fitting together the placental, immune and cardiovascular pieces. The Journal of Pathology. 2010;221(4):363–78.

[83] James JL, Whitley GS, Cartwright JE. Pre-eclampsia: fitting together the placental, immune and cardiovascular pieces. The Journal of Pathology. 2010;221(4):363–78.

[84] Ustuner I, Keskin HL, Tas EE, Neselioglu S, Sengul O, Filiz Avsar A. Maternal serum 25 (OH) D levels in the third trimester of pregnancy during the winter season. The Journal of Maternal-Fetal & Neonatal Medicine. 2011;24(12):1421–6.

[85] Kazemi A, Sharifi F, Jafari N, Mousavinasab N. High prevalence of vitamin D deficiency among pregnant women and their newborns in an Iranian population. Journal of Women's Health. 2009;18(6):835–9.

[86] Shin JS, Choi MY, Longtine MS, Nelson DM. Vitamin D effects on pregnancy and the placenta. Placenta. 2010;31(12):1027–34.

[87] Wei SQ, Audibert F, Hidiroglou N, Sarafin K, Julien P, Wu Y, et al. Longitudinal vitamin D status in pregnancy and the risk of pre-eclampsia. BJOG: An International Journal of Obstetrics & Gynaecology. 2012;119(7):832–9.

[88] Hoeck AD, Pall ML. Will vitamin D supplementation ameliorate diseases characterized by chronic inflammation and fatigue? Medical Hypotheses. 2011;76(2):208–13.

[89] Jablonski KL, Chonchol M, Pierce GL, Walker AE, Seals DR. 25-Hydroxyvitamin D deficiency is associated with inflammation-linked vascular endothelial dysfunction in middle-aged and older adults. Hypertension. 2011;57(1):63–9.

[90] Ngo DT, Sverdlov AL, McNeil JJ, Horowitz JD. Does vitamin D modulate asymmetric dimethylarginine and C-reactive protein concentrations? The American Journal of Medicine. 2010;123(4):335–41.

[91] Peterson CA, Heffernan ME. Serum tumor necrosis factor-alpha concentrations are negatively correlated with serum 25 (OH) D concentrations in healthy women. Journal of Inflammation. 2008;5(1):1.

[92] Wei S-Q, Audibert Fo, Luo Z-C, Nuyt AM, Masse B, Julien P, et al. Maternal plasma 25-hydroxyvitamin D levels, angiogenic factors, and preeclampsia. American Journal of Obstetrics and Gynecology. 2013;208(5):390.e1– e6.

[93] Tarcin O, Yavuz DG, Ozben B, Telli A, Ogunc AV, Yuksel M, et al. Effect of vitamin D deficiency and replacement on endothelial function in asymptomatic subjects. The Journal of Clinical Endocrinology & Metabolism. 2009;94(10):4023–30.

[94] Morisaki N, Esplin MS, Varner MW, Henry E, Oken E. Declines in birth weight and fetal growth independent of gestational length. Obstetrics and Gynecology. 2013;121(1):51.

[95] Bodnar LM, Catov JM, Zmuda JM, Cooper ME, Parrott MS, Roberts JM, et al. Maternal serum 25-hydroxyvitamin D concentrations are associated with small-for-gestational age births in white women. The Journal of Nutrition. 2010;140(5):999–1006.

[96] Staff AC, Benton SJ, von Dadelszen P, Roberts JM, Taylor RN, Powers RW, et al. Redefining preeclampsia using placenta-derived biomarkers. Hypertension. 2013;61(5):932–42.

[97] Christesen HT, Falkenberg T, Lamont RF, JÃ¸rgensen JS. The impact of vitamin D on pregnancy: a systematic review. Acta Obstetricia et Gynecologica Scandinavica. 2012;91(12):1357–67.

[98] Henry HL, Norman AW. Vitamin D: metabolism and biological actions. Annual Review of Nutrition. 1984;4(1):493–520.

[99] Barrera D, Avila E, Hernández G, Méndez I, González L, Halhali A, et al. Calcitriol affects hCG gene transcription in cultured human syncytiotrophoblasts. Reproductive Biology and Endocrinology. 2008;6(3).

[100] Liu NQ, Ouyang Y, Bulut Y, Lagishetty V, Chan SY, Hollis BW, et al. Dietary vitamin D restriction in pregnant female mice is associated with maternal hypertension and altered placental and fetal development. Endocrinology. 2013;154(7):2270–80.

[101] Grundmann M, von Versen-Höynck F. Vitamin D-roles in women's reproductive health? Reproductive Biology and Endocrinology. 2011;9(1):146.

[102] Hirai M, Suzuki S, Hinokio Y, Hirai A, Chiba M, Akai H, et al. Variations in vitamin D-binding protein (group-specific component protein) are associated with fasting plasma insulin levels in Japanese with normal glucose tolerance 1. The Journal of Clinical Endocrinology & Metabolism. 2000;85(5):1951–3.

[103] Ramos-Lopez E, Kahles H, Weber S, Kukic A, Penna-Martinez M, Badenhoop K, et al. Gestational diabetes mellitus and vitamin D deficiency: genetic contribution of CYP27B1 and CYP2R1 polymorphisms. Diabetes, Obesity and Metabolism. 2008;10(8):683–5.

[104] Poel YHM, Hummel P, Lips P, Stam F, Van Der Ploeg T, Simsek S. Vitamin D and gestational diabetes: a systematic review and meta-analysis. European Journal of Internal Medicine. 2012;23(5):465–9.

[105] Norman AW, Frankel JB, Heldt AM, Grodsky GM. Vitamin D deficiency inhibits pancreatic secretion of insulin. Science. 1980;209(4458):823–5.

[106] Sooy K, Schermerhorn T, Noda M, Surana M, Rhoten WB, Meyer M, et al. Calbindin-D28k controls [Ca2+] i and insulin release evidence obtained from calbindin-D28k knockout mice and β cell lines. Journal of Biological Chemistry. 1999;274(48):34343–9.

[107] Dempsey JC, Sorensen TK, Williams MA, Lee IM, Miller RS, Dashow EE, et al. Prospective study of gestational diabetes mellitus risk in relation to maternal recreational physical activity before and during pregnancy. American Journal of Epidemiology. 2004;159(7):663–70.

[108] Allsworth JE, Peipert JF. Prevalence of bacterial vaginosis: 2001-2004 national health and nutrition examination survey data. Obstetrics & Gynecology. 2007;109(1):114–20.

[109] Bodnar LM, Krohn MA, Simhan HN. Maternal vitamin D deficiency is associated with bacterial vaginosis in the first trimester of pregnancy. The Journal of Nutrition. 2009;139(6):1157–61.

[110] Hensel KJ, Randis TM, Gelber SE, Ratner AJ. Pregnancy-specific association of vitamin D deficiency and bacterial vaginosis. American Journal of Obstetrics and Gynecology. 2011;204(1):41.e1–e9.

[111] Davis LM, Chang S-C, Mancini J, Nathanson MS, Witter FR, O'Brien KO. Vitamin D insufficiency is prevalent among pregnant African American adolescents. Journal of Pediatric and Adolescent Gynecology. 2010;23(1):45–52.

[112] Chesney RW. Vitamin D and The Magic Mountain: the anti-infectious role of the vitamin. The Journal of Pediatrics. 2010;156(5):698–703.

[113] Müller K, Diamant M, Bendtzen K. Inhibition of production and function of interleukin-6 by 1, 25-dihydroxyvitamin D 3. Immunology Letters. 1991;28(2):115–20.

[114] Díaz L, Noyola-Martínez N, Barrera D, Hernández G, Avila E, Halhali A, et al. Calcitriol inhibits TNF-α induced inflammatory cytokines in human trophoblasts. Journal of Reproductive Immunology. 2009;81(1):17–24.

[115] Arnson Y, Amital H, Shoenfeld Y. Vitamin D and autoimmunity: new aetiological and therapeutic considerations. Annals of the Rheumatic Diseases. 2007;66(9):1137–42.

[116] Liu N, Kaplan AT, Low J, Nguyen L, Liu GY, Equils O, et al. Vitamin D induces innate antibacterial responses in human trophoblasts via an intracrine pathway. Biology of Reproduction. 2009;80(3):398–406.

[117] Tribe RM. Regulation of human myometrial contractility during pregnancy and labour: are calcium homeostatic pathways important? Experimental Physiology. 2001;86(02):247–54.

[118] Bodnar LM, Simhan HN. The prevalence of preterm birth and season of conception. Paediatric and Perinatal Epidemiology. 2008;22(6):538–45.

[119] Hollis BW, Wagner CL, editors. Randomized controlled trials to determine the safety of vitamin D supplementation during pregnancy and lactation. Fourteenth Workshop on Vitamin D; 2009.

[120] Ford HB, Schust DJ. Recurrent pregnancy loss: etiology, diagnosis, and therapy. Reviews in Obstetrics & Gynecology. 2009;2(2):76–83.

[121] Page JM, Silver RM. Genetic Causes of Recurrent Pregnancy Loss. Clinical Obstetrics and Gynecology. 2016;59(3):498-508.

[122] Mizwicki MT, Norman AW. The vitamin D sterol-vitamin D receptor ensemble model offers unique insights into both genomic and rapid-response signaling. Science Signaling. 2009;2(75):re4–re.

[123] Adams JS, Hewison M. Unexpected actions of vitamin D: new perspectives on the regulation of innate and adaptive immunity. Nature Clinical Practice Endocrinology & Metabolism. 2008;4(2):80–90.

[124] Ota K, Dambaeva S, Han A-R, Beaman K, Gilman-Sachs A, Kwak-Kim J. Vitamin D deficiency may be a risk factor for recurrent pregnancy losses by increasing cellular immunity and autoimmunity. Human Reproduction. 2014;29(2):208–19.

[125] Li Q-P, Qi X, Pramanik R, Pohl NM, Loesch M, Chen G. Stress-induced c-Jun-dependent Vitamin D receptor (VDR) activation dissects the non-classical VDR pathway from the classical VDR activity. Journal of Biological Chemistry. 2007;282(3):1544–51.

[126] Ritterhouse LL, Crowe SR, Niewold TB, Kamen DL, Macwana SR, Roberts VC, et al. Vitamin D deficiency is associated with an increased autoimmune response in healthy individuals and in patients with systemic lupus erythematosus. Annals of the Rheumatic Diseases. 2011;70(9):1569–74.

[127] Chen TC, Chimeh F, Lu Z, Mathieu J, Person KS, Zhang A, et al. Factors that influence the cutaneous synthesis and dietary sources of vitamin D. Archives of Biochemistry and Biophysics. 2007;460(2):213–7.

[128] Kwak JY, Beaman KD, Gilman-Sachs A, Ruiz JE, Schewitz D, Beer AE. Up-regulated expression of CD56+, CD56+/CD16+, and CD19+ cells in peripheral blood lymphocytes in pregnant women with recurrent pregnancy losses. American Journal of Reproductive Immunology. 1995;34(2):93–9.

[129] Erkkola M, Kaila M, Nwaru BI, Kronberg-Kippilä C, Ahonen S, Nevalainen J, et al. Maternal vitamin D intake during pregnancy is inversely associated with asthma and allergic rhinitis in 5-year-old children. Clinical & Experimental Allergy. 2009;39(6):875–82.

[130] Papandreou L, Chasiotis G, Seferiadis K, Thanasoulias NC, Dousias V, Tsanadis G, et al. Calcium levels during the initiation of labor. European Journal of Obstetrics & Gynecology and Reproductive Biology. 2004;115(1):17–22.

[131] Merewood A, Mehta SD, Chen TC, Bauchner H, Holick MF. Association between vitamin D deficiency and primary cesarean section. The Journal of Clinical Endocrinology & Metabolism. 2009;94(3):940–5.

[132] Scholl TO, Chen X, Stein P. Maternal vitamin D status and delivery by cesarean. Nutrients. 2012;4(4):319–30.

[133] Conde-Agudelo A, Villar J, Lindheimer M. Maternal infection and risk of preeclampsia: systematic review and metaanalysis. American Journal of Obstetrics and Gynecology. 2008;198(1):7–22.

[134] Vanek M, Sheiner E, Levy A, Mazor M. Chronic hypertension and the risk for adverse pregnancy outcome after superimposed pre-eclampsia. International Journal of Gynecology & Obstetrics. 2004;86(1):7–11.

[135] Kho AT, Bhattacharya S, Tantisira KG, Carey VJ, Gaedigk R, Leeder JS, et al. Transcriptomic analysis of human lung development. American Journal of Respiratory and Critical Care Medicine. 2010;181(1):54–63.

[136] Barker DJP, Eriksson JG, Forsén T, Osmond C. Fetal origins of adult disease: strength of effects and biological basis. International Journal of Epidemiology. 2002;31(6):1235–9.

[137] Camargo CA, Rifas-Shiman SL, Litonjua AA, Rich-Edwards JW, Weiss ST, Gold DR, et al. Maternal intake of vitamin D during pregnancy and risk of recurrent wheeze in children at 3 y of age. The American Journal of Clinical Nutrition. 2007;85(3):788–95.

[138] Devereux G, Litonjua AA, Turner SW, Craig LCA, McNeill G, Martindale S, et al. Maternal vitamin D intake during pregnancy and early childhood wheezing. The American Journal of Clinical Nutrition. 2007;85(3):853–9.

[139] Altschuler EL. Low maternal vitamin D and schizophrenia in offspring. The Lancet. 2001;358(9291):1464.

[140] Mirzaei F, Michels KB, Munger K, O'Reilly E, Chitnis T, Forman MR, et al. Gestational vitamin D and the risk of multiple sclerosis in offspring. Annals of Neurology. 2011;70(1):30–40.

[141] Hyppönen E, Läärä¤ E, Reunanen A, Järvelin M-R, Virtanen SM. Intake of vitamin D and risk of type 1 diabetes: a birth-cohort study. The Lancet. 2001;358(9292):1500–3.

[142] Zipitis CS, Akobeng AK. Vitamin D supplementation in early childhood and risk of type 1 diabetes: a systematic review and meta-analysis. Archives of Disease in Childhood. 2008;93(6):512–7.

[143] Lucas RM, Ponsonby A-L, Pasco JA, Morley R. Future health implications of prenatal and early-life vitamin D status. Nutrition Reviews. 2008;66(12):710–20.

[144] Mulligan ML, Felton SK, Riek AE, Bernal-Mizrachi C. Implications of vitamin D deficiency in pregnancy and lactation. American Journal of Obstetrics and Gynecology. 2010;202(5):429. e1–e9.

[145] Papandreou D, Malindretos P, Karabouta Z, Rousso I. Possible health implications and low vitamin D status during childhood and adolescence: an updated mini review. International Journal of Endocrinology. 2009;2010.

[146] Kazemijaliseh H, Tehrani FR, Behboudi-Gandevani S, Hosseinpanah F, Azizi F. The prevalence and causes of primary infertility in Iran: a population-based study. Global Journal of Health Science. 2015;7(6):226.

[147] Tehrani FR, Simbar M, Tohidi M, Hosseinpanah F, Azizi F. The prevalence of polycystic ovary syndrome in a community sample of Iranian population: Iranian PCOS prevalence study. Reproductive Biology & Endocrinology. 2011;9(39):39.

[148] Tehrani FR, Rashidi H, Azizi F. The prevalence of idiopathic hirsutism and polycystic ovary syndrome in the Tehran Lipid and Glucose Study. Reproductive Biology & Endocrinology. 2011;9(144):b3.

[149] Tehrani FR, Solaymani-Dodaran M, Hedayati M, Azizi F. Is polycystic ovary syndrome an exception for reproductive aging? Human Reproduction. 2010;25(7):1775–81.

[150] Hosseinpanah F, Barzin M, Erfani H, Serahati S, Ramezani Tehrani F, Azizi F. Lipid accumulation product and insulin resistance in Iranian PCOS prevalence study. Clinical Endocrinology. 2014;81(1):52–7.

[151] Hosseinpanah F, Barzin M, Keihani S, Ramezani Tehrani F, Azizi F. Metabolic aspects of different phenotypes of polycystic ovary syndrome: Iranian PCOS Prevalence Study. Clinical Endocrinology. 2014;81(1):93–9.

[152] Palomba S, Pasquali R, Orio F, Nestler JE. Clomiphene citrate, metformin or both as first step approach in treating anovulatory infertility in patients with polycystic ovary syndrome (PCOS): a systematic review of head-to-head randomized controlled studies and meta-analysis. Clinical Endocrinology. 2009;70(2):311–21.

[153] Wehr E, Trummer O, Giuliani A, Gruber H-Jr, Pieber TR, Obermayer-Pietsch B. Vitamin D-associated polymorphisms are related to insulin resistance and vitamin D deficiency in polycystic ovary syndrome. European Journal of Endocrinology. 2011;164(5):741–9.

[154] Li HWR, Brereton RE, Anderson RA, Wallace AM, Ho CKM. Vitamin D deficiency is common and associated with metabolic risk factors in patients with polycystic ovary syndrome. Metabolism. 2011;60(10):1475–81.

[155] Ott J, Wattar L, Kurz C, Seemann R, Huber JC, Mayerhofer K, et al. Parameters for calcium metabolism in women with polycystic ovary syndrome who undergo clomiphene citrate stimulation: a prospective cohort study. European Journal of Endocrinology. 2012;166(5):897–902.

[156] Rashidi B, Haghollahi F, Shariat M, Zayerii F. The effects of calcium-vitamin D and metformin on polycystic ovary syndrome: a pilot study. Taiwanese Journal of Obstetrics and Gynecology. 2009;48(2):142–7.

[157] Ngo DTM, Chan WP, Rajendran S, Heresztyn T, Amarasekera A, Sverdlov AL, et al. Determinants of insulin responsiveness in young women: impact of polycystic ovarian syndrome, nitric oxide, and vitamin D. Nitric Oxide. 2011;25(3):326–30.

[158] Patra SK, Nasrat H, Goswami B, Jain A. Vitamin D as a predictor of insulin resistance in polycystic ovarian syndrome. Diabetes & Metabolic Syndrome: Clinical Research & Reviews. 2012;6(3):146–9.

[159] Pal L, Berry A, Coraluzzi L, Kustan E, Danton C, Shaw J, et al. Therapeutic implications of vitamin D and calcium in overweight women with polycystic ovary syndrome. Gynecological Endocrinology. 2012;28(12):965–8.

[160] Dunaif A. Insulin resistance and the polycystic ovary syndrome: mechanism and implications for pathogenesis 1. Endocrine Reviews. 1997;18(6):774–800.

[161] Pittas AG, Lau J, Hu FB, Dawson-Hughes B. The role of vitamin D and calcium in type 2 diabetes. A systematic review and meta-analysis. The Journal of Clinical Endocrinology & Metabolism. 2007;92(6):2017–29.

[162] Merhi Z. Advanced glycation end products and their relevance in female reproduction. Human Reproduction. 2013:det383.

[163] Diamantiâ€ Kandarakis E, Katsikis I, Piperi C, Kandaraki E, Piouka A, Papavassiliou AG, et al. Increased serum advanced glycation end-products is a distinct finding in lean women with polycystic ovary syndrome (PCOS). Clinical Endocrinology. 2008;69(4):634–41.

[164] Merhi Z. Advanced glycation end products and their relevance in female reproduction. Human Reproduction. 2014;29(1):135–45.

[165] Merhi Z, Irani M, Doswell AD, Ambroggio J. Follicular fluid soluble receptor for advanced glycation end-products (sRAGE): a potential indicator of ovarian reserve. The Journal of Clinical Endocrinology & Metabolism. 2013;99(2):E226–E33.

[166] Mahmoudi T. Genetic variation in the vitamin D receptor and polycystic ovary syndrome risk. Fertility and Sterility. 2009;92(4):1381–3.

[167] Bagheri M, Abdi Rad I, Hosseini Jazani N, Nanbakhsh F. Vitamin D receptor TaqI gene variant in exon 9 and polycystic ovary syndrome risk. International Journal of Fertility & Sterility. 2013;7(2):116–21.

[168] Maestro Ba, Dávila N, Carranza MC, Calle C. Identification of a Vitamin D response element in the human insulin receptor gene promoter. The Journal of Steroid Biochemistry and Molecular Biology. 2003;84(2):223–30.

[169] Maestro B, Molero S, Bajo S, Davila N, Calle C. Transcriptional activation of the human insulin receptor gene by 1, 25â€ dihydroxyvitamin D3. Cell Biochemistry and Function. 2002;20(3):227–32.

[170] Shoelson SE, Herrero L, Naaz A. Obesity, inflammation, and insulin resistance. Gastroenterology. 2007;132(6):2169–80.

[171] Baird DD, Hill MC, Schectman JM, Hollis BW. Vitamin D and the risk of uterine fibroids. Epidemiology. 2013;24(3):447–53.

[172] Sabry M, Halder SK, Allah ASA, Roshdy E, Rajaratnam V, Al-Hendy A. Serum vitamin D3 level inversely correlates with uterine fibroid volume in different ethnic groups: a cross-sectional observational study. International Journal of Women's Health. 2013;5:93.

[173] Bläuer M, Rovio PiH, Ylikomi T, Heinonen PK. Vitamin D inhibits myometrial and leiomyoma cell proliferation in vitro. Fertility and Sterility. 2009;91(5):1919–25.

[174] Bläuer M, Rovio PH, Ylikomi T, Heinonen PK. Vitamin D inhibits myometrial and leiomyoma cell proliferation in vitro. Fertility and Sterility. 2009;91(5):1919–25.

[175] Leppert PC, Catherino WH, Segars JH. A new hypothesis about the origin of uterine fibroids based on gene expression profiling with microarrays. American Journal of Obstetrics and Gynecology. 2006;195(2):415–20.

[176] Ylikomi T, Laaksi I, Lou Y-R, Martikainen P, Miettinen S, Pennanen P, et al. Antiproliferative action of vitamin D. Vitamins & Hormones. 2002;64:357–406.

[177] World Health O. Prevention and management of osteoporosis: report of a WHO scientific group. Diamond Pocket Books (P) Ltd.; 2003.

[178] Colston KW. Mechanisms implicated in the growth regulatory effects of vitamin D in breast cancer. Endocrine-Related Cancer. 2002;9(1):45–59.

[179] Hill NT, Zhang J, Leonard MK, Lee M, Shamma HN, Kadakia M. 1α,25-Dihydroxyvitamin D3 and the vitamin D receptor regulates ΔNp63α levels and keratinocyte proliferation. Cell Death & Disease. 2015;6(6): e1781.

[180] Vienonen A, Miettinen S, Bläuer M, Martikainen PM, Tomás E, Heinonen PK, et al. Expression of nuclear receptors and cofactors in human endometrium and myometrium. Journal of the Society for Gynecologic Investigation. 2004;11(2):104–12.

[181] Lou Y-R, Laaksi I, SyväläH, Bläuer M, Tammela TLJ, Ylikomi T, et al. 25-hydroxyvitamin D3 is an active hormone in human primary prostatic stromal cells. The FASEB Journal. 2004;18(2):332–4.

[182] Sourial S, Tempest N, Hapangama DK. Theories on the pathogenesis of endometriosis. International Journal of Reproductive Medicine. 2014;2014.

[183] Vigano P, Lattuada D, Mangioni S, Ermellino L, Vignali M, Caporizzo E, et al. Cycling and early pregnant endometrium as a site of regulated expression of the vitamin D system. Journal of Molecular Endocrinology. 2006;36(3):415–24.

[184] Vilarino FL, Bianco B, Lerner TG, Teles JS, Mafra FA, Christofolini DM, et al. Analysis of vitamin D receptor gene polymorphisms in women with and without endometriosis. Human Immunology. 2011;72(4):359–63.

[185] Rudick BJ, Ingles SA, Chung K, Stanczyk FZ, Paulson RJ, Bendikson KA. Influence of vitamin D levels on in vitro fertilization outcomes in donor-recipient cycles. Fertility and Sterility. 2014;101(2):447–52.

[186] Ozkan S, Jindal S, Greenseid K, Shu J, Zeitlian G, Hickmon C, et al. Replete vitamin D stores predict reproductive success following in vitro fertilization. Fertility and Sterility. 2010;94(4):1314–9.

[187] Rudick B, Ingles SA, Stanczyk F, Chung K, Paulson R, Bendikson K. The role of vitamin D levels on IVF outcomes in donor-recipient cycles. Fertility and Sterility. 2011;95(4):S8.

[188] Rudick B, Ingles S, Chung K, Stanczyk F, Paulson R, Bendikson K. IVF 4X more successful for white women with lots of vitamin D – Oct 2012. Fertility and Sterility. 2014.

[189] Evans KN, Nguyen L, Chan J, Innes BA, Bulmer JN, Kilby MD, et al. Effects of 25-hydroxyvitamin D3 and 1, 25-dihydroxyvitamin D3 on cytokine production by human decidual cells. Biology of Reproduction. 2006;75(6):816–22.

[190] Evans KN, Bulmer JN, Kilby MD, Hewison M. Vitamin D and placental-decidual function. Journal of the Society for Gynecologic Investigation. 2004;11(5):263–71.

[191] Tavakoli M, Jeddi-Tehrani M, Salek-Moghaddam A, Rajaei S, Mohammadzadeh A, Sheikhhasani S, et al. Effects of 1, 25 (OH) 2 vitamin D3 on cytokine production by endometrial cells of women with recurrent spontaneous abortion. Fertility and Sterility. 2011;96(3):751–7.

[192] Leffelaar ER, Vrijkotte TGM, van Eijsden M. Maternal early pregnancy vitamin D status in relation to fetal and neonatal growth: results of the multi-ethnic Amsterdam Born Children and their Development cohort. British Journal of Nutrition. 2010;104(01):108–17.

[193] Swamy GK, Garrett ME, Miranda ML, Ashley-Koch AE. Maternal vitamin D receptor genetic variation contributes to infant birthweight among black mothers. American Journal of Medical Genetics Part A. 2011;155(6):1264–71.

[194] Swamy GK, Garrett ME, Miranda ML, Ashley-Koch AE. Maternal vitamin D receptor genetic variation contributes to infant birthweight among black mothers. American Journal of Medical Genetics Part A. 2011;155(6):1264–71.

[195] Enquobahrie DA, Williams MA, Qiu C, Siscovick DS, Sorensen TK. Global maternal early pregnancy peripheral blood mRNA and miRNA expression profiles according to plasma 25-hydroxyvitamin D concentrations. The Journal of Maternal-Fetal & Neonatal Medicine. 2011;24(8):1002–12.

[196] Pasco JA, Wark JD, Carlin JB, Ponsonby A-L, Vuillermin PJ, Morley R. Maternal vitamin D in pregnancy may influence not only offspring bone mass but other aspects of musculoskeletal health and adiposity. Medical Hypotheses. 2008;71(2):266–9.

[197] Martini LA, Wood RJ. Vitamin D status and the metabolic syndrome. Nutrition Reviews. 2006;64(11):479–86.

[198] Shi H, Norman AW, Okamura WH, Sen A, Zemel MB. 1α, 25-Dihydroxyvitamin modulates human adipocyte metabolism via nongenomic action. The FASEB Journal. 2001;15(14):2751–3.

[199] Mahon P, Harvey N, Crozier S, Inskip H, Robinson S, Arden N, et al. Low maternal vitamin D status and fetal bone development: cohort study. Journal of Bone and Mineral Research. 2010;25(1):14–9.

[200] Rochat MK, Ege MJ, Plabst D, Steinle J, Bitter S, Braun-Fahrländer C, et al. Maternal vitamin D intake during pregnancy increases gene expression of ILT3 and ILT4 in cord blood. Clinical & Experimental Allergy. 2010;40(5):786–94.

[201] Chi A, Wildfire J, McLoughlin R, Wood RA, Bloomberg GR, Kattan M, et al. Umbilical cord plasma Umbilical cord plasma 25-hydroxy vitamin concentration and immune function at birth: the Urban Environment and Childhood Asthma study. Clinical & Experimental Allergy. 2011;41(6):842–50.

[202] Chiu C-Y, Huang S-Y, Peng Y-C, Tsai M-H, Hua M-C, Yao T-C, et al. Maternal vitamin D levels are inversely related to allergic sensitization and atopic diseases in early childhood. Pediatric Allergy and Immunology. 2015;26(4):337–43.

[203] Chiu C-Y, Yao T-C, Chen S-H, Tsai M-H, Tu Y-L, Hua M-C, et al. Low cord blood vitamin D levels are associated with increased milk sensitization in early childhood. Pediatric Allergy and Immunology. 2014;25(8):767–72.

[204] Palmer DJ, Sullivan TR, Skeaff CM, Smithers LG, Makrides M. Higher cord blood 25-hydroxyvitamin D concentrations reduce the risk of early childhood eczema: in children with a family history of allergic disease. World Allergy Organization Journal. 2015;8(1):1–7.

[205] Gazibara T, den Dekker HT, de Jongste JC, McGrath JJ, Eyles DW, Burne TH, et al. Associations of maternal and fetal 25-hydroxyvitamin D levels with childhood lung function and asthma. The Generation R Study. Clinical & Experimental Allergy. 2015.

[206] Zosky GR, Hart PH, Whitehouse AJO, Kusel MM, Ang W, Foong RE, et al. Vitamin D deficiency at 16 to 20 week's gestation is associated with impaired lung function and asthma at 6 years of age. Annals of the American Thoracic Society. 2014;11(4):571–7.

[207] Jedrychowski W, Galas A, Pac A, Flak E, Camman D, Rauh V, et al. Prenatal ambient air exposure to polycyclic aromatic hydrocarbons and the occurrence of respiratory symptoms over the first year of life. European Journal of Epidemiology. 2005;20(9):775–82.

[208] Jedrychowski W, Perera F, Maugeri U, Miller RL, Rembiasz M, Flak E, et al. Intrauterine exposure to lead may enhance sensitization to common inhalant allergens in early childhood: a prospective prebirth cohort study. Environmental Research. 2011;111(1):119–24.

[209] Schwalfenberg GK, Genuis SJ. Vitamin D, essential minerals, and toxic elements: exploring interactions between nutrients and toxicants in clinical medicine. The Scientific World Journal. 2015;2015.

[210] Genuis SJ. Sensitivity-related illness: the escalating pandemic of allergy, food intolerance and chemical sensitivity. Science of the Total Environment. 2010;408(24):6047–61.

[211] Camargo CA, Ingham T, Wickens K, Thadhani R, Silvers KM, Epton MJ, et al. Cord-blood 25-hydroxyvitamin D levels and risk of respiratory infection, wheezing, and asthma. Pediatrics. 2011;127(1):e180–e7.

[212] Miyake Y, Sasaki S, Tanaka K, Hirota Y. Dairy food, calcium and vitamin D intake in pregnancy, and wheeze and eczema in infants. European Respiratory Journal. 2010;35(6):1228–34.

[213] Kovacs CS, Kronenberg HM. Maternal-fetal calcium and bone metabolism during pregnancy, puerperium, and lactation 1. Endocrine Reviews. 1997;18(6):832–72.

[214] Sowers M, Zhang D, Hollis BW, Shapiro B, Janney CA, Crutchfield M, et al. Role of calciotrophic hormones in calcium mobilization of lactation. The American Journal of Clinical Nutrition. 1998;67(2):284–91.

[215] Basile LA, Taylor SN, Wagner CL, Horst RL, Hollis BW. The effect of high-dose vitamin D supplementation on serum vitamin D levels and milk calcium concentration in lactating women and their infants. Breastfeeding Medicine. 2006;1(1):27–35.

[216] Wagner CL, Hulsey TC, Fanning D, Ebeling M, Hollis BW. High-dose vitamin D3 supplementation in a cohort of breastfeeding mothers and their infants: a 6-month follow-up pilot study. Breastfeeding Medicine. 2006;1(2):59–70.

[217] Nelson HD, Vesco KK, Haney E, Fu R, Nedrow A, Miller J, et al. Nonhormonal therapies for menopausal hot flashes: systematic review and meta-analysis. JAMA. 2006;295(17):2057–71.

[218] LeBlanc ES, Desai M, Perrin N, Wactawski-Wende J, Manson JE, Cauley JA, et al. Vitamin D levels and menopause-related symptoms. Menopause. 2014;21(11):1197–203.

[219] Lovejoy JC, Champagne CM, De Jonge L, Xie H, Smith SR. Increased visceral fat and decreased energy expenditure during the menopausal transition. International Journal of Obesity. 2008;32(6):949–58.

[220] Ramezani Tehrani F, Behboudi-Gandevani S, Ghanbarian A, Azizi F. Effect of menopause on cardiovascular disease and its risk factors: a 9-year follow-up study. Climacteric. 2014;17(2):164–72.

[221] Shi H, Seeley RJ, Clegg DJ. Sexual differences in the control of energy homeostasis. Frontiers in Neuroendocrinology. 2009;30(3):396–404.

[222] Salehpour A, Hosseinpanah F, Shidfar F, Vafa M, Razaghi M, Dehghani S, et al. A 12-week double-blind randomized clinical trial of vitamin D 3 supplementation on body fat mass in healthy overweight and obese women. Nutrition Journal. 2012;11(1):1.

[223] Lerchbaum E. Vitamin D and menopause – A narrative review. Maturitas. 2014;79(1):3–7.

[224] Group ECW. Perimenopausal risk factors and future health. Human Reproduction Update. 2011;17(5):706–17.

[225] Pludowski P, Holick MF, Pilz S, Wagner CL, Hollis BW, Grant WB, et al. Vitamin D effects on musculoskeletal health, immunity, autoimmunity, cardiovascular disease, cancer, fertility, pregnancy, dementia and mortality – a review of recent evidence. Autoimmunity Reviews. 2013;12(10):976–89.

[226] Wehr E, Pilz S, Schweighofer N, Giuliani A, Kopera D, Pieber TR, et al. Association of hypovitaminosis D with metabolic disturbances in polycystic ovary syndrome. European Journal of Endocrinology. 2009;161(4):575–82.

[227] Greendale GA, Reboussin BA, Hogan P, Barnabei VM, Shumaker S, Johnson S, et al. Symptom relief and side effects of postmenopausal hormones: results from the Postmenopausal Estrogen/Progestin Interventions Trial. Obstetrics & Gynecology. 1998;92(6):982–8.

[228] Felson DT, Cummings SR. Aromatase inhibitors and the syndrome of arthralgias with estrogen deprivation. Arthritis & Rheumatism. 2005;52(9):2594–8.

[229] Glerup H, Mikkelsen K, Poulsen L, Hass E, Overbeck S, Andersen H, et al. Hypovitaminosis D myopathy without biochemical signs of osteomalacia bone involvement. Calcified Tissue International. 2000;66(6):419–24.

[230] Lerchbaum E, Schwetz V, Pilz S, Boehm BO, März W. Association of bone turnover markers with mortality in women referred to coronary angiography: the Ludwigshafen Risk and Cardiovascular Health (LURIC) study. Osteoporosis International. 2014;25(2):455–65.

[231] Pilz S, Kienreich K, Tomaschitz A, Ritz E, Lerchbaum E, Obermayer-Pietsch B, et al. Vitamin D and cancer mortality: systematic review of prospective epidemiological studies. Anti-Cancer Agents in Medicinal Chemistry (Formerly Current Medicinal Chemistry-Anti-Cancer Agents). 2013;13(1):107–17.

[232] Cheng T-YD, LaCroix AZ, Beresford SAA, Goodman GE, Thornquist MD, Zheng Y, et al. Vitamin D intake and lung cancer risk in the Women's Health Initiative. The American Journal of Clinical Nutrition. 2013;98(4):1002–11.

[233] Cauley JA, Chlebowski RT, Wactawski-Wende J, Robbins JA, Rodabough RJ, Chen Z, et al. Calcium plus vitamin D supplementation and health outcomes five years after active intervention ended: the Women's Health Initiative. Journal of Women's Health. 2013;22(11):915–29.

[234] Wharton W, Gleason CE, Olson SRMS, Carlsson CM, Asthana S. Neurobiological underpinnings of the estrogen-mood relationship. Current Psychiatry Reviews. 2012;8(3):247.

[235] Buell JS, Dawson-Hughes B, Scott TM, Weiner DE, Dallal GE, Qui WQ, et al. 25-Hydroxyvitamin D, dementia, and cerebrovascular pathology in elders receiving home services. Neurology. 2010;74(1):18–26.

[236] Eyles DW, Burne THJ, McGrath JJ. Vitamin D, effects on brain development, adult brain function and the links between low levels of vitamin D and neuropsychiatric disease. Frontiers in Neuroendocrinology. 2013;34(1):47–64.

[237] Yoshida M, Kawano N, Yoshida K. Control of sperm motility and fertility: diverse factors and common mechanisms. Cellular and Molecular Life Sciences. 2008;65(21):3446–57.

[238] Hirai T, Tsujimura A, Ueda T, Fujita K, Matsuoka Y, Takao T, et al. Effect of 1, 25-dihydroxyvitamin D on testicular morphology and gene expression in experimental cryptorchid mouse: testis specific cDNA microarray analysis and potential implication in male infertility. The Journal of Urology. 2009;181(3):1487–92.

[239] Jensen MB, Nielsen JE, Jørgensen A, Rajpert-De Meyts E, Kristensen DMb, Jørgensen N, et al. Vitamin D receptor and vitamin D metabolizing enzymes are expressed in the human male reproductive tract. Human Reproduction. 2010;25(5):1303–11.

[240] Aquila S, Guido C, Middea E, Perrotta I, Bruno R, Pellegrino M, et al. Human male gamete endocrinology: 1alpha, 25-dihydroxyvitamin D3 (1, 25 (OH) 2D3) regulates different aspects of human sperm biology and metabolism. Reproductive Biology & Endocrinology. 2009;7(1):140.

[241] Foresta C, Strapazzon G, De Toni L, Perilli L, Di Mambro A, Muciaccia B, et al. Bone mineral density and testicular failure: evidence for a role of vitamin D 25-hydroxylase in human testis. The Journal of Clinical Endocrinology & Metabolism. 2011;96(4):E646–E52.

[242] Walters MR. 1, 25-Dihydroxyvitamin D3 receptors in the seminiferous tubules of the rat testis increase at puberty. Endocrinology. 1984;114(6):2167–74.

[243] Menegaz D, Rosso A, Royer C, Leite LD, Santos ARS, Silva FtRMB. Role of 1,25 (OH) 2 vitamin D 3 on α-[1–14 C] MeAIB accumulation in immature rat testis. Steroids. 2009;74(2):264–9.

[244] Avila E, Díaz L, Barrera D, Halhali A, Méndez I, González L, et al. Regulation of Vitamin D hydroxylases gene expression by 1, 25-dihydroxyvitamin D 3 and cyclic AMP in cultured human syncytiotrophoblasts. The Journal of Steroid Biochemistry and Molecular Biology. 2007;103(1):90–6.

[245] Menegaz D, Barrientos-Duran A, Kline A, Silva FRMB, Norman AW, Mizwicki MT, et al. alpha,25(OH)2-Vitamin D3 stimulation of secretion via chloride channel activation in sertoli cells. The Journal of Steroid Biochemistry and Molecular Biology. 2010;119(3):127–34.

[246] Hamden K, Carreau S, Jamoussi K, Ayadi F, Garmazi F, Mezgenni N, et al. Inhibitory effects of 1,25dihydroxyvitamin D3 and Ajuga iva extract on oxidative stress, toxicity and hypo-fertility in diabetic rat testes. Journal of Physiology and Biochemistry. 2008;64(3):231–9.

[247] Kagi U, Chafouleas JG, Norman AW, Heizmann CW. Developmental appearance of the Ca2+-binding proteins parvalbumin, calbindin D-28K, S-100 proteins and calmodulin during testicular development in the rat. Cell and Tissue Research. 1988;252(2):359–65.

[248] Shahbazi M, Jeddi-Tehrani M, Zareie M, Salek-Moghaddam A, Akhondi MM, Bahmanpoor M, et al. Expression profiling of vitamin D receptor in placenta, decidua and ovary of pregnant mice. Placenta. 2011;32(9):657–64.

[249] Rittenberg V, Seshadri S, Sunkara SK, Sobaleva S, Oteng-Ntim E, El-Toukhy T. Effect of body mass index on IVF treatment outcome: an updated systematic review and meta-analysis. Reproductive Biomedicine Online. 2011;23(4):421–39.

[250] Jensen MB, Bjerrum PJ, Jessen TE, Nielsen JE, Joensen UN, Olesen IA, et al. Vitamin D is positively associated with sperm motility and increases intracellular calcium in human spermatozoa. Human Reproduction. 2011;26(6):1307–17.

[251] Ross AC, Manson JE, Abrams SA, Aloia JF, Brannon PM, Clinton SK, et al. The 2011 report on dietary reference intakes for calcium and vitamin D from the Institute of Medicine: what clinicians need to know. The Journal of Clinical Endocrinology & Metabolism. 2011;96(1):53–8.

Chapter 13

Vitamin D and Female Reproduction

Heba Elhusseini, Daria Lizneva,
Larisa Gavrilova-Jordan, Noura Eziba,
Mohamed Abdelaziz, Soumia Brakta,
Sunil Halder and Ayman Al-Hendy

Additional information is available at the end of the chapter

http://dx.doi.org/10.5772/64502

Abstract

Vitamin D deficiency has an impact on the reproduction of more than 40% of reproductive age women globally. Fibroids are more common among African-American females owing to their decreased milk consumption and reduced absorption of ultraviolet rays, supporting the relation between vitamin D deficiency and fibroid development. Vitamin D has an inhibitory effect on leiomyoma cells by suppression of proliferation cell nuclear antigen (PCNA), BCL-2, BCL-w, CDK1, and catechol-O-methyltransferase (COMT) protein levels. A growing evidence support the relationship between vitamin D deficiency and endometriosis through overexpression of vitamin D recseptor (VDR) and α-hydroxylase enzyme, however, it is still unclear if the endometriosis patients could benefit from vitamin D supplementation. Effect of vitamin D supplementation on the metabolic outcomes of polycystic ovary (PCO) has been studied and reveled that it is negatively correlated with fasting glucose, fasting insulin, triglycerides, C-reactive protein, free androgen index, and Dehydroepiandrosterone (DHEAS) and positively associated with quantitative insulin sensitivity check index (QUICKI), high density lipoprotein cholesterol (HDL-C), and sexual hormone binding globulin (SHBG), whereas its impact on the ovarian function is still unclear. Vitamin D deficiency may worse the obstetrical outcomes, including preeclampsia, gestational diabetes, low birth weight, increased cesarean section rate, neonatal asthma, seizures, and preterm labor. The relationship between serum levels of 25-hydroxy-vitamin D (25(OH) D) and pregnancy rates in ART is still debatable, with the need to conduct more clinical trials toward it. The in vitro antiproliferative and prodifferentiative effect of vitamin D might find a role in control of hyperplastic overactive bladder. Several studies support that vitamin D deficiency constitutes a risk factor for development of many types of cancer such as breast, ovarian, and colorectal.

Keywords: vitamin D deficiency, Female reproduction, Fibroid, Fertility, Overactive bladder

1. Introduction

Over the past decade, a global pandemic of vitamin D deficiency has grown among all racial groups. Based on the National Health and Nutrition Examination Survey analysis, the overall revalence rate of vitamin D deficiency was 41.6%, with the highest rate detected in blacks (82.1%), followed by Hispanics (69.2%) [1].

Vitamin D is a naturally occurring steroidal hormone whose primary role in the human body is calcium homeostasis, keeping bones healthy and strong. However, a recent body of research strongly indicates that vitamin D's relevance and significance extend well beyond just keeping bones healthy and strong but rather plays a more pivotal role in the body's overall health, including its role or lack thereof in chronic diseases such as diabetes, obesity, autoimmune disease(s), cardiovascular disease, and cancer [2]. This is largely attributed to vitamin D's ability to affect different types of cells by turning genes within these cells "on and off"; thereby playing a major role in controlling cellular growth, function, and death [3].

Many studies have recently investigated the relationship between levels of vitamin D levels and cancer [3]. Low levels of vitamin D have been associated with a 30–50% increased risk of colon, prostate, and breast cancer [4]. Indeed, vitamin D's growing role in human immunity might provide a logical explanation for these disease manifestations. However, data that support a definitive causal relationship between vitamin D deficiency and these cancers, as well as, further elucidate the associated benefits of vitamin D supplementation are extremely limited [4].

In female reproduction, the importance of vitamin D was initially appreciated *in vivo*, as mice who were either deficient in vitamin D or lacked the vitamin D receptor (VDR), suffered from underdeveloped uteri and an inability to form normal mature eggs, which in turn lead to infertility [5]. In humans, VDR, a member of the nuclear receptor family, is expressed in many female organs, including the ovaries (granulosa cells), uterus (endometrium and myometrium), and placenta [6]. These receptors are targeted by the active form of vitamin D (calcitriol = 1, 25 dihydroxy vitamin D) and produce an array of effects in female reproduction. For example, calcitriol regulates genes involved in estrogen synthesis [6]. It also controls several genes involved in embryo implantation [7].

Poor vitamin D status has been associated with a wide array of obstetrical complications and gynecological diseases [6, 7]. Furthermore, vitamin D has also played a progressive role in assisted reproductive techniques such as *in vitro* fertilization [8, 9]. In this chapter, we critically summarize the most recent data regarding the impact of vitamin D deficiency on female reproduction and related disorders.

http://dx.doi.org/10.5772/64502

2. Vitamin D and uterine fibroids

Uterine fibroids (AKA: leiomyoma) are the most common hormone-dependent gynecologic tumors, affecting up to 70% of reproductive aged women. They arise from the proliferation of smooth muscle cells, forming a mass surrounded by a pseudocapsule of compressed muscle fibers [10, 11].

They are often asymptomatic, discovered incidentally in routine bimanual pelvic and/or ultrasound examination. Nevertheless, some leiomyomas may be complicated by a variety of symptoms including abnormal uterine bleeding, pelvic pressure, and pain, increased urinary incontinence, bowel disturbance, and are associated with infertility and recurrent abortion [12, 13]. Consequently, surgery represents the main treatment modality for symptomatic cases [11].

Hysterectomy is usually an option for women who have completed childbearing; however, many women may prefer to be treated with other conservative therapies as myomectomy to preserve her future fertility [14]. No definitive medical treatment has been established and can be used for short-term therapy. Evolving agents might have a role in the near future, such as vitamin D, green tea extract, and elagolix (oral GnRH antagonist). Furthermore, agents, such as selective E receptor modulators (SERMs) and gestrinone, can be used to decrease leiomyoma size with minimal side effects [15].

Sabry and Al-Hendy [15] have studied the potential effect of epigallocatechin-3-gallate (EGCG), one of the major green tea components, on the human leiomyoma cells. They found that EGCG inhibits the proliferation of these cells and induces apoptosis.

The pathophysiology behind the development of uterine fibroids is still not completely understood, growing evidence has supported the fact that both estrogen and progesterone play major roles in fibroid growth [16]. Al-Hendy group studied the Med12 gene somatic mutations in females with symptomatic uterine fibroids from the southern United States. They found four novel somatic mutations in the Med12 gene in uterine fibroids in this population, whereas, no mutations were identified in the Med12 gene in normal myometrium in these women [17].

Several studies revealed a two- to threefold higher incidence of uterine fibroids in African-American females as compared with other racial types, including Caucasians, Hispanics, and Asians [18]. This is supported by finding of uterine leiomyomas in 75% of hysterectomies performed on African-American women [19].

Analysis of leiomyoma phenotype performed by Baird et al. revealed that 73% of African-American females had multiple leiomyoma on ultrasound, whereas only 45% of Caucasian females pretended this phenotype [18]. The cumulative incidence of development of fibroid among African-American females is 80% by age 50 years, with annual incidence of 3% during their reproductive period [20, 21].

There are several dietary sources of vitamin D, such as fatty fish, fish oils, fortified foods, and vitamin supplements; however, sunlight exposure remains the main source of vitamin D [22]. High melanin concentrations in African-Americans have largely contributed to decreases the absorption of ultraviolet rays from the sun. Furthermore, decreased milk consumption due to lactose intolerance diminishes the levels of vitamin D as well [23, 24].

Al-Hendy et al. have addressed a correlation between lower serum vitamin D levels and an increased risk of uterine leiomyoma in 2013 in a cohort of black and white females from North Africa. In addition, they revealed a significant inverse association between vitamin D serum levels and the severity of fibroids among African-American females [16].

These findings were supported by Baird et al. when they determine that women with sufficient levels of vitamin D were less likely to develop uterine fibroids and found that levels of vitamin D was 10% of African-Americans and 50% of Caucasians, with an adjusted odds ratio of 0.68 [25]. Also, Paffoni et al. found that women with vitamin D deficiency were more likely to have uterine fibroids, with an adjusted odds ratio of 2.4 [26].

The exact mechanism of uterine leiomyoma development is still unclear; however, there are several contributing factors, including clonal smooth muscle cell proliferations, chromosomal abnormalities, hormonal deregulation, and growth and angiogenic factors [27–30]. Catherino et al. have postulated that protein encoding genes from the extracellular matrix (ECM) were overexpressed in leiomyomas. Consequently, analysis of the ECM in leiomyoma tissue revealed a disturbed orientation of collagen fibril with reduction of its binding protein, which is called dermatopontin. Nevertheless, the latter was associated with an increase in transforming growth factor (TGF)-b3 messenger RNA levels [31].

Currently, TGF-b3 represents the only growth factor found to be overexpressed in leiomyoma samples during the secretory phase [32]. Recent COMT (catechol-O-methyltransferase) and ER-α (E receptor-α) polymorphism analyses in women from different ethnic groups was performed by Al-Hendy et al. and concluded that females with a high expression genotype for COMT were 2.5 times more likely to develop leiomyomas than females with other genotypes. That points to the vital role of submicroscopic genetic anomalies in formation of leiomyoma in African-American females [33].

Blauer et al. studied the role of vitamin D3 in the regulation of uterine leiomyomas growth and demonstrated that bioactive 1a, 25(OH) 2D3 inhibits the growth of both leiomyoma and myometrial cells derived from human tissues of premenopausal females undergoing hysterectomy. This growth inhibition was found to be a concentration dependent, being a concentration of 100 nM-the physiological level, sufficient to produce that inhibition [34, 35]. Baird et al. addressed that women with uterine leiomyomas have lower levels of serum vitamin D3 compared to their healthy counterpart women [25]. Moreover, serum levels of vitamin D3 are inversely proportional to leiomyoma sizes, supporting that vitamin D3 deficiency could be a potential risk factor for the development of uterine leiomyoma [26].

Al-Hendy group studied the mechanism of action of vitamin D on human uterine leiomyoma cell proliferation. Cells were treated with vitamin D3, followed by measurement of proliferation cell nuclear antigen (PCNA), BCL-2, BCL-w, CDK1, and COMT protein levels. They found a downregulation of PCNA, CDK1, and BCL-2 and suppression of COMT expression in human leiomyoma cells, favoring that vitamin D3 inhibits growth and induces apoptosis in cultured leiomyoma cells. In the following study, they tested the effect of vitamin D3 on TGF-b3–

induced fibrosis-related protein expression in human cells and concluded the suppressant effect of vitamin D3 onTGF-b3 in human leiomyoma cells [34].

Wei et al. conducted a study to verify the ethnic differences in tumorigenic factors of uterine leiomyomas. They identified selective genes by performing tissue microarray analyses and specific immunohistochemistry procedures involved in the development of leiomyomas and compared the results with matched myometrial tissue. They revealed that P receptor PR-A was upregulated in fibroid tissue of African-American females in comparison to other ethnic groups. Moreover, the E receptor, ER-a, was elevated in both the normal myometrial and leiomyoma tissues of African-American females when compared with other ethnic groups [35].

Recently, Al-Hendy et al. assessed the effect of vitamin D3 on leiomyoma growth in the Eker Rat model of uterine fibroids. They found that treatment with vitamin D3 significantly minimize leiomyoma size by inhibiting cell growth, proliferation-related genes (PCNA, cyclin D1 [Ccnd1], c-Myc, CDK1, CDK2, and CDK4), antiapoptotic genes (BCL2 and BCL-xl), and E receptor ER-a, and P receptors PR-A and PR-B [36]. Similarly, they found that paricalcitol, an analog of 1, 25-dihydroxyvitamin D3, significantly decreased fibroid tumor size in female nude mice as compared with placebo [37].

3. Vitamin D and endometriosis

Endometriosis is a chronic gynecological disorder affecting 5–10% of female population of reproductive age, with increased prevalence up to 30–40% among infertile women [38]. It can be defined as the presence of endometrial tissue in ectopic locations including ovaries, bladder, and bowel. The most common symptoms are dysmenorrhea, dyspareunia, chronic pelvic pain, and infertility [39].

The exact pathogenesis of the endometriosis is still questionable; however, several theories have been suggested. One of the most supporting theories is the development of immune system dysfunction, which leads to a state of chronic inflammation [40, 41].

A series of immunologic changes have been reported leading to endometriosis development, including a reduction in T -cell cytotoxicity, a functional deficit of natural-killer lymphocytes and higher concentration of activated macrophages in the peritoneal fluid, which consequently trigger a cascade of cytokines and vascular endothelial growth factors promoting proliferation of endometrial cells and angiogenesis [42, 43]. Genetic predisposition may play a role in incidence of endometriosis. It has been reported that first degree relatives have a three- to fivefold increased risk of endometriosis development [44].

Furthermore, there are several recognized endometriosis susceptibility genes, which are associated with steroid hormone action, immune response, oxidative stress, glucose homeostasis, vascular and tissue remodeling, and apoptosis [45, 46].

Several investigators studied the potential correlation between endometriosis and vitamin D. Viganò et al. addressed that the endometrium expresses the VDR and 1α -hydroxylase enzyme irrespective of the menstrual cycle. Furthermore, they found that 1α-hydroxylase is expressed both in the eutopic and in the ectopic endometrial cells of women affected by endometriosis and that the enzyme expression is higher in the proliferative phase of the menstrual cycle [43].

Agic et al. supported these results and found an elevation of 24-hydroxylase in patients with endometriosis, indicated a very active metabolic process of vitamin D in the endometrium. These studies propose a local paracrine action of vitamin D, that could be involved either in the regulation of the immune system activity and in the cytokine production [47].

However, it is very hard to determine whether the endometriosis patients may benefit from Vitamin D supplementation, as the relationship between vitamin D and endometriosis seems to be more complicated. Hartwell and colleagues tested for the first time the metabolism of vitamin D in 42 women with endometriosis. They discovered that levels of 25(OH) D in the serum were normal, whereas the levels of 1, 25(OH) 2D3 were increased compared to the control group [48]. Lasco et al. conducted a prospective study, to examine the effect of a singleloading dose of cholecalciferol (300,000 IU) on primary dysmenorrhea; they found a significant reduction of pain in the supplemented group compared with the placebo group ($P < 0.001$) [49].

Somigliana et al. showed that the levels of 25(OH) D were significantly increased in the group of women with endometriosis, whereas the levels of 1, 25(OH) 2D3 and calcium were the same compared to the control group [50].

A prospective study was conducted by Harris et al. They reviewed 70,556 women, including 1385 with endometriosis and 69,171 matched controls regarding age, season, race, geographical region, alcohol intake, and physical activity. They found an inverse association between serum values of 25(OH) D and endometriosis: women in the highest predicted 25(OH) D quintile and highest intake of vitamin D from food had respectively a 24% and 21% lower risk of endometriosis compared with those in the lowest quintile. These results support the hypothesis that low levels of vitamin D are associated with an increased risk of endometriosis [51].

4. Vitamin D in polycystic ovary syndrome

Polycystic ovary syndrome (PCOS) is the most common endocrinopathy in the reproductive age women, with a prevalence of 6–19% in the general population [52–56]. The etiology of the syndrome remains largely unknown. Key characteristics of PCOS include ovulatory dysfunction, hyperandrogenism, and polycystic ovaries [57–59].

Insulin resistance (IR), yet another key feature of PCOS, plays a significant role in the development of metabolic complications such as type 2 diabetes mellitus, dyslipidemia, abdominal obesity, and increased risk of cardiovascular disease (CAD) [60, 61].

Current evidence supports the important role of vitamin D in energy metabolism and homeostasis. Animal studies have demonstrated that vitamin D signaling is directly involved in transcriptional activation of the insulin receptor gene [62] and inhibits pro-inflammatory cytokines. Thus, the metabolic disarrangements observed in IR among PCOS patients may have a connection to the suboptimal Vitamin D level [63].

A recent systematic review and meta-analysis summarized the relationship between serum level of vitamin D and metabolic outcomes in women with PCOS, as well as determined the effects of vitamin D supplementation [64]. A pooled estimate of five observations revealed no significant difference in 1, 25(OH) 2D levels among PCOS patients as compared to the controls (SMD: 0.18; 95%CI: −0.10 to 0.45) [64].

Interestingly, vitamin D deficient PCOS patients were found to have lower HDL-C, higher fasting glucose, fasting insulin, HOMA-IR, HOMA-β, and FAI [64]. Serum levels of 25(OH) D were negatively correlated with fasting glucose, fasting insulin, triglycerides, C-reactive protein, free androgen index, and DHEAS among PCOS patients [62]. Moreover, vitamin D was found to be positively associated with QUICKI, HDL-C, and SHBG [64].

Studies addressing pretreatment and post-treatment with vitamin D supplementation have shown only significant decreases in triglyceride levels −0.45 (−0.73, −0.17) [64]. Furthermore, supplementation with vitamin D did not demonstrate any significant difference in metabolic parameters, androgen levels, and serum levels of 25(OH) D as compared to placebo [64].

In summary, clinical data support that vitamin D status is related to metabolic dysfunctions in PCOS. Moreover, vitamin D deficiency may worsen existing metabolic disarrangement in PCOS [64]. However, limited clinical evidence found no improvement of those disarrangements with a standard vitamin D supplementation. Over the past several years, there has been significant interest to vitamin D's effect on ovarian function in PCOS [65–71].

Clinical data demonstrate that 25OHD3 deficiency can be negative predictor of follicle development with clomiphene citrate stimulation [65]. However, the exact mechanism of vitamin D action on mammalian ovaries is not clearly understood. A vitamin D receptor has being identified in ovarian granulosa cells [66, 67], and animal studies suggest that the promoter for Anti-Müllerian hormone (AMH) is under vitamin D downregulation [67].

It is also well-known now that excessive ovarian production of AMH, secreted by growing follicles is an important feature of PCOS [68]. Taken into account that vitamin D signaling can modify the expression of AMH in ovaries, it is highly probable that vitamin D supplementation may also affect ovarian physiology in PCOS and possibly improve folliculogenesis. Surprisingly, there have been very few studies published. Thys-Jacobs et al. reported a study with 13 oligomenorrheic normocalcemic PCOS women who received vitamin D and calcium. Two months of treatment resulted in normalized menstrual cyclicity for 7 of 13 women [68].

In a more recent small clinical trial, 60 infertile PCOS patients were randomized into three groups: [1] vitamin D with calcium; [2] metformin only; and [3] combination of both [69]. The combination group demonstrated the higher number of dominant follicles at second- and third-month follow-up visits [69]. In another study of 67 vitamin D deficient women with and without PCOS, vitamin D supplementation was associated with decreased serum levels of AMH in PCOS patients, which suggests a possible improvement in ovarian physiology [70].

In conclusion, the results of basic research and several small clinical studies suggest that vitamin D has a positive effect on ovarian function in PCOS women; however, further clinical trials are needed.

5. Vitamin D and pregnancy

The role of vitamin D in pregnancy outcomes has recently gathered much attention. Pregnancy is a state of increased calcium demand and fetal/neonatal vitamin D status is dependent on the maternal level of vitamin D [71, 72].

Observational studies have shown consistent associations between obstetrical clinical outcomes and poor maternal vitamin D status, including preeclampsia, gestational diabetes, low birth weight, increased cesarean section rate, neonatal asthma, seizures, and preterm labor [73–75].

Although vitamin D deficiency in pregnancy is documented to be common by CDC (Center of Disease Control and Prevention) and the WHO (World Health Organization), a clear consensus for screening and management of vitamin D deficiency in pregnancy has not yet been adopted, owing to a paucity of research data regarding the role of vitamin D in pregnancy biology and limited clinical trials on the use of vitamin D supplementation to improve obstetrical outcomes [73, 76].

Studies have showed that vitamin D supplementation impacts inflammatory markers, contractile-associated proteins, estrogen receptor α, and progesterone receptor A/B ratio in human uterine smooth muscle cells. Vitamin D prevents inflammation-induced changes in myometrial cells mediated through the nuclear factor (NF)-κB pathway [77]. Deficient or low levels (hypovitaminosis) of vitamin D during pregnancy might be a risk factor for preterm birth [78, 79].

There is increasing evidence that corticotropin releasing hormone (CRH) plays a pivotal role in the control of human pregnancy and parturition. During human pregnancy, the placenta and fetal membranes produce large amounts of CRH, which steadily increases in concentration with advancing pregnancy [80]. It has been shown that CRH promotes myometrium quiescence during most of pregnancy, whereas it facilitates myometrial contractility after the onset of parturition. However, the mechanisms by which CRH exerts such dual effects remain unclear [81].

Vitamin D and CRH might interact during pregnancy through a number of pathways. The activation of CRH receptors initiate a variety of subsequent signals including protein kinase C (PKC) pathway that increase vitamin D receptor expression [82]. CRH synthesis is partly regulated through a noncanonical NF-κB pathway, which might be affected by vitamin D status [82]. A poor understanding of spontaneous preterm birth (PTB) or its risk factors and a lack of reliable biomarkers contribute to the difficulty in prevention, early diagnosis, and treatment of PTB [83].

6. Vitamin D and reproduction

Evidence from both animal and human studies strongly suggests a potential role of vitamin D in human reproduction. Assisted reproductive technology (ART) has been presented as a valuable model to study the effect of vitamin D deficiency on specific aspects of human fertility as it allows the separate evaluation of the various step of the reproductive process, including sperm function, folliculogenesis, and embryo implantation [84].

In vitro studies, it has been shown that vitamin D receptors are expressed in murine endometrium and ovary throughout the estrous cycle, whereas knockout experiments have shown that vitamin D receptor null mice experience uterine hypoplasia and impaired folliculogenesis [85].

Whereas, *in vivo* data supporting a role for vitamin D in female fertility and embryo implantation are still not conclusive. Some studies have revealed findings showing that maternal vitamin D deficiency is associated with lower pregnancy rates and others demonstrating that vitamin D deficiency does not affect the final reproductive outcome. Multiple studies investigated the association between serum levels of 25-hydroxy-vitamin D (25(OH) D) and pregnancy rates in ART with controversial results [84].

It is observed that serum 25 (OH)D levels were significantly related to implantation, clinical pregnancy, and live birth rates, although opposite trends were found according to patients ethnicity being critical in non-Hispanic whites but not in Asian ethnicity. In a second study, the same authors examined serum 25(OH) D concentration among recipients of oocyte donation, finding a positive association between vitamin D status and clinical pregnancy rate and suggesting the specific effect of 25(OH)D levels on ART outcomes to be mediated by endometrial receptivity rather than by ovarian stimulation or embryo parameters [85]. Interestingly, both cyclic and early pregnancy endometrium represent an extrarenal site of vitamin D synthesis; thus, the effect of vitamin D at the uterine level is thought to be exerted via the vitamin D receptor (VDR) through either the regulation of target genes or the hormonal effects on the local immune response [84].

A recent study has evaluated the influence of vitamin D deficiency on pregnancy rates among women undergoing In Vitro Fertilization/Intracytoplasmic Sperm Injection (IVF/ICSI) and day 5 single embryo transfer (SET). A total of 368 consecutive infertile women treated within a period of 15 months were included in the study. Serum vitamin concentration was measured retrospectively in all included patients. They found that clinical pregnancy rates were significantly lower in women with vitamin D deficiency compared with those with higher vitamin D values. Finally, even when restricting the analysis to women undergoing elective SET, vitamin deficiency was again independently associated with pregnancy rates [84].

In a cross-sectional analysis, a cohort of 1072 women with a mean age of 36.3 attending an academic infertility center were used to examine serum 25-hydroxy-vitamin D 25(OH)D) levels in relation to demographic characteristics, seasons, and general health risk factors. They found that median 25(OH) D concentration was below 30 ng/ml for 89% of the entire year. Over the whole year, 6.5% of patients had 25(OH) D levels </- 10 ng/ml, 40.1% </-20 ng/ml, and 77.4%

</-30 ng/ml. Global solar radiation was weakly correlated with 25(OH) D levels. Multivariate data analysis reveals that, 25(OH) D levels were inversely associated with basal metabolic rate (BMI); conversely, 25(OH) D levels were positively associated with height and endometriosis history. Serum 25(OH) D levels are highly deficient in women seeking medical help for couple's infertility. Levels are significantly associated with body composition, seasonal modification, and causes of infertility. Importantly, this deficiency status may last during pregnancy with more severe consequences [84].

In another cross-sectional study, Paffoni et al. investigated the IVF outcome in women with deficient 25(OH) D serum levels <20 mg/ml. They included 154 women with serum 25(OH)D <20 ng/ml and 181 women with levels of >/-20 ng/ml. They found that the clinical pregnancy rates were 20% and 31%, respectively, with an adjusted odds ratio of 2.15 for clinical pregnancy in women with vitamin D >/-20 ng/ml. Furthermore, a subgroup analysis revealed that the group with the highest serum level of vitamin D (>30 ng/ml) resulted in the highest chances of pregnancy [85].

More recently, Dressler et al. conducted a retrospective cohort study at two centers in Germany to investigate the prevalence of vitamin D deficiency among women with impaired fertility and to identify the risk factors associated. They found that 98.2% of women at center 1 and 81.3% of women at center 2 had deficient or insufficient vitamin D levels. Moreover, they found that overweight BMI and limited exposure to sun (winter, spring, and autumn trimester) were associated with vitamin D deficiency [86].

In an observation case-control study, Al-Jaroudi et al. compared the dietary vitamin D and calcium intake among subfertile women and pregnant (control) women to determine vitamin D levels. They found that vitamin D levels were significantly higher in the subfertile group compared to the control group (59.0% vs 40.4%; P<0.01) [87].

In Contrast, Franasiak et al. showed that vitamin D status was unrelated to pregnancy outcomes in women undergoing euploid blastocyst transfer [88].

They attempted to characterize the relationship between serum 25-hydroxy vitamin D (25-OH D) levels and implantation and clinical pregnancy outcomes in 517 women undergoing a euploid blastocyst embryo transfer. They concluded that serum vitamin D ranges and pregnancy outcomes did not correlate. However, their results may not apply to the patients who do not undergo extended embryo culture, blastocyst biopsy for comprehensive chromosome screening and euploid embryo transfer [88]. In a prospective cohort study, vitamin D (25OH-D) serum and follicular fluid levels were analyzed in 82 infertile women undergoing ART. They found that fertilization rate decreased significantly and the implantation rate increased (not significantly) with increasing levels of 25OH-D [89]. Using the same approach, Farzadi et al. reported a correlation between follicular fluid 25(OH) D concentration and assisted reproductive outcomes in an Iranian population [90].

In a retrospective study, serum and follicular levels of 25-OH vitamin D were collected from 80 infertile female candidates for IVF/ICSI to investigate the possible association of vitamin D with assisted reproductive outcome. They found a statically significant positive correlation between 25-OH vitamin D levels with patient age and implantation rate [91].

Although optimization of vitamin D levels is encouraged for the general reproductive health but more research is needed to understand the impact on reproductive potential. Most studies have small sample sizes, heterogeneous experimental design, and great confounders, such as obesity. Prospective studies are needed to confirm causal relationship and to investigate the potential therapeutic benefits of vitamin D supplementation in this population. Vitamin D deficiency has been shown to impair pregnancy rates in women undergoing single blastocyst transfer. Future prospective confirmatory studies are needed; preferably randomized controlled trials of vitamin D supplementation with an appropriate assessment of pregnancy outcomes.

7. Vitamin D and overactive bladder

Overactive bladder syndrome (OBS) is a highly prevalent condition, affecting 17% of the population worldwide, with more than 17 million people affected in the United States and more than 22 million adults affected in Europe [92–94].

Being a recently defined syndrome, its risk factors have not been determined yet; however, it is believed to be multifactorial. Zhang et al. concluded that the contributing risk factors are advanced age, menopause, parity >2, constipation, Hx of episiotomy, and high basal metabolic rate (BMI) [95].

Different drugs directed toward the central, peripheral sympathetic, parasympathetic, or sensory nervous pathways, as well as the detrusor muscle itself, have been studied and described [96]. Antimuscarinics have become the standard therapy for OAB, but their tolerability is limited by several adverse events, often leading to poor compliance and drug discontinuation [97].

Consequently, because the need for new drugs that provide similar or even greater clinical efficacy but with fewer side effects is evolving, those patients with OAB consistently require long-term therapy to control their symptoms [97, 98]. OBS shares epidemiological and pathophysiological features with preterm birth. Our recently published work suggests that vitamin D deficiency is a novel risk factor for preterm birth, a condition about four times more prevalent in African-Americans; who also have higher prevalence of vitamin D deficiency; as compared to their Caucasian counterparts. Furthermore, we have shown that vitamin D elicits a robust anti-inflammatory response in human myometrial cells [99]. Coyne et al. has demonstrated the prevalence of OBS in the total United States population and found that black women had a higher prevalence (32.6%), compared to Hispanic (29%) and white women (29.4%) [100].

This finding was supported by many studies as well [101, 102]. Vitamin D3 functions through the nuclear vitamin D receptor (VDR) and acts on VDR target genes. This vitamin D-mediated gene activation requires a VDR/retinoid X receptor heterodimer complex [102]. Recent studies have examined the expression of vitamin D receptors (VDRs) in the human bladder [102]. Bladder cell overgrowth and smooth muscle overactivity have been implicated in the initial steps of bladder decompensation and lower urinary tract symptoms (LUTS) [103].

The hyperplastic overactive bladder could represent an ideal candidate for treatment with paricalcitol in view of its antiproliferative and prodifferentiative effects on bladder cells in culture, which probably contribute to the control of smooth muscle cell overactivity, as well as considering the strong association of a high dietary intake of vitamin D with a decreased risk of overactive bladder [104].

Schroder et al. used a rat model of partial bladder outflow obstruction and found that VDR agonists reduced the incidence of spontaneous bladder contractions during filling through the inhibition of RhoA/Rho-kinase activity [105].

8. Vitamin D deficiency and risk of gynecological cancer in women

A number of studies have shown its association with risk of several types of cancers [106, 107]. Higher prevalence of vitamin D deficiency, together with the increased risks of certain types of cancer in those who are deficient in vitamin D3, suggesting that vitamin D deficiency may account for several thousand premature deaths from colorectal [108], breast [109], ovarian [110], and prostate [111] cancer annually [112]. People exposed to sunlight were noted to be less likely developed cancer. Thus, these findings inspired us for ensuring adequate vitamin D intake in order to reduce the risk of several gynecological cancers such as breast, ovarian, endometrial, and cervical cancers.

8.1. Vitamin D and breast cancer risk

Several case-control and laboratory tests have demonstrated an important role of vitamin D in the prevention of breast cancer. Low vitamin D intake is associated with increased risk of breast cancer in premenopausal women [106].

Daily vitamin D intake of greater than 500 IU had been shown significantly reduced breast cancer risk than those were consumed less of vitamin D [107, 113].

In a study by Lin et al. showed that higher intake of vitamin D and calcium were able to reduce the risk of premenopausal breast cancer [114]. 1,25(OH)2D3 exerts its antiproliferative effects on breast cancer cells by a number of ways, including by altering the expression of oncogenes and tumor suppressor genes, several cyclins, and cyclin-dependent kinase inhibitors p21WAF-1/CIP-1 and p27kip1 [114, 115]. 1,25(OH)2D3 also induce apoptosis in breast cancer cells by stimulating Ca2+ release from intracellular stores that result in rising cytosolic Ca2+ which triggers calpain-mediated caspase-independent programmed cell death [115]. This synergistic actions of calcium and vitamin D are probably the cause why high intake of low-fat dairy products is associated with a reduced risk of breast cancer in premenopausal women [116, 117].

Studies have shown that breast cancer death rates tend to be higher in low winter sunlight levels, whereas it is lower in sunny areas [118]. Women who are regularly exposed to sunlight and ingest sufficient amounts of vitamin D had significantly lower prevalence of breast cancer [119]. It has also been shown that women in the lowest quartile of serum levels of 1,25(OH)2D3 had a five times higher risk of breast cancer than those in the highest quartile [120]. Low levels of 1,25(OH)2D3 were also associated with faster progression of metastatic breast cancer [121]. Studies showed that high intake of vitamin D and calcium markedly reduced the incidence of mammary cancer in experimental mice and rats that were given high-fat diets [122–123]. Furthermore, high levels of vitamin D and calcium intake was able to reduce the incidence of mammary cancer in rats [123].

8.2. Vitamin D and ovarian cancer risk

Ovarian cancer is the fifth leading cause of cancer death among women in the United States [124]. Low levels of serum vitamin D was reported in ovarian cancer patients [125–127], and that low concentrations of 25(OH)D3 was associated with lower overall survival rate, whereas higher 25(OH)D3 concentrations significantly associated with longer survival among women with ovarian cancer [128].

This observation indicates that severe vitamin D deficiency may play a role in the development of more aggressive ovarian cancer. Several epidemiological studies have identified higher mortality rates of ovarian cancer in areas of higher latitude and lower levels of solar irradiation [110, 128, 130].

Most of these studies have also shown a lower mortality rate of premenopausal ovarian cancer in sunny regions [128, 129]. These findings have been supported by observational studies of dietary intake of vitamin D [131] and of pre-diagnostic serum 25(OH) D3 [132]. A study had shown that the lower level of pre-diagnostic serum 25(OH) D3 was associated with high risk of ovarian cancer in overweight women, whereas that was not the case for thinner women [132].

It is also recommended that serum 25(OH) D3 measurement could be a standard procedure that might be helpful to diagnose ovarian cancer patients with worse prognosis. In addition, supplementation of vitamin D3 at moderate doses achieving 25(OH) D3 serum concentrations of 30–80 ng/ml could be beneficial for reducing the risk of developing ovarian cancer.

8.3. Vitamin D and cervical cancer risk

An independent study in China and France also showed inverse correlation between solar UVB indices and cervical cancer incidence rate [133, 134].

In addition, a case-control study in Japan showed significant reduction of cervical cancer risk with increasing oral vitamin D intake [135]. Moreover, a recent case report indicated that patient suffering with abdominal pain due to cervical cancer-related treatment was improved after vitamin D replacement therapy [136].

These studies suggest that vitamin D deficiency might play risk for cervical cancer development that can be prevented by oral intake of vitamin D.

8.4. Vitamin D and endometrial cancer risk

A case–control study from seven cohorts evaluated the inverse association between serum concentrations of vitamin D3 and the risk of development of endometrial cancer [137–139].

It is recommended to measure serum concentrations of vitamin D3 that permits to estimate the risk of association with endometrial cancer, and proper levels of vitamin D intake could reduce the risk of development of endometrial cancer.

8.5. Vitamin D and cancer prevention

A number of epidemiological studies have demonstrated the association between vitamin D deficiency and risk of several types of cancers. Strong evidence also indicates that vitamin D3 intake is associated with reduced incidence and death rates of colon, breast, prostate, and ovarian cancers [140].

Evidence also proved that vitamin D3 intake of 2000 IU/day would lead to the reduction of breast cancer and colon cancer incidence [140]. Thus, vitamin D3 supplementation could address the high prevalence of vitamin D deficiency and could prevent many deaths from breast and colorectal cancers in the United States [140].

The measurement of serum concentrations of vitamin D3 is important to assay the risk of various cancers, and an intake of recommended levels of vitamin D3 per day could be very beneficial to prevent many deaths from cancers in the United States.

9. Conclusions and future directions

In strictly seasonal breeders, an increase in photoperiod (longer day time, more sun) affects pineal gland, which in turn alter melatonin secretion, affect gonadotropin secretion, and finally place animal in or out of estrus [141].

Humans are continuous but probably still partial seasonal breeder. The sunshine hormone (vitamin D) may have something to do with it. As female human emerge from the cloudy/rainy winters, vitamin D rises and female reproduction is optimized (effects on ovary, egg quality, endometrium (implantation), and myometrium), beside potential central effects of vitamin D on hypothalamic-pituitary axis [142].

Conception occurs soon and delivery takes place about 9 months later by early autumn, which is also optimal as it is harvest season and there is abundance of food to support the nursing mom to take care of the hungry baby. In fact, CDC life birth rates support such model and consistently show highest rates in late summer and early fall month [143]. Clearly, such innate breeding pattern in human has been influenced and largely muted by various ever evolving religious, cultural, and social traditions and etiquettes in human civilization on this planet.

The future research focusing on translational applications of various fundamental observation described in this review will likely to have major positive impact on women reproductive health. Both pharma and academia have synthesized many highly potent and safe VDR agonists that will soon undergo rigorous preclinical and subsequent clinical evaluation for utility in various female reproductive disorders. Furthermore, more clinical research should examine the possible association of vitamin D deficiency with additional adverse reproductive

http://dx.doi.org/10.5772/64502

outcomes. Clearly, future effort will be utilized in patients' counseling regarding screening for vitamin D status and appropriate vitamin D supplementation when indicated for overall health benefits, including bone health, reproductive health, and chronic disease risk reduction.

Author details

Heba Elhusseini[1], Daria Lizneva[1], Larisa Gavrilova-Jordan[1], Noura Eziba[1], Mohamed Abdelaziz[1,2], Soumia Brakta[1], Sunil Halder[1] and Ayman Al-Hendy[1*]

*Address all correspondence to: aalhendy@gru.edu

1 Deparment of Obstetrics and Gynecology, Medical College of Georgia, Georgia Regents University, Augusta, GA, USA

2 Department of Obstetrics and Gynecology, Mansoura Faculty of Medicine, Mansoura University Hospital, Mansoura, Egypt

References

[1] Kimberly YZF and Wendy LS. Prevalence and correlates of vitamin D deficiency in US adults. Nutr Res 2011(31): 48–54.

[2] Michaelsson, K; Baron, JA; Snellman, G; Gedeborg, R; Byberg, L; Sundstrom, J; Berglund, L; Arnlov, J et al. Plasma vitamin D and mortality in older men, a community based prospective cohort study. The Am J Clin Nutr 2010;92(4):841–848.

[3] Yoshizawa T, Handa Y, Uematsu Y, Takeda S, Sekine K, Yoshihara Y, et al. Mice lacking the vitamin D receptor exhibit impaired bone formation, uterine hypoplasia and growth retardation after weaning. Nat Genet 1997;16:391–396.

[4] Halloran BP and DeLuca HF. Effect of vitamin D deficiency on fertility and reproductive capacity in the female rat. J Nutr 1980;110:1573–1580.

[5] Johnson LE, DeLuca HF. Vitamin D receptor null mutant mice fed high levels of calcium are fertile. J Nutr 2001;131:1787–1791.

[6] Vigano P, Lattuada D, Mangioni S, Ermellino L, Vignali M, Caporizzo E, Panina-Bordignon P, Besozzi M, Di Blasio AM. Cycling and early pregnant endometrium as a site of regulated expression of the vitamin D system. J Mol Endocrinol 2006;36(3):415–424.

[7] Bodnar LM, Catov JM, Simhan HN, Holick MF, Powers RW, Roberts JM. Maternal vitamin D deficiency increases the risk of preeclampsia. J Clin Endocrinol Metab 2007;92(9):3517–3522.

[8] Ozkan S, Jindal S, Greenseid K, Shu J, Zeitlian G, Hickmon C, Pal L. Replete vitamin D stores predict reproductive success following *in vitro* fertilization. Fertil Steril 2010;94(4):1314–1319.

[9] Rudick B, Ingles SA, Chung K, Stanczyk F, Paulson R, Bendikson K. Characterizing the influence of vitamin D levels on IVF outcomes. Human reproduction 2012; 27 (11): 3321–3327.

[10] Cook JD, Walker CL. Treatment strategies for uterine leiomyoma: the role of hormonal modulation. Semin Reprod Med 2004;22: 105–111.

[11] Eltoukhi HM, Modi MN, Weston M, Armstrong AY, Stewart EA.: The health disparities of uterine fibroid tumors for African American women: a public health issue. Am J Obstet Gynecol 2014; 210:194–199.

[12] Hart R, Khalaf Y, Yeong CT, Seed P, Taylor A, Braude P. A prospective controlled study of the effect of intramural uterine fibroids on the outcome of assisted conception. Hum Reprod 2001;16:2411–2417.

[13] Surrey ES, Lietz AK, Schoolcraft WB. Impact of intramural leiomyomata in patients with a normal endometrial cavity on *in vitro* fertilization-embryo transfer cycle outcome. Fertil Steril 2001;75:405–410.

[14] Donna Day Baird, Michael C. Hill, Joel M. Schectman, and Bruce W. Hollis. Vitamin D and the risk of uterine fibroids. Epidemiology 2013;24(3):447–453.

[15] Sabry M, Al-Hendy A. Innovative oral treatments of uterine leiomyoma. Obstet Gynecol Int 2012; 2012: 943635.

[16] Mohamed Sabry, Sunil K Halder, Abdou S Ait Allah, Eman Roshdy, Veera Rajaratnam, Ayman Al-Hendy. Serum vitamin D3 level inversely correlates with uterine fibroid volume in different ethnic groups: a cross-sectional observational study. Int J Women's Health 2013:5 93–100.

[17] Halder SK, Laknaur A, Miller J, Layman LC, Diamond M, Al-Hendy A. Novel MED12 gene somatic mutations in women from the Southern United States with symptomatic uterine fibroids. Mol Genet Genomics 2015;290: 505–511.

[18] Baird DD, Dunson DB, Hill MC, Cousins D, Schectman JM. High cumulative incidence of uterine leiomyoma in black and white women: ultrasound evidence. Am J Obstet Gynecol 2003;188: 100–107.

[19] Davis BJ, Haneke KE, Miner K, Kowalik A, Barrett JC, Peddada S, et al. The fibroid growth study: determinants of therapeutic intervention. J Womens Health 2009;18: 725–732.

[20] Wise LA, Palmer JR, Stewart EA, Rosenberg L. Age-specific incidence rates for self-reported uterine leiomyomata in the Black Women's Health Study. Obstet Gyncol 2005;105: 563–568.

[21] Holick MF. Vitamin D: a millenium perspective. J Cell Biochem 2003; 88: 296–307.

[22] Holick MF. Too little vitamin D in premenopausal women: why should we care? Am J Clin Nutr 2002;76: 3–4.

[23] Holick MF. Sunlight and vitamin D for bone health and prevention of autoimmune diseases, cancers, and cardiovascular disease. Am J Clin Nutr 2004; 80(6 Suppl): 1678S-1688S.

[24] Clemens TL, Adams JS, Henderson SL, et al. Increased skin pigment reduces the capacity of skin to synthesise vitamin D3. Lancet 1982;1:74–76.

[25] Baird DD, Hill MC, Schectman JM, et al. Vitamin d and the risk of uterine fibroids. Epidemiology 2013; 24: 447–453.

[26] Paffoni A, Somigliana E, Vigano P, et al. Vitamin D status in women with uterine leiomyomas. J Clin Endocrinol Metab 2013;98: E1374-E1378.

[27] Flake GP, Andersen J, Dixon D. Etiology and pathogenesis of uterine leiomyomas: a review. Environ Health Perspect 2003; 111: 1037–1054.

[28] Arslan AA, Gold LI, Mittal K, et al. Gene expression studies provide clues to the pathogenesis of uterine leiomyoma: new evidence and a systematic review. Hum Reprod 2005; 20:852–863.

[29] Shushan A, Ben-Bassat H, Mishani E, et al. Inhibition of leiomyoma cell proliferation *in vitro* by genistein and the protein tyrosine kinase inhibitor TKS050. Fertil Steril 2007; 87:127–135.

[30] Catherino WH, Leppert PC, Stenmark MH, et al. Reduced dermatopontin expression is a molecular link between uterine leiomyomas and keloids. Genes Chromosomes Cancer 2004; 40:204–217.

[31] Arici A, Sozen I. Transforming growth factor-beta3 is expressed at high levels in leiomyoma where it stimulates fibronectin expression and cell proliferation. Fertil Steril 2000;73:1006–1111.

[32] Al-Hendy and Salama SA. Ethnic distribution of estrogen receptor-alpha polymorphism is associated with a higher prevalence of uterine leiomyomas in black Americans. Fertil Steril 2006;86: 686–693.

[33] Sharan C, Halder SK, Thota C, et al. Vitamin D inhibits proliferation of human uterine leiomyoma cells via catechol-O-methyltransferase. Fertil Steril 2011;95:247–253.

[34] Halder SK, Goodwin JS, Al-Hendy A; 1,25-Dihydroxyvitamin D3 reduces TGF-beta3-induced fibrosis-related gene expression in human uterine leiomyoma cells. J Clin Endocrinol Metab 2011;96: E754-E762.

[35] Wei JJ, Chiriboga L, Arslan AA, et al. Ethnic differences in expression of the dysregulated proteins in uterine leiomyomata. Hum Reprod 2006; 21:57–67.

[36] Halder SK, Sharan C, Al-Hendy A. 1,25-dihydroxyvitamin D3 treatment shrinks uterine leiomyoma tumors in the Eker rat model. Biol Reprod 2012;86: 116.

[37] Halder SK, Sharan C, Al-Hendy O: Paricalcitol, a vitamin d receptor activator, inhibits tumor formation in a murine model of uterine fibroids. Reprod Sci 2014;21:1108–1119.

[38] Meuleman C, Vandenabeele B, Fieuws S. High prevalence of endometriosis in infertile women with normal ovulation and normospermic partners. Fertil Steril 2009; 92: 68–74.

[39] de Ziegler D, Borghese B, Chapron C. Endometriosis and infertility: Pathophysiology and management. Lancet 2010; 376: 730–738.

[40] Bischoff FZ and Simpson JL. Heritability and molecular genetic studies of endometriosis. Hum Reprod Update 2000; 6: 37–44.

[41] McLaren J, Prentice A, Charnock-Jones DS, et al. Vascular endothelial growth factor (VEGF) concentrations are elevated in peritoneal fluid of women with endometriosis. Hum Reprod 1996; 11(1):220–223.

[42] Osuga Y, Koga K, Hirota Y, et al. Lymphocytes in endometriosis. Am J Reprod Immunol 2011; 65(1):1–10.

[43] Viganó P, Lattuada D, Mangioni S, et al. Cycling and early pregnant endometrium as a site of regulated expression of the vitamin D system. J Mol Endocrinol 2006; 36(3): 415–424.

[44] Montgomery GW, Nyholt DR, Zhao ZZ, et al. The search for genes contributing to endometriosis risk. Hum Reprod Update 2008; 14: 447–457.

[45] Braun DP, Ding J, Shaheen F, et al. Quantitative expression of apoptosis-regulating genes in endometrium from women with and without endometriosis. Fertil Steril 2007; 87: 263–268.

[46] Sourial S, Tempest N, Hapangama DK. Theories on the pathogenesis of endometriosis. Int J Reprod Med 2014; 2014:1–9.

[47] Agic A, Xu H, Altgassen C, et al. Relative expression of 1,25-dihydroxyvitamin D3 receptor, vitamin D 1 alphahydroxylase, vitamin D 24-hydroxylase, and vitamin D 25-hydroxylase in endometriosis and gynecologic cancers. Reprod Sci 2007; 14(5):486–497.

[48] Hartwell D, Rødbro P, Jensen SB, Thomsen K, Christiansen C. Vitamin D metabolites-relation to age, menopause and endometriosis. Scand J Clin Lab Invest 1990 50(2):115–121.

[49] Lasco A, Catalano A, Benvenga S. Improvement of primary dysmenorrhea caused by a single oral dose of vitamin D: results of a randomized, double-blind, placebo-controlled study. Arch Intern Med 2012 27; 172(4):366–367.

[50] Somigliana E, Panina-Bordignon P, Murone S, et al. Vitamin D reserve is higher in women with endometriosis. Hum Reprod 2007;22(8):2273–2278.

[51] Harris HR, Chavarro JE, Malspeis S, et al. Dairy-food, calcium, magnesium, and vitamin D intake and endometriosis: a prospective cohort study. Am J Epidemiol 2013;177(5): 420–430.

[52] Asuncion M, Calvo RM, San Millan JL, Sancho J, Avila S, Escobar-Morreale HF. A prospective study of the prevalence of the polycystic ovary syndrome in unselected Caucasian women from Spain. J Clin Endocrinol Metab 2000; 85:2434–2438.

[53] Azziz R, Woods KS, Reyna R, Key TJ, Knochenhauer ES, Yildiz BO. The prevalence and features of the polycystic ovary syndrome in an unselected population. J Clin Endocrinol Metab 2004; 89:2745–2749.

[54] Diamanti-Kandarakis E, Kouli CR, Bergiele AT, Filandra FA, Tsianateli TC, Spina GG, et al. A survey of the polycystic ovary syndrome in the Greek island of Lesbos: hormonal and metabolic profile. J Clin Endocrinol Metab 1999; 84:4006–4011.

[55] March WA, Moore VM, Willson KJ, Phillips DI, Norman RJ, Davies MJ. The prevalence of polycystic ovary syndrome in a community sample assessed under contrasting diagnostic criteria. Hum Reprod 2010; 25:544–551.

[56] Yildiz BO, Bozdag G, Yapici Z, Esinler I, Yarali H. Prevalence, phenotype and cardiometabolic risk of polycystic ovary syndrome under different diagnostic criteria. Hum Reprod 2012, 27: 3067–3073.

[57] Rotterdam ESHRE/ASRM-Sponsored PCOS Consensus Workshop Group. Revised 2003 consensus on diagnostic criteria and long-term health risks related to polycystic ovary syndrome. Fertil Steril 2004; 81:19–25.

[58] Rotterdam ESHRE/ASRM-Sponsored PCOS Consensus Workshop Group. Revised 2003 consensus on diagnostic criteria and long-term health risks related to polycystic ovary syndrome. Hum Reprod 2004; 19: 41–47.

[59] Diamanti-Kandarakis E & Dunaif A. Insulin resistance and the polycystic ovary syndrome revisited: an update on mechanisms and implications. Endocr Rev 2012; 33: 981–1030.

[60] Anderson SA, Barry JA, Hardiman PJ. Risk of coronary heart disease and risk of stroke in women with polycystic ovary syndrome: a systematic review and meta-analysis. Int J Cardiol 2014; 176:486–489.

[61] Calle C, Maestro B, García-Arencibia M. Genomic actions of 1,25-dihydroxyvitamin D3 on insulin receptor gene expression, insulin receptor number and insulin activity in the kidney, liver and adipose tissue of streptozotocin-induced diabetic rats. BMC Mol Biol 2008; 18: 9–65.

[62] Teegarden, D.; Donkin, S.S. Vitamin D. Emerging new roles in insulin sensitivity. Nutr Res Rev 2009; 22: 82–92.

[63] He C, Lin Z, Robb SW, Ezeamama AE. Serum vitamin D levels and polycystic ovary syndrome: a systematic review and meta-analysis. Nutrients 2015; 7:4555–4577.

[64] Ott J, Wattar L, Kurz C, Seemann R, Huber JC, Mayerhofer K, et al. Parameters for calcium metabolism in women with polycystic ovary syndrome who undergo clomiphene citrate stimulation: a prospective cohort study. Eur J Endocrinol 2012;166:897–902.

[65] Durlinger AL, Kramer P, Karels B, de Jong FH, Uilenbroek JT, Grootegoed JA, Themmen AP. Control of primordial follicle recruitment by anti-mullerian hormone in the mouse ovary. Endocrinology 1999;140:5789–5796.

[66] Wojtusik J, Johnson PA. Vitamin D regulates anti-mullerian hormone expression in granulosa cells of the hen. Biol Reprod 2012;86:91.

[67] Iliodromiti S, Kelsey TW, Anderson RA, Nelson SM J. Can anti-Mullerian hormone predict the diagnosis of polycystic ovary syndrome? A systematic review and meta-analysis of extracted data. Clin Endocrinol Metab 2013;98:3332–3340.

[68] Thys-Jacobs S, Donovan D, Papadopoulos A, Sarrel P, Bilezikian JP. Vitamin D and calcium dysregulation in the polycystic ovarian syndrome. Steroids 1999;64:430–435.

[69] Rashidi B, Haghollahi F, Shariat M, Zayerii F. The effects of calcium-vitamin D and metformin on polycystic ovary syndrome: a pilot study. Taiwan J Obstet Gynecol 2009;48:142–147.

[70] Irani M, Minkoff H, Seifer DB, Merhi Z, Vitamin D. Increases serum levels of the soluble receptor for advanced glycation end products in women with PCOS. J Clin Endocrinol Metab 2014;99:E886-E890.

[71] Bodnar LM, Catov JM, Zmuda JM, Cooper ME, Parrott MS, Roberts JM, et al. Maternal serum 25-hydroxyvitamin D concentrations are associated with small-for-gestational age births in white women. J Nutr 2010;140(5):999–1006.

[72] Charney VA, Bertin FR, Kritchevsky J, Miller MA. Pathology in practice. Myometrial leiomyoma in a Vietnamese potbellied pig. J Am Veter Med Assoc 2014;244(5):541–543.

[73] Karras SN, Anagnostis P, Annweiler C, Naughton DP, Petroczi A, Bili E, et al. Maternal vitamin D status during pregnancy: the Mediterranean reality. Eur J Clin Nutr 2014;68(8):864–869.

[74] Reeves IV, Bamji ZD, Rosario GB, Lewis KM, Young MA, Washington KN. Vitamin D deficiency in pregnant women of ethnic minority: a potential contributor to preeclampsia. J Perinatol 2014;34(10):767–773.

[75] Tan YL, Naidu A. Rare postpartum ruptured degenerated fibroid: a case report. J Obstet Gynaecol Res 2014;40(5):1423–1425.

[76] Thota C, Laknaur A, Farmer T, Ladson G, Al-Hendy A, Ismail N. Vitamin D regulates contractile profile in human uterine myometrial cells via NF-kappaB pathway. Am J Obstet Gynecol 2014;210(4):347; 1–10.

[77] Thota C, Farmer T, Garfield RE, Menon R, Al-Hendy A. Vitamin D elicits anti-inflammatory response, inhibits contractile-associated proteins, and modulates Toll-like receptors in human myometrial cells. Reprod Sci 2013;20(4):463–475.

[78] Trivedi S, Joachim M, McElrath T, Kliman HJ, Allred EN, Fichorova RN, et al. Fetal-placental inflammation, but not adrenal activation, is associated with extreme preterm delivery. Am J Obstet Gynecol 2012;206(3):236 e1–8.

[79] You X, Gao L, Liu J, Xu C, Liu C, Li Y, et al. CRH activation of different signaling pathways results in differential calcium signaling in human pregnant myometrium before and during labor. J Clin Endocrinol Metab 2012;97(10):E1851–E1861.

[80] Jin L, Chen C, Guo R, Wan R, Li S. Role of corticotropin-releasing hormone family peptides in androgen receptor and vitamin D receptor expression and translocation in human breast cancer MCF-7 cells. Eur J Pharmacol 2012;684(1–3):27–35.

[81] Wang WS, Liu C, Li WJ, Zhu P, Li JN, Sun K. Involvement of CRH and hCG in the induction of aromatase by cortisol in human placental syncytiotrophoblasts. Placenta 2014;35(1):30–36.

[82] Holick MF. Vitamin D deficiency. N Engl J Med 2007;357(3):266–281.

[83] Pagliardini, Luca et al. High prevalence of vitamin D deficiency in infertile women referring for assisted reproduction. Nutrients. 2015;7:9972–9984.

[84] Colonese, Francesca et al. The pleinotropic effects of vitamin D in gynaecological and obstetric diseases: an overview on a hot topic. BioMed Res Int. 2015;2015:1–11.

[85] Paffoni, Alesio et al. Vitamin D deficiency and infertility: insights from *in vitro* fertilization cycles. J Clin Endocrinol Metab. 2014;99(11):E2372–E2376.

[86] Dressler, N et al. BMI and season are associated with vitamin D Deficiency in women with impaired fertility: a two-center analysis. Gynecol Endocrinol Reprod Med. 2016;293(4):907–914.

[87] Al-Jaroudi et al. Vitamin D deficiency among subfertile women: case-control study. Gyncol Endocrinol. 2016;32(4): 272–275.

[88] Franasiak et al. Vitamin D levels do not affect IVF outcomes following the transfer of euploid blastocysts. Am J Obstet Gynecol. 2015. 212(3):315.e1–6.

[89] Firouzabadi et al. Value of follicular fluid vitamin D in predicting the pregnancy rate in an IVF program. Arch Gynecol Obstet. 2014;289(1):201–206.

[90] Farzadi, L. et al. Correlaton between follicular fluid 25-OH vitamin D and assisted reproductive outcomes. Iran J Reprod Med 2015;13:361–366.

[91] Aleyasin et al. Predictive value of the level of vitamin D in follicular fluid on the outcome of assisted reproductive technology. Eur J Obstet Gynecol Reprod Biol. 2011; 159(1):132–137.

[92] Haylen BT, de Ridder D, Freeman RM, Swift SE, Berghmans B, Lee J, et al. An International Urogynecological Association (IUGA)/International Continence Society (ICS) joint report on the terminology for female pelvic floor dysfunction. Int Urogynecol J 2010;21(1):5–26.

[93] Stewart WF, Van Rooyen JB, Cundiff GW, Abrams P, Herzog AR, Corey R, et al. Prevalence and burden of overactive bladder in the United States. World J Urol 2003;20(6):327–336.

[94] Wein AJ. Pharmacological agents for the treatment of urinary incontinence due to overactive bladder. Expert Opin Investig Drugs 2001;10(1):65–83.

[95] Zhang W, Song Y, He X, Huang H, Xu B, and Song J. Prevalence and Risk Factors of Overactive Bladder Syndrome in Fuzhou Chinese Women. Neurourol Urodyn 2006;(25):717–721.

[96] Sears CL, Lewis C, Noel K, Albright TS, Fischer JR. Overactive bladder medication adherence when medication is free to patients. J Urol 2010;183(3):1077–1081.

[97] Athanasopoulos A, Cruz F. The medical treatment of overactive bladder, including current and future treatments. Expert Opin Pharmacother 2011;12(7):1041–1055.

[98] D'Souza AO, Smith MJ, Miller LA, Doyle J, Ariely R. Persistence, adherence, and switch rates among extended-release and immediate-release overactive bladder medications in a regional managed care plan. J Manag Care Pharm 2008;14(3):291–301.

[99] Freedman LP. Multimeric Coactivator Complexes for Steroid/Nuclear Receptors. Trends Endocrinol Metab: TEM 1999;10(10):403–407.

[100] Coyne KS, Sexton CC, Bell JA, Thompson CL, Dmochowski R, Bavendam T, Chen CI and Clemens JQ. Re: the prevalence of lower urinary tract symptoms (LUTS) and overactive bladder (OAB) by racial/ethnic group and age: results from OAB-POLL. Neurourol Urodyn 2013;32:230–237.

[101] Downes E, Sikirica V, Gilabert-Estelles J, Bolge SC, Dodd SL, Maroulis C, et al. The burden of uterine fibroids in five European countries. Eur J Obstet Gynecol Reprod Biol 2010;152(1):96–102.

[102] Crescioli C, Morelli A, Adorini L, Ferruzzi P, Luconi M, Vannelli GB, et al. Human bladder as a novel target for vitamin D receptor ligands. J Clin Endocrinol Metab 2005;90(2):962–972.

[103] Dallosso HM, McGrother CW, Matthews RJ, Donaldson MM, Leicestershire MRCISG. Nutrient composition of the diet and the development of overactive bladder: a longitudinal study in women. Neurourol Urodyn 2004;23(3):204–210.

[104] Morelli A, Squecco R, Failli P, Filippi S, Vignozzi L, Chavalmane AK, et al. The vitamin D receptor agonist elocalcitol upregulates L-type calcium channel activity in human and rat bladder. Am J Physiol Cell Physiol 2008;294(5):C1206–C1214.

[105] Schroder A, Colli E, Maggi M, Andersson KE. Effects of a vitamin D(3) analogue in a rat model of bladder outlet obstruction. BJU Int 2006;98(3):637–642.

[106] Lipkin M, and Newmark HL. Vitamin D, calcium and prevention of breast cancer: a review. J Am Coll Nutr 1999;18(5 Suppl):392S–397S.

[107] Hansen CM, Binderup L, Hamberg KJ, and Carlberg C. Vitamin D and cancer: effects of 1,25(OH)2D3 and its analogs on growth control and tumorigenesis. Front Biosci 2001;6:D820–D848.

[108] Garland CF, and Garland FC. Do sunlight and vitamin D reduce the likelihood of colon cancer? Int J Epidemiol 1980;9(3):227–231.

[109] Garland FC, Garland CF, Gorham ED, and Young JF. Geographic variation in breast cancer mortality in the United States: a hypothesis involving exposure to solar radiation. Prevent Med 1990;19(6):614–622.

[110] Lefkowitz ES, and Garland CF. Sunlight, vitamin D, and ovarian cancer mortality rates in US women. Int J Epidemiol 1994;23(6):1133–1136.

[111] Schwartz GG, and Hulka BS. Is vitamin D deficiency a risk factor for prostate cancer? (Hypothesis). Anticancer Res 1990;10(5A):1307–1311.

[112] Grant WB. An estimate of premature cancer mortality in the U.S. due to inadequate doses of solar ultraviolet-B radiation. Cancer 2002;94(6):1867–1875.

[113] Shin MH, Holmes MD, Hankinson SE, Wu K, Colditz GA, and Willett WC. Intake of dairy products, calcium, and vitamin d and risk of breast cancer. J Natl Cancer Inst 2002;94(17):1301–1311.

[114] Lin J, Manson JE, Lee IM, Cook NR, Buring JE, and Zhang SM. Intakes of calcium and vitamin D and breast cancer risk in women. Arch Intern Med 2007;167(10):1050–1059.

[115] Deeb KK, Trump DL, and Johnson CS. Vitamin D signalling pathways in cancer: potential for anticancer therapeutics. Nat Rev Cancer 2007;7(9):684–700.

[116] Colston KW, and Hansen CM. Mechanisms implicated in the growth regulatory effects of vitamin D in breast cancer. Endocr Relat Cancer 2002;9(1):45–59.

[117] Mathiasen IS, Sergeev IN, Bastholm L, Elling F, Norman AW, and Jaattela M. Calcium and calpain as key mediators of apoptosis-like death induced by vitamin D compounds in breast cancer cells. J Biol Chem 2002;277(34):30738–30745.

[118] Gorham ED, Garland CF, and Garland FC. Acid haze air pollution and breast and colon cancer mortality in 20 Canadian cities. Can J Public Health = Revue canadienne de sante publique 1989;80(2):96–100.

[119] John EM, Schwartz GG, Dreon DM, and Koo J. Vitamin D and breast cancer risk: the NHANES I Epidemiologic follow-up study, 1971–1975 to 1992. National Health

and Nutrition Examination Survey. Cancer Epidemiol Biomarkers Prev 1999;8(5): 399–406.

[120] Janowsky EC, Lester GE, Weinberg CR, Millikan RC, Schildkraut JM, Garrett PA, and Hulka BS. Association between low levels of 1,25-dihydroxyvitamin D and breast cancer risk. Public Health Nutr 1999;2(3):283–291.

[121] Mawer EB, Walls J, Howell A, Davies M, Ratcliffe WA, and Bundred NJ. Serum 1,25-dihydroxyvitamin D may be related inversely to disease activity in breast cancer patients with bone metastases. J Clin Endocrinol Metab 1997;82(1):118–122.

[122] Newmark HL. Vitamin D adequacy: a possible relationship to breast cancer. Adv Exp Med Biol 1994;364;109–114.

[123] Carroll KK, Jacobson EA, Eckel LA, and Newmark HL. Calcium and carcinogenesis of the mammary gland. Am J Clin Nutr 1991;54(1 Suppl):206S-208S.

[124] Siegel R, Ma J, Zou Z, and Jemal A. Cancer statistics, 2014. CA: Cancer J Clinicians 2014;64(1):9–29.

[125] Webb PM, de Fazio A, Protani MM, Ibiebele TI, Nagle CM, Brand AH, Blomfield PI, Grant P, Perrin LC, Neale RE, et al. Circulating 25-hydroxyvitamin D and survival in women with ovarian cancer. Am J Clin Nutr 2015;102(1):109–114.

[126] Granato T, Manganaro L, Petri L, Porpora MG, Viggiani V, Angeloni A, and Anastasi E. Low 25-OH vitamin D levels at time of diagnosis and recurrence of ovarian cancer. Tumour Biol: J Int Soc Oncodevelop Biol Med 2015.

[127] Garland CF, Mohr SB, Gorham ED, Grant WB, and Garland FC. Role of ultraviolet B irradiance and vitamin D in prevention of ovarian cancer. Am J Prev Med 2006;31(6): 512–514.

[128] Freedman DM, Dosemeci M, and McGlynn K. Sunlight and mortality from breast, ovarian, colon, prostate, and non-melanoma skin cancer: a composite death certificate based case-control study. Occup Environ Med 2002;59(4):257–262.

[129] Grant WB. Ecologic studies of solar UV-B radiation and cancer mortality rates. Recent results in cancer research. 2003;164(371–7.

[130] Grant WB. An ecological study of cancer incidence and mortality rates in France with respect to latitude, an index for vitamin D production. Dermato-endocrinology 2010;2(2):62–67.

[131] Salazar-Martinez E, Lazcano-Ponce EC, Gonzalez Lira-Lira G, Escudero-De los Rios P, and Hernandez-Avila M. Nutritional determinants of epithelial ovarian cancer risk: a case-control study in Mexico. Oncology 2002;63(2):151–157.

[132] Tworoger SS, Lee IM, Buring JE, Rosner B, Hollis BW, and Hankinson SE. Plasma 25-hydroxyvitamin D and 1,25-dihydroxyvitamin D and risk of incident ovarian cancer. Cancer Epidemiol Biomarkers Prev 2007;16(4):783–788.

[133] Chen W, Clements M, Rahman B, Zhang S, Qiao Y, and Armstrong BK. Relationship between cancer mortality/incidence and ambient ultraviolet B irradiance in China. Cancer Causes Control: CCC 2010;21(10):1701–1709.

[134] Grant WB, and Garland CF. The association of solar ultraviolet B (UVB) with reducing risk of cancer: multifactorial ecologic analysis of geographic variation in age-adjusted cancer mortality rates. Anticancer Res 2006;26(4A):2687–2699.

[135] Hosono S, Matsuo K, Kajiyama H, Hirose K, Suzuki T, Kawase T, Kidokoro K, Nakanishi T, Hamajima N, Kikkawa F, et al. Association between dietary calcium and vitamin D intake and cervical carcinogenesis among Japanese women. Eur J Clin Nutr 2010;64(4): 400–409.

[136] Whitehurst JL, and Reid CM. Vitamin D deficiency as a cause of chronic pain in the palliative medicine clinic: two case reports. Palliat Med 2014;28(1):87–89.

[137] Helzlsouer KJ, and Committee VS. Overview of the Cohort Consortium Vitamin D Pooling Project of Rarer Cancers. Am J Epidemiol 2010;172(1):4–9.

[138] Zeleniuch-Jacquotte A, Gallicchio L, Hartmuller V, Helzlsouer KJ, McCullough ML, Setiawan VW, Shu XO, Weinstein SJ, Weiss JM, Arslan AA, et al. Circulating 25-hydroxyvitamin D and risk of endometrial cancer: Cohort Consortium Vitamin D Pooling Project of Rarer Cancers. Am J Epidemiol 2010;172(1):36–46.

[139] Gallicchio L, Helzlsouer KJ, Chow WH, Freedman DM, Hankinson SE, Hartge P, Hartmuller V, Harvey C, Hayes RB, Horst RL, et al. Circulating 25-hydroxyvitamin D and the risk of rarer cancers: Design and methods of the Cohort Consortium Vitamin D Pooling Project of Rarer Cancers. Am J Epidemiol 2010;172(1):10 20.

[140] Garland CF, Gorham ED, Mohr SB, and Garland FC. Vitamin D for cancer prevention: global perspective. Ann Epidemiol 2009;19(7):468–483.

[141] Lehman M. N., Goodman R. L., Karsch F.J., Jackson G. L, Berriman S. J., Jansen H.T. The GnRH System of Seasonal Breeders: Anatomy and Plasticity. Brain Res Bull. 44(4): 445–57.

[142] Heidari B and Mirghassemi MB. Seasonal variations in serum vitamin D according to age and sex. Caspian J InternMed 2012; 3(4): 535–540.

[143] NVSR Monthly Provisional Reports-January 2014-June 2015: http://www.cdc.gov/nchs/products/nvsr/monthly_provisional_notice.htm.